Enrique Herrera-Viedma, Gabriella Pasi, Fabio Crestani (Eds.)

Soft Computing in Web Information Retrieval

Studies in Fuzziness and Soft Computing, Volume 197

Editor-in-chief
Prof. Janusz Kacprzyk
Systems Research Institute
Polish Academy of Sciences
ul. Newelska 6
01-447 Warsaw
Poland
E-mail: kacprzyk@ibspan.waw.pl

Further volumes of this series can be found on our homepage: springer.com

Vol. 181. Nadia Nedjah, Luiza de Macedo Mourelle
Fuzzy Systems Engineering, 2005
ISBN 3-540-25322-X

Vol. 182. John N. Mordeson, Kiran R. Bhutani, Azriel Rosenfeld
Fuzzy Group Theory, 2005
ISBN 3-540-25072-7

Vol. 183. Larry Bull, Tim Kovacs (Eds.)
Foundations of Learning Classifier Systems, 2005
ISBN 3-540-25073-5

Vol. 184. Barry G. Silverman, Ashlesha Jain, Ajita Ichalkaranje, Lakhmi C. Jain (Eds.)
Intelligent Paradigms for Healthcare Enterprises, 2005
ISBN 3-540-22903-5

Vol. 185. Spiros Sirmakessis (Ed.)
Knowledge Mining, 2005
ISBN 3-540-25070-0

Vol. 186. Radim Bělohlávek, Vilém Vychodil
Fuzzy Equational Logic, 2005
ISBN 3-540-26254-7

Vol. 187. Zhong Li, Wolfgang A. Halang, Guanrong Chen (Eds.)
Integration of Fuzzy Logic and Chaos Theory, 2006
ISBN 3-540-26899-5

Vol. 188. James J. Buckley, Leonard J. Jowers
Simulating Continuous Fuzzy Systems, 2006
ISBN 3-540-28455-9

Vol. 189. Hans Bandemer
Mathematics of Uncertainty, 2006
ISBN 3-540-28457-5

Vol. 190. Ying-ping Chen
Extending the Scalability of Linkage Learning Genetic Algorithms, 2006
ISBN 3-540-28459-1

Vol. 191. Martin V. Butz
Rule-Based Evolutionary Online Learning Systems, 2006
ISBN 3-540-25379-3

Vol. 192. Jose A. Lozano, Pedro Larrañaga, Iñaki Inza, Endika Bengoetxea (Eds.)
Towards a New Evolutionary Computation, 2006
ISBN 3-540-29006-0

Vol. 193. Ingo Glöckner
Fuzzy Quantifiers: A Computational Theory, 2006
ISBN 3-540-29634-4

Vol. 194. Dawn E. Holmes, Lakhmi C. Jain (Eds.)
Innovations in Machince Learning, 2006
ISBN 3-540-30609-9

Vol. 195. Zongmin Ma
Fuzzy Database Modeling of Imprecise and Uncertain Engineering Information, 2006
ISBN 3-540-30675-7

Vol. 196. James J. Buckley
Fuzzy Probability and Statistics, 2006
ISBN 3-540-30841-5

Vol. 197. Enrique Herrera-Viedma, Gabriella Pasi, Fabio Crestani (Eds.)
Soft Computing in Web Information Retrieval, 2006
ISBN 3-540-31588-8

Enrique Herrera-Viedma
Gabriella Pasi
Fabio Crestani
(Eds.)

Soft Computing in Web Information Retrieval

Models and Applications

Springer

Professor Enrique Herrera-Viedma
Department of Computer Science and A.I
E.T.S.I. Informatica
University of Granada
C/Periodista Daniel
Saucedo Aranda s/n
Granada, Spain
E-mail: viedma@decsai.ugr.es

Professor Fabio Crestani
Department of Computer
and Information Sciences
University of Strathclyde
Livingstone Tower
26 Richmond Street
Glasgow G1 1XH
Scotland, UK
E-mail: F.Crestani@cis.strath.ac.uk

Professor Gabriella Pasi
Università degli Studi di Milano Bicocca
Department of Informatics
Systems and Communication (DISCo)
Via Bicocca degli Arcimboldi
8 (Edificio U7)
20126 Milano (ITALY)
E-mail: pasi@disco.unimib.it

ISBN 978-3-642-06854-6 e-ISBN 978-3-540-31590-2

ISSN print edition: 1434-9922
ISSN electronic edition: 1860-0808

Springer is a part of Springer Science+Business Media
springer.com

Printed in The Netherlands

Preface

The *World Wide Web*, or simply the Web, is a popular and interactive medium to collect, disseminate, and access an increasingly huge amount of information. Nowadays, *information access on the Web* is the main problem of the so called Web Information Retrieval (IR). The Web represents a new framework that is rather different from with respect to the traditional IR framework and sheds new and difficult challenges.

The Web presents particular characteristics that limit the existing IR technologies and determine the need to design new information access technologies:

- the Web is possibly the biggest and dynamic information resource existing.
- the Web presents a structure of linked pages.
- the Web is growing and updating at a very high rate.
- the Web is very heterogeneous.
- last but not least, imprecision and vagueness characterize several tasks in Web IR, such as assessing the relevance of Web pages, dealing with the multimedia nature of information, identifying spam problem, discovering deception, etc.

Furthermore, due to this complexity, any major advance in the field of information access on the Web requires the application of intelligent techniques. In fact, several authors suggest to proceed towards the *Web Intelligence* by incorporating and embedding some form of intelligence (such as learning capabilities, and tolerance to uncertainty, vagueness and imprecision) in Web technologies.

Soft Computing (SC) techniques constitute a synergy of methodologies (fuzzy logic, neural networks, probabilistic reasoning, rough-set theory, evolutionary computing and parts of machine learning theory) useful for solving problems requiring some form of intelligence. The basis of SC is its tolerance to imprecision, uncertainty, partial truth, and approximation. Because of these properties SC can provide very powerful tools for modelling the different activities related with the information access problem. In a previous book edited in this series, titled "*Soft Computing in Information Retrieval,*

Techniques and Applications", F. Crestani and G. Pasi (Eds.), collected a selection of SC-based approaches to the traditional IR. The present edited volume focuses on the use of the SC techniques for improving information access in Web IR.

This book presents some recent works on the application of SC techniques in information access on the Web. The book comprises 15 chapters from internationally known researchers. It is divided in four parts reflecting the areas of research of the presented works.

The first part focuses on the use of SC in *Document Classification*. The chapter by Bordogna and Pasi proposes a hierarchical fuzzy clustering algorithm for dynamically supporting document filtering that performs a fuzzy hierarchical categorization of documents allowing updating as new documents are fed. The chapter by de Campos, Fernández-Luna, and Huete presents a theoretical framework for classifying Web pages in a hierarchical directory using the Bayesian Network formalism which is able to perform multi-label text categorization in a category tree in a Web framework. The chapter by Loia and Senatore describes a customized system for information discovery based on fuzzy clustering of RDF-based documents which are classified in terms of the semantics of their metadata. The chapter by Zhang, Fan, Chen, Fox, Gonçalves, Cristo, and Pável Calado defines a Genetic Programming Approach for Combining Structural and Citation-Based Evidence for Text Classification in Web Digital Libraries.

The second part presents experiences on the development of the *Semantic Web* using SC techniques. The chapter by Ceravolo, Damiani, and Viviani describes a complete approach for developing a Trust Layer service, aimed at improving the quality of automatically generated semantic Web-style metadata and based on non-intrusive collection of user feedback. The chapter by Herrera-Viedma, Peis, and Morales-del-Castillo defines a model of a Web fuzzy linguistic multi-agent system that combines the use of Semantic Web technologies together with the application of user profiles to carry out its information access processes. The chapter by Barriocanal and Sicilia proposes a first fuzzy approach for the design of ontology-based browsers. The chapter by Loiseau, Boughanem, and Prade presents an evaluation method of term-based queries using possibilistic ontologies which allows to retrieve information containing terms that may not match exactly those of the query. This tool can be applied in both textual information retrieval and data base management.

The third part shows different SC approaches to the *Web Information Retrieval*. The chapter by Dominich, Skrop, and Tuza presents a unified formal framework for three major methods used for Web retrieval tasks: PageRank, HITS, I^2R. It is based on the Artificial Neural Networks and the generic network equation. It was shown that the PageRank, HITS and I^2R methods can be formally obtained from the generic equation as different particular cases by making certain assumptions reflecting the corresponding underlying paradigm. The chapter by Losada, Díaz-Hermida, and Bugarín carries out an empirical study that demonstrates the usefulness of the semi-fuzzy quantifiers for improving the query languages in information retrieval. In particular, it is shown that fuzzy quantifiers-based IR models are competitive with respect to models such as the vector-space model. The chapter by Martín-Bautista, Sánchez, Serrano, and Vila describes a query refinement technique based on fuzzy association rules that helps the user to search information and improve the Web information retrieval. The chapter by Valverde-Albacete proposes a formal model of the batch retrieval phase of a Web retrieval interaction or any other batch retrieval task which is designed using hard techniques like Concept Formal Analysis and soft techniques like Rough-Set Theory.

The fourth part reports a selection of *Web Applications* developed by means of SC techniques. The chapter by Cristo, Ribeiro-Neto, Golgher, and de Moura presents an analysis on key concepts and variables related to search advertising, both in the commercial and in the technology fronts. It studies some SC approaches to the Web topic as the case of content-targeted advertising based on Bayesian Networks. The chapter by Domingo-Ferrer, Mateo-Sanz, and Sebé provides a fast method for generating numerical hybrid microdata in a way that preserves attribute means, variances and covariances, as well as (to some extent) record similarity and subdomain analyses. Finally, the chapter by Sobrino-Cerdeiriña, Fernández-Lanza,and Graña-Gil proposes a general model for implementing large dictionaries in natural language processing applications which is able to store a considerable amount of data relating to the words contained in these dictionaries. Additionally, it shows how this model can be applied to implement and transform a Spanish dictionary of synonyms into a computational framework able to represent relations of synonymy between words.

Ultimately, the goal of this book is to show that Web IR can be a stimulating area of research where SC technologies can be applied satisfactorily. This book is a proof of this and we think that it will not be the last one.

Granada
October, 2005

Enrique Herrera-Viedma
Gabriella Pasi
Fabio Crestani

Acknowledgments

We would like to thank the authors of the papers, that with their effort showed that it is possible to improve the performance of the Web technologies through SC techniques and made possible the apparition of this book.

Our gratitude also goes to Ricardo Baeza-Yates for his foreword, and to the reviewers (Miyamoto, Kraft, Sobrino, Losada, Huete, Domingo-Ferrer, Damiani, Martín-Bautista, Dominich, Ribeiro-Neto, Bordogna, Olsina, Mich, Fan, Olivas, Peis, Loia, Sicilia). Without their help and collaboration we could not have assured the high quality of this book (we received 22 contributions and each paper was reviewed by at least three referees).

Finally, thanks to Janusz Kacprzyk, the series editor of Studies in Fuzziness and Soft Computing, for accepting our proposal of this volume.

Foreword

The Web currently is the largest repository of data available, comprising a maremagnum of different media over more than ten billion interconnected pages. However, volume is not necessarely the main challenge, as content and link spamming makes information retrieval even harder. Indeed, searching for information in the Web has been called "adversarial Web retrieval". Hence, the main challenges are:

- to keep an up-to-date index of all the pages of the Web by crawling it,
- to assess how much we can trust the content of a given page (this is a key issue also for the semantic Web),
- to compute the relevance of the page with respect to the user query, and
- to give a personalized answer.

These challenges imply several sub-challenges and several related problems such as what advertising can be shown in the answer page or how to use different sources of information to rank a page (content, links, usage, etc.)

Soft-computing can help in many of the challenges above, specially in off-line tasks where we can preprocess data to build additional data structures that are fast enough for on-line use. Important examples are new retrieval models, categorization of documents, link analysis, trust models, creation of pseudo-semantic resources, fuzzy search, linguistic processing, adaptive interfaces, etc. This book contains several of the problems and applications mentioned above and it is one step forward on the fascinating research path that lies in front of us.

Barcelona, Spain
October 2005

Ricardo Baeza-Yates

Contents

Part I

Document Classification

A Dynamic Hierarchical Fuzzy Clustering Algorithm for Information Filtering

Gloria Bordogna[1], Marco Pagani[1], and Gabriella Pasi[2]

[1] CNR – IDPA, Sez. di Milano, Gruppo di Georisorse, via Pasubio 5, c/o POINT, 24044 Dalmine (BG) Italy
gloria.bordogna@idpa.cnr.it

[2] Dip. Di Informatica, Sistemistica e Comunicazione Univ. Degli Studi di Milano Bicocca p.le Ateneo Nuovo, 1, Milano, Italy
gabriella.pasi@itc.cnr.it

Summary. In this contribution we propose a hierarchical fuzzy clustering algorithm for dynamically supporting information filtering. The idea is that document filtering can draw advantages from a dynamic hierarchical fuzzy clustering of the documents into overlapping topic categories corresponding with different levels of granularity of the categorisation. Users can have either general interests or specific ones depending on their profile and thus they must be feed with documents belonging to the categories of interest that can correspond with either a high level topic, such as sport news, or a subtopics, such as football news, or even a very specific topics such as football matches of their favourite team. The hierarchical structure of the automatically identified clusters is built so that each level corresponds with a distinct level of overlapping of the clusters in it, so that in climbing the hierarchy this value increases since the topics represented in the upper levels are more general, i.e., fuzzier. The hierarchy of fuzzy clusters is used to support the filtering criteria that are personalized based on user profiles. Since a filter monitors one or more continuously feed document streams, the clustering must be able both to generate a fuzzy hierarchical classification of a collection of documents and to update the hierarchy of existing categories by either including newly found documents or detecting new categories when such new documents have contents that are different from those represented by the existing clusters. The fuzzy clustering algorithm is based on a generalization of the fuzzy C-means algorithm that is iteratively applied to each hierarchical level to identify clusters of the higher level. In order to apply this algorithm in document filtering it has been extended so as to use a cosine similarity instead of the usual Euclidean distance, and to automatically estimate the number of the clusters to detect at each hierarchical level. This number is identified based either on an explicit input that specifies the minimum percentage of common index terms that the clusters of the level can share (that is equivalent to indicate a tolerance for overlapping between the topics dealt with in each fuzzy cluster) or on a statistical analysis of the cumulative curve of overlapping degrees between all pairs of clusters of the level. This way the problem of application of the fuzzy C means that requires the specification of the desired number of the clusters is overcome.

G. Bordogna et al.: *A Dynamic Hierarchical Fuzzy Clustering Algorithm for Information Filtering*, StudFuzz **197**, 3–23 (2006)
www.springerlink.com

1 Introduction

Content-based filtering is an automatic process that has the objective of monitoring a series of document streams and of feeding users only with appropriate documents matching their interests represented in their personal profiles. This process is performed based on a filtering model that compares the user's preferences represented in their personal or group profile with the available information about document contents, i.e., the documents' representation based generally on meta-data and content keywords [3, 4, 12, 22, 29].

One potential problem with standard filtering systems is that all reasoning is done online. With impatient users waiting for quick responses, the search for similar items must be very time-efficient. This time restriction results in fewer possibilities when trying to improve or extend the content-based filtering strategies.

In order to improve both speed and effectiveness, current approaches to building filtering systems often try to perform some of the reasoning offline using clustering techniques [4, 17, 28, 30]. Clustering can be employed in the filtering task either to group users having potential similar interests, this technique is known as collaborative filtering, or to group the documents in topic categories, or even for both tasks. The clustering module is run off-line periodically so as to identify a reasonable number of documents' representatives (centroids) that can directly be matched with the user interests in the profiles thus reducing the number of the matching operations necessary to identify the interesting documents and achieving a speed up of the filtering task and an improvement of its effectiveness. Another reason to employ clustering techniques is the fact that content based filtering runs the risk of only recommending items almost identical to the ones the user has appreciated before.

The most obvious solution is to categorize the items in the system. A novel approach in highly dynamic contexts where data change frequently, is called category-based filtering [28]. Category-based filtering can be seen mainly as an extension to existing filtering strategies. In a category-based filtering system, user preferences reflect attitudes not towards single items, but categories of similar items. Its main characteristic is that selection of information is based on category ratings instead of single item ratings, in contrast to other content-based filtering strategies. To function, category-based filtering requires categorization of every item, either manually or by an automated process. Selecting items based on category ratings instead of ratings of individual items is especially suited for domains where there is a constant flow of new information (e.g. news and adverts), provided that effective categorization is possible.

Several problems are involved in the application of such techniques in document filtering. First of all, the interests of users can have different level of granularity, which means that they can reflect either general topics corresponding with high level categories, or more specific topics corresponding with subcategories of topics, for instance news stories. A user can be interested in general "*sport*" news, while another one may be interested only in "*football*" news or even in a more specific subtopic such as "*football matches of his/her favorite team*". Thus a feasible solution is to support category-based filtering though a hierarchical clustering technique able to identify categories structured in a hierarchy reflecting different levels of granularity of the topics.

The second problem is that documents and categories of topics often cannot be partitioned into well defined disjoint groups, but they may be associated with several groups based on their content at the same time. For example a document reporting on the participation of a sport man to the film festival in Venice could be related with both the categories of "*sport*" news and "*entertainement*". For this reason fuzzy clustering techniques are well suited to the purpose of document classification into overlapping categories.

The most popularly applied fuzzy clustering algorithm is the fuzzy C-means [11]. However this algorithm does not provide a hierarchical categorization and suffers from the need to specify some parameters that drastically influence the results, such as the number of the clusters that one is willing to identify and the seeds where to start the grouping.

Finally, this algorithm is non dynamic, but it is static in the sense that it does not provide the possibility to update existing clusters that is a mandatory requirement when applying clustering in a dynamic context such as filtering.

Several proposals of modification of the fuzzy C-means have been defined that tackle one of the above deficiencies, but, as far as we know, up to date there is not a proposal that faces all of them.

In this contribution we propose a dynamic hierarchical fuzzy clustering algorithm for information filtering that performs a fuzzy hierarchical categorization of documents allowing updating as new documents are fed.

In the next section we describe the basic notions of clustering techniques, in Section 3. the characteristics of the proposed method are illustrated and in Section 4 the description of the algorithm is provided. Section 5 illustrates the first experiment with the application of the proposed algorithm and the conclusions summarize the main achievements and open issues.

2 Categorization of Documents in IR Based on Clustering Techniques

The problem of categorization of a collection of documents has some similarities with the problem of knowledge discovery. Finding out topic categories is in fact a way to organize materials and thus to know more about its content.

One common approach to document categorization and more generally to knowledge discovery is clustering.

Two alternative families of clustering methods can be adopted: partitioning clustering methods and hierarchical clustering methods [9, 14].

In order to apply a clustering algorithm to a document collection D one needs to represent each document $d_i \in D$ as a vector in an N dimensional space of indexes like in the Vector Space model [22]:

> $d_i = (tf_{1i}, \ldots, tf_{Ni})$ in which tf_{ki} is a numeric value expressing the significance of the index term t_k in synthesising the content of document d_i.

tf_{ki} can be automatically computed as the frequency of the term in the document text normalized with respect to the document length, or it can be the classic index term weight defined as the product of the normalized term frequency (TF) and the Inverse Document Frequency (IDF) [2, 24].

In the following subsections we will introduce the main approaches to clustering.

2.1 Partitioning Clustering Method

The clustering algorithms based on the partitioning method create a flat non hierarchical partition of the documents into C clusters [8]. The approach consists in partitioning the document collection into C containers (clusters) identifying topic categories, such that documents dealing with similar contents (i.e., represented by similar feature vectors) are grouped into the same container. Each container corresponds then to a given category. This is a well known approach that has been applied since long time at indexing phase for efficient retrieval purposes, or for implementing associate retrieval techniques based for instance on either local or global analysis aimed at expanding the set of documents retrieved by an initial query [5, 33].

Similar to machine learning algorithms, clustering can be completely supervised or unsupervised. In supervised clustering, labelled documents are grouped into known pre-specified topic categories [6, 23], while in unsupervised clustering both the number of the categories and the categories themselves are unknown and documents are assigned to a group using some similarity measure (distance or density function). This last approach is the most feasible one since in IR a complete labelled collection is generally not available.

Basically the most common algorithm based on the partitioning method is the crisp C-means algorithm that needs as input the number C of desired clusters that one wants to generate. It basically performs the following steps:

1. C points in the document space are selected as initial centroids of the clusters;
2. all documents vectors are assigned to the closest centroid based on the computation of their similarity to the centroids;
3. the centroids of the clusters are recomputed;
4. steps 2 and 3 are repeated until the centroids stabilize

This algorithm is relatively efficient and scalable, and its complexity is linear to the number of documents in the collection.

However, the application of clustering methods in IR suffers from some drawbacks mainly the fact that the topic categories are not well defined and disjoint ones [33]. Some topics can have some overlapping so that it becomes sometimes cumbersome to completely associate a document to a single category. This is the reason that has motivated the application of fuzzy clustering techniques to document categorization. Topics that characterise a given knowledge domain or application area are somehow associated with each other. Those topics may also be related to topics of other domains or areas. Therefore, documents may contain information that is relevant to different domains to some degree. With fuzzy clustering methods documents are associated with several clusters at the same time and thus, useful relationships between knowledge domains or application areas may be discovered, which would otherwise remain implicit when applying hard clustering methods.

In fuzzy clustering each document can belong to more than a single category (cluster) to a degree in [0,1] according to some similarity measure used by the clustering algorithm.

The fuzzy C-means algorithm (FCM) is based on the extension of the classic partitional C-means clustering technique [11, 21].

It is based on the minimization of the J function defined as follows:

$$J(M, C) = \sum_{i=1}^{|C|} \sum_{k=1}^{|D|} \mu_{ik}^{m} \|d_k - c_i\|^2 \tag{1}$$

in which:

M is the matrix of the membership degrees of documents to the clusters, i.e., $\mu_{ik} \in [0, 1]$ is the membership degree of document k, represented by vector d_k, to the i-th cluster, represented by the centroid vector $c_i = (c_{1i}, \ldots, c_{Ni})$;
$m > 1$ is the fuzzification parameter, (for $m = 1$ we have the crisp clustering)
$|C|$ and $|D|$ are the number of the desired clusters (this is supplied as input to the FCM), and the number of documents in the collection D respectively;
$\| * \|^2$ is the squared Euclidean distance measure in the N dimensional space of the indexes.

The optimisation of function J is achieved by an iterative process that computes the representations of the $|C|$ clusters at each step by applying the following:

$$c_i = \frac{\sum_{k=1}^{|D|} \mu_{ik}^{m} d_k}{\sum_{k=1}^{|D|} \mu_{ik}^{m} d_k} \quad \text{for} \quad i = 1, \ldots, |C| \tag{2}$$

and successively updates the membership degrees of documents to the clusters by applying the following:

$$\mu_{ik} = \frac{\|d_k - c_i\|^2}{\sum_{i=1}^{|C|} \|d_k - c_i\|^2} \quad \text{for} \quad i = 1, \ldots, |C| \quad \text{and} \quad k = 1, \ldots, |D| \tag{3}$$

The iteration stops at step r when the value of function J does not substantially change, i.e. when $(J(M^r, C^r) - J(M^{r-1}, C^{r-1})) < \varepsilon$ at the iteration step r. The application of the FCM algorithm to the purpose of document categorization in IR suffers from several drawbacks.

One of the problems is relative to the use of the Euclidean distance function to determine the clusters. It is well known that the Euclidean distance is not appropriate in IR, while the cosine similarity coefficient is widely used in IR. For this reason in [19] a modified Fuzzy C-Means (FCM) algorithm for clustering textual documents has been defined which replaces the Euclidean distance with the cosine similarity coefficient. The modified algorithm works with normalised n-dimensional data vectors that lie in hyper-sphere of unit radius and hence has been named Hyper-spherical Fuzzy c-Means (H-FCM). The experiments they made with the H-FCM algorithm for document clustering have shown that it outperforms the original FCM algorithm as well as the hard-Means algorithm [10, 11].

However the H-FCM algorithm requires the selection of the desired number of clusters $c = |C|$. In IR the optimum number c is not known a priori. A typical approach to find the best c is to run the clustering algorithm for a range of c values and then apply validity measures to determine which c leads to the best partition of the data set. The validity of individual clusters is usually evaluated based on their compactness and density [32].

In low-dimensional spaces it is acceptable to assume that valid clusters are compact, dense and well separated from each other. However, text documents are typically represented as high-dimensional sparse vectors. In such a space, the similarity between documents and cluster centroids is generally low and hence, compact clusters are not expected. Furthermore, in the filtering context one cannot expect to have all the clusters with more or less the same density and compactness. Generally when a new story is started to be reported in the news it is expected that the number of news about it grows for a period of time and then decreases when the story becomes obsolete, thus increasing and then decreasing the density of the cluster associated with it.

By taking into account these considerations we have modified the FCM algorithm so as to automatically identify a criterion for the instantiation of c depending on the degree of overlapping of the representations of documents' regarded as fuzzy subsets of terms on the space of indexes. The proposed approach is introduced in Section 3.

Another issue that is very relevant in the IR context is the possibility to identify categories that correspond with different levels of granularity of the topics dealt with in a collection of documents. This means that the higher the number of clusters the more specific will be the topics covered by the documents in those clusters. What is relevant is the possibility to identify a hierarchy between the clusters so as to be able to analyse the categories at different levels of granularity. This is another reason that motivated to development of our modified hierarchical fuzzy C-means algorithm.

2.2 Hierarchical Clustering Method

Hierarchical clustering methods yield a nested sequence of partitions, with a single root cluster at the top and singleton clusters of individual points at the bottom [20]. Each intermediate level can be regarded as combining two clusters from the next lower level (or splitting the cluster of the next higher level). The result of a hierarchical clustering can be depicted as a tree, called a dendrogram. There are two basic approaches to generate a hierarchy of clusters.

The **divisive** approaches start with one cluster including all the documents and, at each step, split a cluster until only singleton clusters, i.e., individual documents remain. In this case it is necessary to decide when to perform the splitting. Methods in this category usually suffer from their inability to perform adjustment once a merge or split has been done. This inflexibility often lowers the clustering accuracy and makes it inappropriate when one has to update the hierarchy with new points, i.e., documents.

The **agglomerative** approaches start with the points (documents in our context) as individual clusters and, at each step merge the most similar or closest clusters. This requires the adoption of a similarity or distance measure. The agglomerative approaches are more common and have been used in information retrieval [25]. The basic algorithm of these approaches is the following:

- Computation of the similarity between all pairs of clusters, i.e., calculate a similarity matrix S whose entry s_{ij} is the degree of similarity between document (or cluster) i and document (or cluster) j.
- Merging of the most similar (closest) clusters.
- Updating of the similarity matrix s to reflect the pairwise similarity between the new clusters
- Repeat steps 2 and 3 until a single cluster remains.

The disadvantage of crisp agglomerative clustering is that the built hierarchy is rigid allowing only the membership of a document to a single cluster. Secondly, since each cluster merges two clusters of the next lower level the generated hierarchy does not reflect the implicit structure of topics and subtopics relationships. Further its complexity is $O(n^2)$ that could get infeasible for very large document sets.

2.3 Incremental Clustering Method

The two approaches seen so far assume that the set of documents to be categorized is static. In many applications, such as document filtering, these approaches are not feasible since they demand a waste of time at every run and this cannot be done efficiently when we have large collections that are updated with a few documents [28, 30].

When the collection is dynamic, incremental clustering algorithms are more appropriate [15, 16]. These algorithms generally generate a flat crisp partition of the objects that come available on the stream, since they work by assigning objects to their respective clusters as they arrive. Problems faced by such algorithms include how to find the best cluster to assign for the next object, how to deal with insertion of new clusters, and how to reassign objects to newly created clusters that were not present when the object was first introduced.

Basically these algorithms process documents sequentially, as they arrive, and compare such document to all existing clusters. If the similarity between the document and any cluster is above a given threshold, then the document is added as a member of the closest cluster; otherwise it forms its own cluster. The incremental approaches differ in the criterion used to compute the similarity between a new element and the existing clusters. The **single-Pass** clustering algorithm works by computing the similarity between a document and a cluster as the average of the similarity degrees between the document vector and the vectors representing the documents in the considered cluster (i.e., the similarity between the cluster centroid vector and the document vector). The **K-nearest neighbour** clustering algorithm computes the similarity of any new document to every other document and chooses the top k documents that exhibit the k-th highest similarity degrees. The new document is then assigned to the cluster containing the majority of the selected k documents.

3 The Rationale of the Proposed Approach

In Information Retrieval, clustering methods have also been applied to support the filtering tasks. Differently than in ad hoc retrieval where generally the collection is static and what varies more frequently is the user query, in the filtering task the collection of documents is dynamic, a continuous stream of documents, and the system does not have to match the document representation against a query but against a user profile representing the topics of potential interest to a given user. Filtering is at the basis of recommender systems, automatic e-mail dispatchers, automatic news wire deliverable.

In this context the application of clustering methods in order to select documents that have some relation with the user profile have to take into account several features specific of the filtering task, mainly the high dynamism of the collection that demands continuous updating; the efficiency of the task that must provide users with new documents as soon as they are available; the possibility to set up flexible selection criteria so as not to miss potentially relevant documents. These characteristics have turned out into the following design specifications:

- The clustering algorithm must produce a **fuzzy partition** of the set of documents so as to reflect the ambiguity of the categorization of the topics dealt with in a document [18]. For example a *news about a film on Albert Einstein reporting his opinion on the research relative to the atomic bomb during the second world war* can be categorized both as related to *scientific literature*, *history* and *entertainment* as well. A fuzzy classification makes it possible to rank the news within any of the above categories so that a user potentially interested in any of them does not run the risk to miss it. This has suggested the application of the fuzzy C means algorithm (FCM).
- It should produce a **fuzzy hierarchy** of clusters so as to represent the subject categories at different levels of granularity, from the most specific ones corresponding with the clusters of the lowest hierarchical level (the deepest level in the tree structure), to the most general ones, corresponding with the clusters of the top level. For example, general news categories can be *scientific news, politics* and there may be overlapping subcategories such as *news about scientists* and *news on science and politics.* Since topics categories do not have crisp boundaries but may overlap one another, the hierarchy should be as flexible as possible, i.e., a fuzzy hierarchy allowing each cluster of a level to belong with distinct degrees to every cluster in the next upper level.
- This has suggested the definition of a fuzzy agglomerative clustering algorithm that is based on the recursive application of the FCM. The algorithm works bottom up in building the levels of the fuzzy hierarchy. Each level corresponds with an allowed degree of overlapping between the clusters in it. This means that at each level the topics represented by the identified clusters can have an overlapping and this overlapping increases as we climb the hierarchy since the topics become more general. The overlapping degree can be intuitively related to the number of common index terms that the clusters of the level can share. Once the centroids of the clusters in a level of the hierarchy have been identified, the FCM algorithm is re-applied to the cluster centroids of that level for identifying the fuzzy clusters of the next upper level. The FCM is applied at each level so as to detect a number of clusters that reflect topics with a given degree of specificity (or granularity), so that, in going up the hierarchy, the granularity decreases, thus identifying more general topics and as a consequence a smaller number of clusters. By proceeding this way the number of the clusters in each upper level reduces with respect to the number of clusters of the lower level. The process ends when a stopping rule is met, i.e., when the number of the clusters automatically computed for a level remains unvaried at the next upper level or a single cluster is built.
- One must be able to update the hierarchy of the clusters as new documents are available (**incremental dynamic clustering algorithm**). This may possibly increase the number of the categories already identified, and thus may require to compute the association of the old documents to the new categories. These operations must be performed through an efficient algo-

rithm so as to be able to run it on-line as new documents come available through the stream. These requirements made necessary to extend the agglomerative fuzzy clustering algorithm so as to be able to run it in update mode thus performing an incremental dynamic clustering. In the incremental modality, as a new document is submitted in input to the clustering module it is compared to every cluster centroid of the lowest hierarchical level. If the similarity degree between the new document and a cluster overcomes the minimum membership degree of the level, the new document is assigned to the cluster. If this condition is not satisfied for any cluster of the level, the new document is instantiated as a new fuzzy cluster of the lowest hierarchical level and every pre-existing document with a similarity with it above the minimal membership degree of the level is assigned to this cluster. Finally the centroids of the clusters must be updated. This process is iteratively applied until the modified centroids reach a stable position.
- Finally, when one wants to influence the grouping so as to reflect as far as possible the interests of all the users of the filtering system some further characteristics that the clustering algorithm should meet have been identified. This feature is particularly useful in collaborative filtering where users having similar interests are grouped and represented by a common profile. These group-profiles are more stable that single user profiles and thus can constitute a starting base for categorizing documents. To this end we have explored the application of **semi-supervised fuzzy clustering techniques** [1, 26]. Semi-supervised clustering techniques consist in clustering data by taking into account available knowledge. Generally semi-supervised clustering techniques exploit a priori classified documents to influence the process of grouping and at the same time to improve the accuracy of unsupervised clustering. The goal is to relate clusters to some predefined classes, in our application the user interests. Our approach tries to influence the fuzzy hierarchical incremental clustering by starting the process not by random points in the term space but by appropriately chosen seeds that are the initial prototypes of the clusters corresponding with the vectors representing the users interests [21]. This choice also copes with the well-known problem of instability of the results produced by the FCM when the seeds are identified randomly.

Besides these main features of the proposed clustering method there are other more specific features that have been considered.

- Since **documents are represented by large and sparse vectors** of weights, in order to be managed in an efficient way, the proposed clustering algorithm has been developed by adopting appropriate methods for dealing with sparse data.
- Since the optimal **number of clusters that must be generated at each level of the hierarchy is not known** in advance, a criterion has been defined to guess its most appropriate value so as to drive the aggregation based on the FCM at each level. This number is identified based either

on a completely automatic process or on an explicit input. In the second case the system administrator can specify the percentage of common terms that index each cluster in a level. This value is interpreted as the minimum overlapping degree that pairs of fuzzy clusters of the next lower level must have in order to be merged into a single cluster of the current level. Clearly, in climbing the hierarchy the allowed minimum percentage of shared terms between fuzzy clusters increases since the objective is to identify clusters corresponding with more general topics. The minimum overlapping degree is used to determine the number of clusters to identify for the level.

In the next section we formally introduce the proposed approach.

4 The Dynamic Fuzzy Hierarchical Clustering Algorithm

First of all let us introduce the notation used to formalize the algorithm.

A document $d_i \in D$ is represented as a vector of weights:

$d_i = (tf_{1i}, \ldots, tf_{Ni})$ in an N dimensional space of indexes $t_1, \ldots t_N \in T$ with T the set of indexes.

4.1 Documents Indexing Criteria for Clustering Purposes

The index term weight tf_{ki} associated with a term-document pair (t_k, d_i) is a numeric value expressing the significance of the index term t_k in synthesising the content of document d_i. It is computed during the indexing process before the application of the clustering algorithm. Some considerations on its possible definition when applying clustering to support filtering are due. In the simplest case tf_{ki} can be automatically computed as the frequency of the term in the document text normalized with respect to the document length. This definition is independent on the term properties of the collections.

On the contrary, the commonly used definition of tf_{ki} based on the traditional $tf * idf$ scheme would need a continuous updating of the inverse document frequency computation as new documents come available in the filtering task. This would greatly degrade the efficiency of the filtering. For this reason, it is generally preferred the adoption of the simple weighting scheme based solely on term frequency tf, motivated by both the need to maintain as far as possible simple and efficient the generation of the documents' vectors and the observation that previous experimentations of crisp clustering of textual documents have shown that its adoption does not substantially penalize the results produced with respect to the adoption of the $tf * idf$ weighting scheme [2, 24].

Further, in the literature it has been shown that the results produced by a crisp clustering of documents does not vary too much in considering 100 or 300 or even greater numbers of significant index terms. These indications suggest to apply indexing criteria that try to reduce as much as possible the cardinality of T [10].

For example, the terms that index one single document of the collection D are not useful to determine a positive similarity between documents as well as the index terms that appear in all the documents of the collection since they do not provide any discrimination capacity.

Based on these considerations, after adopting a common indexing procedure that applies lexicographic analysis (ex. Porter's stemming algorithm), removal of stop words (ex a stop word list from the SMART IR engine *english.stop* from ftp://ftp.cs.cornell.edu/pub/smart/), elimination of terms with ranks above or below the upper and lower cut-offs on the frequency/rank curve (Zipfian curve), a final selection can be applied based on the computation of the discrimination power [22] of the indexes to try to further reduce the document space dimensionality. The term discrimination power is defined as the term's ability in decreasing the density of the term space. This is a topological interpretation and can be computed as:

$$\Delta j = \mathrm{Den} - \mathrm{Den}_j \tag{4}$$

in which:

$$\mathrm{Den}_j = \sum_{i=1}^{N} \sum_{j=1 \wedge i \neq j}^{N} \frac{\mathrm{sim}(d_i, d_j)}{N(N-1)} \tag{5}$$

is the density of the space when adding index term t_j and **Den** is the value of the space density before the extension of the space dimension with the new term. An index term t_j is selected as a good index for clustering if its discrimination power $\Delta j > 0$, otherwise it is rejected . However there are other possible definitions that consider the term discrimination power in relation with the ability of the term to decrease the entropy of the retrieval, i.e., the uncertainty in the computation of the Retrieval Status Value [7].

4.2 Input of the Clustering Algorithm

Once the vectors of weights representing a set of documents have been generated by the indexing procedure they are supplied in input to the clustering algorithm in the form of a matrix I of dimensions $M \times N$ where $M = |D|$ is the cardinality of the starting collection of documents. The value of M can vary from 100 to for instance 10 M documents while we try to keep limited the value of N corresponding with the number of index terms.

Notwithstanding the fact that generally $N \ll M$, the I matrix is sparse since each document is indexed only by a small subset of all the selected indexes.

4.3 The Fuzzy Hierarchy of Documents

A fuzzy cluster of the hierarchical level L is indicated by C_{iL} and has three different representations. First, C_{iL}, with $L > 1$, is represented as a fuzzy subset of clusters of the next lower level $L-1$ as follows:

$$C_{iL} = \{(\mu_{ik}^{L-1}/C_{k(L-1)}), k = 1, N_{(L-1)}\} \tag{6}$$

where $N_{(L-1)}$ is the number of clusters of the level $(L-1)$, and $\mu_{ik}^{L-1} \in [0,1]$ is the membership degree of cluster $C_{k(L-1)}$ to the upper level cluster C_{iL}. Notice that for $L = 1$ C_{k0} is document d_k.

Second, each cluster is represented as a fuzzy subset of documents:

$$C_{iL} = \{(\mu_{iLk}/d_k), k = 1, M\} \tag{7}$$

In the case of the hierarchical level $L = 1$ this representation is directly obtained by the application of the FCM algorithm, and in this case $\mu_{i1k} = \mu_{ik}^0$. For any other level $L > 1$ the membership degree μ_{iLk} of document node d_k to the cluster C_{iL} can be derived based on a recursive and iterative procedure that traverses all the possible paths p existing between document d_k and the cluster $C_{iL} p \in \text{PATHS}(k, iL)$. On each path p let us indicate by $path(p, L, J)$ the set of membership degrees associated with its edges. Then the minimum of these membership degrees μ_{ab}^J associated with each edge between C_{aJ+1} and C_{bJ} with $J = 0, L-1$ of the traversed path p is selected: $\mu_{iLk}^{\text{path}\, p} = \min_{a,b \in \text{path}(p,L,J), J=0,L-1} (\mu_{ab}^J)$ and then the final membership degree μ_{iLk} is computed as the maximum of these values: $\mu_{iLk} = \max_{p \in \text{PATHS}(k,iL)} (\mu_{iLk}^{\text{path}\, p})$

For example, in Figure 1, the membership degree μ_{22k} of d_k to cluster C_{22} is given by the membership degree μ_{3k}^0 associated with the edge between d_k and C_{31} (notice that the thicker the edge the greater is the membership degree associated with it):

$$\mu_{22k} = \max(\min(\mu_{2k}^0, \mu_{22}^1), \min(\mu_{3k}^0, \mu_{23}^1))$$

The rationale of this criterion is to consider each path as a chain made of concatenated rings whose global strength is given by its weakest ring; the strongest chain is defined to represent the strength of the link between a cluster and a document.

Finally, associated with a fuzzy cluster C_{iL} there is a centroid:

$$c_{iL} = (tf_{1iL}, \ldots, tf_{NiL}) \tag{8}$$

in the N dimensional space of indexes $t_1, \ldots t_N \in T$, where $tf_{kiL} \in [0,1]$ is the membership degree of term t_k to the fuzzy cluster C_{iL}. The semantics of tf_{kiL} is to represent the significance of t_k in qualifying the common subjects dealt with in all documents of cluster C_{iL}.

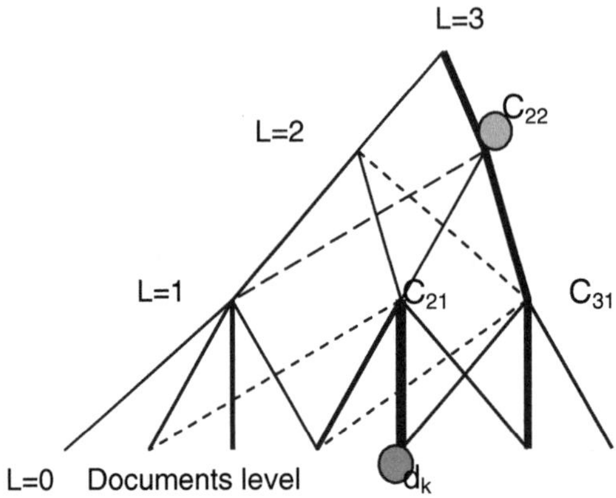

Fig. 1. Representation of a fuzzy hierarchy.

4.4 Generation of the Fuzzy Clusters

To generate the fuzzy clusters in a level L of the hierarchy, the FCM algorithm is applied.

For the first level $L = 1$ the I matrix of the documents vectors constitute the input of the algorithm.

In order to influence the grouping of the documents at level $L = 1$ we start the FCM with suitable seeds corresponding with points in the document space that identify document vectors representing collective user interests. In this way we influence the algorithm to grow fuzzy clusters around interesting regions of the document space and avoid the problem of instability that occurs when starting the FCM with random seeds.

For the levels greater than 1 a matrix I_L of the fuzzy cluster centroids of that level is built in which each row is a vector of weights, i.e. the membership degrees of the index terms to the centroids (see definition (8)).

By considering that previous experimentations on documents clustering have shown that the most appropriate definition of the distance between documents (or centroids) to be used by the FCM in formulae (1) and (3) is the complement of the cosine similarity measure we have modified the classic FCM to compute in formulae (1) and (3) the following distance measure:

$$\|c_{k(L-1)} - c_{iL}\| = 1 - \frac{\sum_{j=1}^{N} tf_{jk(L-1)} * tf_{jiL}}{\sqrt{\sum_{j=1}^{N} tf_{jk(L-1)}^2} * \sqrt{\sum_{j=1}^{N} tf_{jiL}^2}} \tag{9}$$

in which $c_{k(L-1)}$ is document d_k for level $L = 2$.

Other required input parameters are the following:

- the fuzziness parameter m (see formula (1)) indicating the desired fuzziness of the clusters to generate. A value of $m = 1$ generates a crisp partition, a greater value a fuzzy partition. An appropriate value of m must be estimated depending on the collection. For example on the *Open Directory Project* collection it has been set $m = 2$ [19].

- The desired number of clusters at each level N_L. The algorithm has been extended so as to automatically estimate an appropriate value of N_L based on a statistical criterion.
  - The value N_L must be smaller than the number of the clusters of the next lower level $N_L < N_{L-1}$;
  - To determine N_L we first compute the degree of overlapping of each pair of fuzzy clusters $C_{i(L-1)}$ $C_{j(L-1)}$ of the next lower level $L-1$ defined by the fuzzy Jaccard coefficient:

$$\text{overlap}(C_{iL}, C_{jL}) = \frac{\sum_{k=1}^{N_{L-1}} \min(tf_{ki(L-1)}, tf_{kj(L-1)})}{\sum_{k=1}^{N_{L-1}} \max(tf_{ki(L-1)}, tf_{kj(L-1)})} \tag{10}$$

In this definition we consider the clusters represented as fuzzy sets of index terms through formula (8). Second, we build the cumulative **Histogram** curve of the cardinalities of the sets containing the clusters having an overlapping degree greater than s with *at least* k other clusters (11):

$$\text{Histogram}(s) = N_s \quad s \in [0,1] \quad \text{and} \quad N_s := \sum_{i=1}^{N_{L-1}}$$

$$\left[\textit{at least } k \left(\sum_{j \neq i \wedge j=1}^{N_{L-1}} (\text{overlap}(C_{j(L-1)}, C_{i(L-1)}) > s) \right) \right] \tag{11}$$

in which *at least* k is a fuzzy linguistic quantifier defined as a crisp threshold function, i.e., *at least* k $(x) = 1$ if $x \geqslant k$, 0 otherwise [2]. This definition is crisp, we could even use a fuzzy definition that instead of producing a zero degree for the values $x < k$ penalises the result producing decreasing values as x is smaller. By choosing greater k values we want to be more demanding on the number of the clusters of level $L-1$ that must overlap each others in order to be grouped at the next upper level L.

Third we determine the value σ of overlapping corresponding with the greatest variation of the **Histogram** curve trend. Notice that the **Histogram** curve is not increasing with s and the null second derivative identifies the greatest concavity or convexity (see Figure 2 that represents the cumulative Histogram curve for level $L = 1$ on the collection *odp* [19]). σ is considered as a critical overlapping value that corresponds with the best partition of the fuzzy clusters that deal with common subject matters (identified by common set of index terms and that consequently must be merged in a single cluster at the upper level L) with respect to those that are almost disjoint one another and thus deal with different topics. The value of N_L is chosen as:

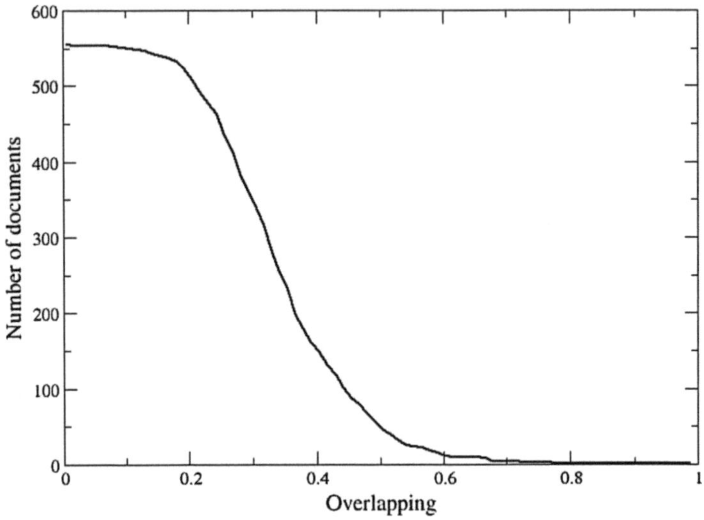

Fig. 2. Histogram curve of the cardinalities of the sets of documents (in the *odp* collection) as a function of their overlapping degree with other 3 documents.

$$N_L = \text{Histogram } (\sigma) \tag{12}$$

N_L is the cardinality of the set of clusters of level $L-1$ that have an overlapping degree greater than σ with *at least* k other clusters. From the preliminary testing we have carried out we can observe that while the automatic computation of σ based on the analysis of the overlapping of the documents works well for determining N_1 it is not so reliable for the upper levels since in these cases the statistical analysis is performed on a limited number of clusters.

Alternatively the value of N_L can be determined based on an input x that specified the minimum percentage of common terms that two clusters of level L must share. The x value is interpreted as the overlapping degree between two clusters to be merged and thus N_L is derived by computing $N_L = \mathbf{Histogram}(x)$.

When $N_L = N_{L-1}$ the fuzzy C means is not reapplied any more and we consider that the final level of the hierarchy has been reached. So, the built hierarchy does not necessarily have a single node as root, but can have several nodes at the top level.

4.5 Updating the Fuzzy Hierarchy with New Documents

In the filtering task we may need to categorized the new documents that are made available through the continuous stream. However, it is not feasible to reapply the entire process to the increased collection that may grow to thousands of documents in a week or two. Specifically, the cost of applying the fuzzy hierarchical clustering algorithm we have described in Subsection 4.4 is mainly due to the processing of the documents for the generation of the clusters of the first level. So we have designed an incremental clustering algorithm that can be applied to update an existing fuzzy hierarchy by possibly adding new clusters as new documents become available.

The input to this phase is a matrix $\mathrm{I}_{\mathrm{update}}$ in which each row is a new document vector and, differently that in the reset mode described in the previous subsection, in update mode the existing hierarchy must be updated with the new set of documents. This is achieved by processing one document vector d_k at a time and by categorizing it with respect to the existing clusters of the first level.

- First of all the similarity of the document vector d_k is computed with respect to the every cluster's centroid of the first level $L = 1$: $\mathrm{sim}(d_k, c_{i1})$ for $i = 1, N_1$, defined as the cosine similarity measure.
- If for at least a cluster this value is above the minimal membership degree of any document to any cluster of the level the document is assigned to the cluster with a membership degree equal to its similarity value, otherwise the document vector is instantiated as a centroid of a new fuzzy cluster of the level and every document in the collection is associated with this cluster if its similarity with respect to the new centroid is above the minimal similarity degree of the level.
- The centroids are then updated by averaging the contributions of their assigned documents. This process is repeated until the centroids do not vary their position consistently.

Once the clusters of the first levels have been updated with all the new documents, for the updating of the next upper levels clusters the fuzzy C means is reapplied by starting with the new clusters centroids of the first level. In this case since the number of the clusters is limited with respect to the number of documents the procedure is efficient.

5 Preliminary Experiments

The hierarchical fuzzy clustering algorithm has been implemented in C++ under windows/xp operating system.

Preliminary testing has been conducted on a collection of 556 documents that represents an extraction of the data provided by the *Open Directory Project* (http://dmoz.org). The documents illustrate the content of web sites and have metadata that classify them into five categories: **sport** (59 documents), **safety** (99 documents), **game** (205 documents), **math** (84 documents), **lego** (187 documents). Some documents are classified in two categories at the same time ("**math** and **game**", and "**sport** and **game**") thus reflecting the ambiguity of their topics.

For representing the documents we have used the 620 indexes provided by the *OPD* together with the documents themselves.

We have run the algorithm several times with the automatic criterion for the estimation of the number of clusters to generate at each level and random vectors for starting the grouping. The critical value of overlapping σ from which we derive the number of documents has been identified on the **Histogram** curve (see Figure 2 for the bottom level) obtained by imposing that each document overlaps *at least 4* other documents.

First of all, we have observed that the algorithm behaviour is greatly dependent on the random vectors that are used. However this problem can be overcome by starting the grouping from predefined seeds, for example by using the vectors representing the user interests. By setting the initial vectors, we have obtained 33 clusters at the bottom level and 13 at the top level. The number of clusters so generated is higher than the number of the a priori categories, thus the unsupervised classification has a finer granularity than the a priori categorization.

We started the analysis of the results by comparing the a priori categorization provided with the data with the fuzzy clusters generated at the bottom level of the hierarchy. To this end, from the fuzzy partition we derived a crisp partition by associating each document to a single fuzzy cluster of the bottom level: a document is instantiated as a full member of the fuzzy cluster to which it has the greatest membership degree. This way we observed that 32 out of the 33 fuzzy clusters remain active with documents associated with them while a cluster remains empty. To figure out this situation we can depict the bottom level fuzzy clusters as either saturated clouds or faint clouds. A plausible interpretation of this fact is that the clustering algorithm performs a classification at a given level of resolution (33 clusters) that is too fine with respect to the main topics dealt with in the collection of documents. Thus the algorithm is forced to identify not just the main topics, corresponding with the active clusters (32), but also secondary topics (1), the faint clusters. So a document can be associated both with main topics (with higher membership degree) and secondary topics (with smaller membership degree). By comparing the a priori categorization and the bottom level classification so generated we measured an average precision of 95% and an average recall of 20%. The low recall value is not surprising since we have many clusters corresponding with the same a priori category, while we can observe that the generated clusters are homogeneous with respect to the categories.

We have also noticed that the cardinalities of the crisp clusters are very different: some clusters have associated only one element while there are big clusters, the biggest one having 148 elements all labelled **lego** class. This preliminary testing, while encouraging on the consistency of the grouping have shown the need to refine the automatic criterion for determining the proper number of clusters to generate at the upper levels.

By forcing the number to 5 clusters and by generating the crisp active clusters as we did for the bottom level we obtained an average precision of

95%. This reveals that the reapplication of the FCM algorithm to generate the new clusters of the next upper level works well by grouping clusters that are homogeneous with respect to their a priori categories.

6 Conclusions

We have proposed a new hierarchical fuzzy clustering algorithm for supporting category-based filtering. This proposal is particularly suited for highly dynamic environments such as newswires filtering.

The novelty of our approach is basically twofold: the generation of a fuzzy hierarchy of clusters reflecting distinct levels of specificity of the topic categories, and the possibility to updating the hierarchy with new documents.

More technical innovations concern the extension of the FCM algorithm so as to automatically determine the proper number of the clusters to generate, and the use of the cosine similarity measure to drive the grouping.

The first testing of the algorithm described in the paper is encouraging and was useful to outline some weak points of the proposal, mainly the need to refine the criterion to automatically estimate the proper number of the clusters to generate at the upper levels of the hierarchy.

Acknowledgements

The present work has been supported by the European STREP project PENG: *PErsonalised News content programminG*, Contract no.: 004597, 01-09-2004.

References

1. Basu S., Banerjee A., Mooney R.J., Semi-supervised Clustering by Seeding, in Proc. 19th Int. Conf. On Machine Learning (ICML-2002). Sydney, 2002.
2. Bordogna G., Pasi G., Personalised Indexing and Retrieval of Heterogeneous Structured Documents, Information Retrieval Journal, 8, 301–318, 2005.
3. Claypool M., Gokhale A., Miranda T., Murnikov P., Netes D., Sartin M., Combining Content-based and Collaborative Filters in an Online Newspaper, in Proc. ACM SIGIR'99 Workshop on Recommender Systems-Implemenation and Evaluation, Berkeley CA, 1999.
4. Connor M., Herlocker J., Clustering for Collaborative Filtering, in Proc. of ACM SIGIR Workshop on Recommender Systems, Berkeley CA, 1999.
5. Cutting D.R., Karger D.R., Pedersen J.O., Tukey J.W., Scatter/Gather: A Cluster-based Approach to Browsing Large Document Collections, in Proc. of 15th Ann In. SIGIR'92., 1992.
6. Debole F., Sebastiani F., Supervised Term Weighting for Automated Text Categorization. In Proc. SAC-03, 18th ACM Symposium on Applied Computing, 2003.

7. Dominich S., Goth J., Kiezer T., Szlavik Z., Entropy-based interpretation of Retrieval Status Value-based Retrieval. Journal of the American Society for Information Science and Technology. John Wiley & Sons, 55(7), 613–627, 2004.
8. Estivill-Castro V., Why so Many Clustering Algorithms: a Position Paper, ACM SIGKDD Explorations Newsletter, 4 (1), 2002.
9. Everitt B.S., Cluster Analysis, 3rd edition. Edward Arnold /Halsted Press, London, 1992.
10. Grossman D.A., Information retrieval, Algorithms and Heuristics, Kluwer Academic Publishers, 1998.
11. Hathaway, R.J., Bezdek, J.C. and Hu Y., Generalized Fuzzy C-Means Clustering Strategies Using Lp Norm Distances, IEEE Transactions on Fuzzy Systems, 8(5), 576–582, 2000.
12. Herrera-Viedma E., Herrera F., Martinez L., Herrera J.C., Lopez A.G., Incorporatine Filtering Techniques in a Fuzzy Linguistic Multi-Agent Model for Information Gathering on the Web, Fuzzy sets and Systems, 148, 61–83, 2004.
13. http://www.newsinessence.com.
14. Jain A.K., Murty M.N., Flynn P.J., Data Clustering: a Review, ACM Computing Surveys, 31(3), 264–323, 1999.
15. Jung, SungYoung, Taek-Soo Kim, An Incremental Similarity Computation Method in Agglomerative Hierarchical Clustering, in Proc. Of the 2nd International Symposium on Advanced Intelligent Systems, Daejeon, Korea, August 25, 2001
16. Khaled M. Hammouda, Mohamed S. Kamel: Incremental Document Clustering Using Cluster Similarity Histograms. 597–601, 2003.
17. Kraft D., Chen J., Martin–Bautista M.J., Vila M.A., Textual Information Retrieval with User Profiles using Fuzzy Clustering and Inferencing, in Intelligent Exploration of the Web, Szczepaniak P., Segovia J., Kacprzyk J., Zadeh L.A., Studies in Fuzziness and Soft Comp. Series, 111, Physica Verlag, 2003.
18. Lin K., Kondadadi Ravikuma, A Similarity-Based Soft Clustering Algorithm for Documents, in Proc. of the 7th International Conference on Database Systems for Advanced Applications, 40–47, 2001.
19. Mendes Rodrigues M.E.S. and Sacks L., A Scalable Hierarchical Fuzzy Clustering Algorithm for Text Mining, in Proc. of the 4th International Conference on Recent Advances in Soft Computing, RASC'2004, 269–274, Nottingham, UK, 2004.
20. Murtagh. F. A Survey of Recent Advances in Hierarchical Clustering Algorithms which Use Cluster Centres. Computer Journal, 26, 354–359, 1984.
21. Pedrycz W., Clustering anf Fuzzy Clustering, chapter 1, in Knowledge-based clustering, J. Wiley and Son, 2005.
22. Salton G., and McGill M.J., Introduction to modern information retrieval. McGraw-Hill Int. Book Co. 1984.
23. Sebastiani F., Text Categorization. In *Text Mining and its Applications*, Alessandro Zanasi (ed.), WIT Press, Southampton, UK, 2005.
24. Sparck Jones, K. A., A Statistical Interpretation of Term Specificity and its Application in Retrieval., Journal of Documentation, 28(1), 11–20, 1972.
25. Steinbach M., Karypis G., Kumar V., A Comparison of Document Clustering Techniques, In Proc. of KDD Workshop on Text Mining, 2000.
26. Tang N., Vemuri V.R., Web-based Knowledge Acquisition to Inpute Missing Values for Classification, in Proc. of the 2004 IEEE/WI/ACM Int. Joint Conf.

On the Web Intelligence and Intelligent Agent Tech. (WI/IAT-2004). Beijing, China, 2004.
27. The Ordered Weighted Averaging Operators: Theory and Applications, R.R. Yager and J. Kacprzyk eds., Kluwer Academic Publishers, 1997.
28. Ungar, L.H., Foster, D.P.: Clustering Methods for Collaborative Filtering. Proceedings of the Workshop on Recommendation Systems, AAAI Press, Menlo Park California, 1998.
29. van Rijsbergen, C. J. Information Retrieval. London, England, Butterworths & Co., Ltd., 1979.
30. Wai-chiu Wong, Ada Wai-chee Fu, Incremental Document Clustering for Web Page Classification, in Proc. 2000 Int. Conf. on Information Society in the 21st Century: Emerging Technologies and New Challenges (IS2000), Aizu-Wakamatsu City, Fukushima, Japan November 5–8, 2000.
31. Walls F., Jin H., Sista S., Schwartz R., Topic detection in Broadcast News, Proc. of the DARPA Broadcast News Workshop, Feb 28–Mar 3, 1999.
32. Xuejian Xiong, Kian Lee Tan, Similarity-driven cluster merging method for unsupervised fuzzy clustering, in Proc. of the 20th ACM International Conference on Uncertainty in artificial intelligence, 611–618, 2004.
33. Zhao Y., Karypis G., Criterion Functions for Document Clustering: Experiments and Analysis. Machine Learning, 2003.
34. Zhao Y., Karypis G., Empirical and Theoretical Comparisons of Selected Criterion functions for document clustering. Machine Learning, 55, 311–331, 2004.

A Theoretical Framework for Web Categorization in Hierarchical Directories using Bayesian Networks

Luis M. de Campos, Juan M. Fernández-Luna, and Juan F. Huete

Departamento de Ciencias de la Computación e Inteligencia Artificial, E.T.S.I. Informática. Universidad de Granada, C.P. 18071, Granada (Spain)
{lci,jmfluna,jhg}@decsai.ugr.es

Summary. In this paper, we shall present a theoretical framework for classifying web pages in a hierarchical directory using the Bayesian Network formalism. In particular, we shall focus on the problem of multi-label text categorization, where a given document can be assigned to any number of categories in the hierarchy. The idea is to explicitly represent the dependence relationships between the different categories in the hierarchy, although adapted to include the category descriptors. Given a new document (web page) to be classified, a Bayesian Network inference process shall be used to compute the probability of each category given the document. The web page is then assigned to those classes with the highest posterior probability.

1 Introduction

Information available on Internet can be accessed by at least two different methods: searching and browsing. When searching, users submit a query to a search engine and obtain a list of web pages which are ranked according to their relevance in terms of the query. When browsing, users navigate through a hierarchy of interlinked concepts or classes until they find relevant documents. A user's information need is less defined in the second method than it is in the first.

This paper shall explore the second method of satisfying information needs. Web pages are organized into a directory or catalog[1] which contains a wide set of categories or classes. These classes are usually arranged hierarchically. Each class contains a group of web pages, the main subject of which is the topic represented by the class to which they have been assigned. Yahoo! [17] is a typical example of a directory which is constructed manually (i.e. by human experts, who are also responsible for web document allocation).

[1] In this paper, we shall use the terms *directory* and *catalog* interchangeably.

L.M. de Campos et al.: *A Theoretical Framework for Web Categorization in Hierarchical Directories using Bayesian Networks*, StudFuzz **197**, 25–43 (2006)
www.springerlink.com

As the number of web pages on Internet is constantly growing, the task of determining a category for each new web document is very slow, one reason why techniques related to text classification [13] are being developed and applied to this framework. The main differences between general text classification and web classification methods is that the first is usually applied to a flat set of categories, while in the second, the categories belong to a well-defined structure. Automatic web categorization by means of a directory therefore implies finding the most appropriate class or category in the hierarchy for a given web page, using a computer program, taking into account the content of that candidate document, in order to insert it into that class.

As mentioned above, directories are hierarchies of categories. These are normally structured according to an inclusion relation, or in other words, an IS-A relation [12], usually forming a tree-like structure. The top-most category (placed in the root node) is supposedly the most general, with an increasing level of specificity as the category approaches the leaf nodes (the most specific ones). There are several organizations of the directory, according to the taxonomy presented in [15]. In the model that we propose in this paper, we shall deal with two types of these[2]: *Virtual Category tree*, where categories are the tree nodes, and documents are assigned to the leaf nodes (the most specific classes); and *Category tree*, where web pages could also be placed in internal nodes.

In this paper, we shall study how to apply the formalism of Bayesian Networks [11] for the classification of web pages in a hierarchical directory. In particular, our objective is to perform *multi-label text categorization* in a *category tree* in a web framework. In order to build the classifier, our inputs are the hierarchical catalog on one side, and a set of documents (each represented as a vector of terms) that have been previously classified under these categories on the other. Figure 1 presents an example of the input data, where for instance $d2$ might be a document about a multimedia presentation of the comedy "The Pink Panther". The hierarchy is usually completed with the inclusion of a new (dummy) general category containing all the classes.

The idea is to construct a Bayesian Network which represents the dependence relationships between the input data (the catalog and the set of pre-classified documents), although adapted to the unique characteristics of the problem. Given a new document (web page) to be classified, the a posteriori probability of relevance for each category is computed using an efficient Bayesian Network inference process. The web page is then assigned to those classes with a higher posterior probability.

Since multi-label categorization might be allowed, the classifier is committed to achieving the greatest possible specificity. If, on the one hand, the document is about how to construct a computer, explaining its components (CPU, hard disks, monitors, etc.), as for instance document $d4$ in Figure 1,

[2] The remaining taxonomy structures present the most general topology of a directed acyclic graph.

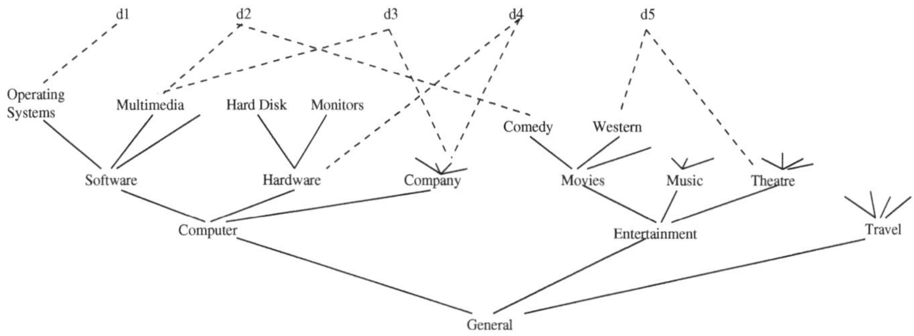

Fig. 1. Documents and hierarchical catalogs.

then it is convenient to classify the document under the label "Hardware" and not to classify it under each different subclass. On the other hand, if a document is about the new release of RedHat Linux, such as for example document $d1$ in Figure 1, then the classifier must select the "Operating System" category. Moreover, and as a general rule that must be fulfilled, a document cannot be classified under two different categories in the same branch of the tree, for example document $d2$ cannot be classified as "comedy" and "movie" simultaneously.

In order to describe the previously mentioned model, this paper is divided into the following sections: Section 2 presents a brief introduction to Bayesian Networks; Section 3 explores related work in hierarchical categorization; Section 4 describes the qualitative and quantitative components of the model; Section 5 shows how the inference can be performed efficiently; and finally, Section 6 includes the paper's conclusions and proposals for future research.

2 Introduction to Bayesian Networks

Bayesian networks are graphical models which are capable of efficiently representing and manipulating n-dimensional probability distributions [11]. A Bayesian network uses two components to codify qualitative and quantitative knowledge:

- A *Directed Acyclic Graph* (DAG), $G = (V, E)$, where the nodes in V represent the random variables from the problem we want to solve, and the topology of the graph (the arcs in E) encodes conditional (in)dependence relationships between the variables (by means of the presence or absence of direct connections between pairs of variables);
- A set of conditional probability distributions drawn from the graph structure: for each variable $X_i \in V$, we have a family of conditional probability distributions $P(X_i|pa(X_i))$, where $pa(X_i)$ represents any combination of the values of the variables in $Pa(X_i)$, and $Pa(X_i)$ is the parent set of X_i in G.

From these conditional distributions, we can recover the joint distribution over V:

$$P(X_1, X_2, \ldots, X_n) = \prod_{i=1}^{n} P(X_i|pa(X_i)) \tag{1}$$

This decomposition of the joint distribution results in important savings in storage requirements. It also allows probabilistic inference (propagation) to be performed in many cases, i.e. the posterior probability to be computed for any variable given some evidence about the values of other variables in the graph [7, 11]. The dependence/independence relationships enabling this decomposition are graphically encoded (through the d-separation criterion [11]) by means of the presence or absence of direct connections between pairs of variables. Bayesian networks can efficiently perform reasoning tasks: the independences represented in the graph reduce changes in the state of knowledge to local computations.

3 Related Work on Hierarchical Categorization

In [13], there is a very thorough review of the different methods related to general text classification. In this section, however, we shall focus our attention on certain papers related to hierarchical classification.

A first attempt to solve the problem was to ignore the structure and to consider the problem as a flat set of independent categories. However, the common philosophy found in literature when the hierarchical structure of the classes is taken into account is that different classifiers are learnt which deal with smaller problems, and these distinguish first at higher levels, and later in more specific classes.

This methodology is followed by Koller and Sahami [9], who take advantage of the hierarchical nature of the problem, dividing the original classification problem into smaller problems, which could only be solved with a small set of features, and allowing more complex Bayesian classifiers for such a task. More precisely, they build a hierarchy of classifiers which are applied successively.

Mladenić [10] put a similar idea into practice in order to use Yahoo! to classify documents. She used the structure to carry out a feature selection process as well as to learn a set of independent Naive Bayesian classifiers.

Another example can be found in [20], where Weigend et al. use an approach based on a two-level classification scheme on Reuters-22173 dataset: the first stage attempts to classify the new document into main categories (called "meta-topics") which group all the classes comprising the hierarchy. Once the meta-topic has been decided, a second stage attempts to assign a more specific category. In both processes, neural networks are employed.

Support Vector Machines have also been applied to the hierarchical classification of web content, as in [4]. These only consider top levels of the hierarchy, and base their work on learning some classifiers and combining the resulting scores. A second paper where Support Vector Machines are applied is [15].

Another interesting paper was written by Frommholz [5]. This researcher uses a completely different approach in that he uses a non-hierarchical classifier at a first stage. A hierarchical classifier is later used to improve the results obtained by the first, using the classes closest to the class that was selected in the first attempt. Another interesting innovation of his model is that it can assign documents to internal classes.

Finally, [12] is another example that puts this idea into practice, but, in this case, with neural networks.

To sum up, we can say that in all these approaches, classification in performed in a top-down manner. This fact allows the models to scale well with the number of characteristics used [18]. One of the problems of this approach appears when a document is wrongly classified at high levels and cannot be passed to the classifiers at lower levels. Sun et al. [16] introduce the blocking factor measure as a new kind of classifier-centric performance measure to tackle this problem but involving more classifiers and with some degradation in precision.

4 Representing Hierarchical Web Directories using a Bayesian Network

In this section, we shall describe the Bayesian network classifier. In order to do so, we shall proceed incrementally with a step-by-step explanation of the topology of a particular component of the classifier, i.e. we shall identify the set of variables[3] and how they are related. Finally, when the topology has been completed, we shall describe how the quantitative values (probability distributions) might be assessed.

- *Hierarchical catalog:* In this case, each category in the catalog will be considered a random variable and therefore represented by a node in the Bayesian Network. For convenience in terms of inference, as we shall explain later, categories are divided into two groups: *basic categories*, which are the most specific and do not contain other classes: $\mathcal{C}_b = \{B_1, B_2, \ldots, B_m\}$; and *complex categories*, those that are composed of other basic or complex categories: $\mathcal{C}_c = \{S_1, S_2, \ldots, S_n\}$. Therefore, $\mathcal{C} = \mathcal{C}_b \cup \mathcal{C}_c$. Each node B_j or S_k has an associated binary random variable, which can take its values from the sets $\{b^-, b^+\}$ or $\{s^-, s^+\}$ (the category is not related or is related to a given document, respectively, in the sense that its description matches the document content). In this paper, we shall use C_i to represent a general category (basic or complex) and $\{c^-, c^+\}$ to denote the particular values it takes.

[3] In this paper, the random variable and its associated node in the graph will be noted identically.

Regarding the topology of the model, there is an arc from one category to the particular category it belongs to, expressing the fact that the relevance of a given category will depend on the relevance values of the different elements that comprise it. With this criterion, we shall obtain a tree-like topology, more specifically a polytree, since a node can have more than one parent. Figure 2 represents the Bayesian Network associated with the catalog.

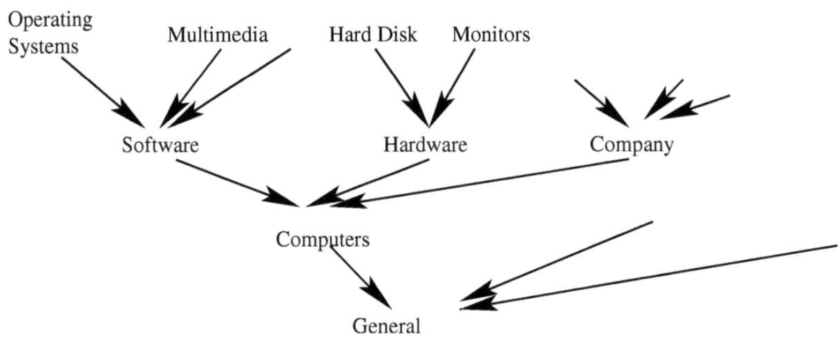

Fig. 2. Bayesian Network based catalog.

- *Relating documents and categories:* For each document D_j, we know the set of basic or complex categories where it has been pre-classified, for instance by a human expert. We have two different problems: the first consists in determining which characteristics of those documents relating to category C_i will be used for categorization purposes, and the second consists in modeling how these characteristics should be related to the class. From a Bayesian network perspective, these problems are the selection of the set of variables of interest and the construction of the network topology.
  – Characteristic Selection: The set of characteristics that will be considered relevant to C_i is a subset of the set of terms contained in those documents pre-classified under C_i. We use $\mathcal{T}_j$ to denote the subset of terms summarizing the content of a document D_j. These terms are obtained after "stop word" elimination and the stemming processes. In order to reduce the high dimensionality of the term space, a term selection process should also be applied, for instance by considering only the terms with the highest discrimination power or by using methods based on Information-Theoretical functions [19].

  We have also decided to consider the proper set of documents as relevant variables. We think that this approach might be useful in web domains where the set of documents is highly (inter)connected by means of links. For instance, from this information we might determine the quality of a document, in the sense that it might give a better description of the class than others.

The Bayesian Network must therefore include variables representing the set of pre-classified documents, $\mathcal{D}$, and the set of terms obtained after dimensionality reduction, $\mathcal{T}$. The first will be represented by the set $\mathcal{D} = \{D_1, D_2, \ldots, D_n\}$, while the second will be denoted as $\mathcal{T} = \{T_1, T_2, \ldots, T_l\}$. As before, each node represents a bivaluated variable taking its values from the sets $\{d^-, d^+\}$ and $\{t^-, t^+\}$(the document/term is not related or is related to the document), respectively.

– Relating characteristics and classes: In order to represent the relationships between terms and documents, we shall include an arc from each term T_i to those documents that have been indexed with T_i. A similar approach could have been used to relate documents and class, i.e. to include an arc between a given document and all the categories which it belongs to. In this case, we would be expressing the fact that the relevance of a category will depend on the relevance values of the different documents classified under it. For basic categories, this criterion seems to be coherent but considering the previous Bayesian network representing the catalog, complex categories should be related to both documents and categories. In order to give a coherent semantics to complex category nodes[4], we therefore propose that a virtual node, S'_k, be included and which represents the complex category S_k, gathering the information supported by those documents preclassified under category S_k. It should be noted that virtual nodes will act as basic categories, and in terms of notation, the set of basic categories is therefore extended in the following way: $\mathcal{C}_b = \{B_1, B_2, \ldots, B_f, S'_1, \ldots S'_n\}$, where $S'_1, \ldots S'_n$ represent the virtual nodes.

Finally, an arc from a virtual category to the particular complex category it represents will be added, and this expresses the fact that the relevance of the complex category will also depend on the relevance values of the documents belonging to this category. Figure 3 represents the topology of the classifier in this step, where virtual category nodes are represented with filled boxes and their relationships with the associated complex category nodes are represented with dashed lines.

It should be noted that in the Bayesian network, root nodes are terms and the leaf node represents the general category. It is also worth mentioning that the hierarchical structure of the model determines that each category $C \in \mathcal{C}$ has only one category as its child, the single category containing C (except for the most general category, which has no child). We shall use $Ch(C)$ to denote the single child node associated with node C (with $Ch(C) = null$ if C is the general node).

Once the model has been completed, regardless of the assessment of the probability values (see Section 4.2), it can be used to perform classification tasks. Thus, given a new unseen document, D, we must be able to compute the posterior probabilities of relevance of all the categories $C \in \mathcal{C}$, $p(c^+|D)$,

[4] This fact will be essential when assessing the probability distributions.

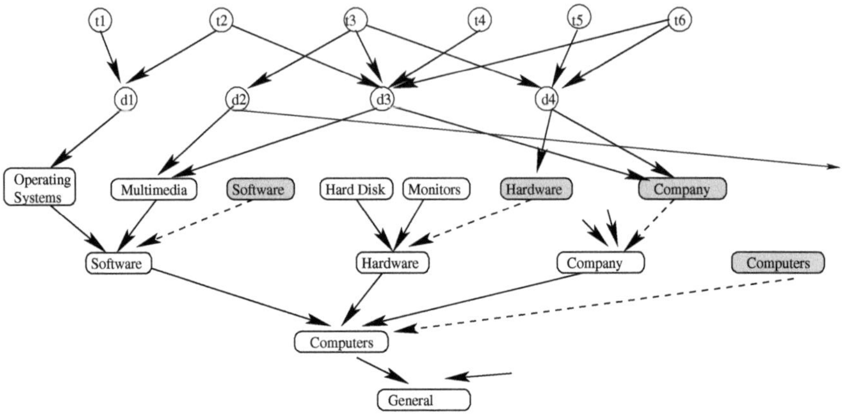

Fig. 3. Bayesian Network based classifier.

representing our belief that D can be categorized under C. In order to compute these probability values, we must instantiate all the terms indexing the document to the related value, t^+, and propagate this information through the network (in Section 5 we show how this process can be performed efficiently). Once we have computed all the probabilities $p(c^+|D)$, the document should then be classified under the category with the maximum value in the case of single label categorization or those categories with higher posterior probability values in the case of multi-label categorization, checking that the document has not been classified under two different categories in the same branch of the catalog. It should be noted that for the set of categories, the probabilities $p(c^+|D)$ might be considered as a kind of measure of vector closeness in the term space.

Perhaps the main advantage of this proposal with respect to classical methods (see Section 3) is that in our case there is no need for a training stage (since the topology is fixed, learning the classifier comes down to computing the probability distributions). This advantage is all the more important when we consider the Web categorization task because the sets $\mathcal{D}$ and $\mathcal{C}$ might not be completely available from the start. New pre-classified documents can therefore be incorporated into the system and new categories can be added and obsolete ones deleted easily. Moreover, our model scales well with the dimensionality of the term space and the dimensionality of the catalog. Another important feature of our model is that we can naturally perform multi-label categorization, explicitly exploiting the hierarchical structure of the category, whereas those methods using a divide and conquer-type approach (which decompose the classification problem into a number of smaller classification problems) have mainly been designed for single label categorization since they select the best branch of the hierarchy at each step.

4.1 Improving the Basic Model

In order to motivate our approach, let us only consider the set $\mathcal{C}_b$ that includes basic categories and virtual nodes (as a representation of the complex categories). In this case, by using $p(c_i^+|D), \forall C_i \in \mathcal{C}_b$, we can perform a flat (non-hierarchical) classification measuring the closeness of document D to the centroid of the positive documents pre-classified under C_i.

A typical error made by flat classifiers is that a large proportion of the false positives (incorrect categories with high probabilities) were on topics which are semantically related to the proper one (mainly because the vocabulary they use is connected, but only incidentally, with the topic). For example, while the term "LINUX" is a good predictor of the category "Operating System", it also tends to appear in documents relating to "Multimedia", for instance considering how to set up a particular application in this operating system. The problem worsens when we consider virtual nodes (representing complex classes), where they should have many terms in common with the subclasses which they comprise. The key is that these terms will rarely appear in documents relating to other branches of the catalog. For instance, it will be difficult to find the term "LINUX" in connection with the entertainment category. Although this discussion has only been made for basic categories, it can easily be extended to the remaining categories in the hierarchy.

In this section, we shall therefore present an approach to tackle this problem, and which basically consists in also taking the information supported by the set of *related categories* into account when making decisions. This idea has also been considered in different text classification approaches such as Schapire et al. [14] for flat categorization and Weingend et al. [20] for hierarchical categorization. In our case, and considering the catalog hierarchy, the set of categories related to C_i will be defined as those sibling categories in the catalog plus the category it belongs to. This definition is equivalent to considering those categories represented by nodes with the same child in the Bayesian network. At this point, we must note that the information supported by the set of related categories of C_i is gathered by the complex category it belongs to, $Ch(C_i)$. It therefore seems natural that the final decision will depend on both the relevance of C_i and the relevance of $Ch(C_i)$.

In order to model this fact in the Bayesian Network, a new set of variables, $\mathcal{A}$, will be included representing the *adequacy* of a category to the new document. In particular, for each category variable C_i (excluding the general class and the set of virtual nodes), a new adequacy variable A_i is added, i.e. $\mathcal{A} = \{A_1, A_2, \ldots, A_{m+n}\}$ with $A_i \in \{a_i^-, a_i^+\}$ representing the fact that the i-th category is or is not appropriate for the document. The model will be completed by adding two new arcs for each adequacy node: the first arc connects node C_i with A_i, and the second connects node $Ch(C_i)$ also with A_i, expressing the fact that the adequacy of a category to the document will depend on the relevance of C_i and $Ch(C_i)$. Section 4.2 discusses how we can measure the strength of these relationships.

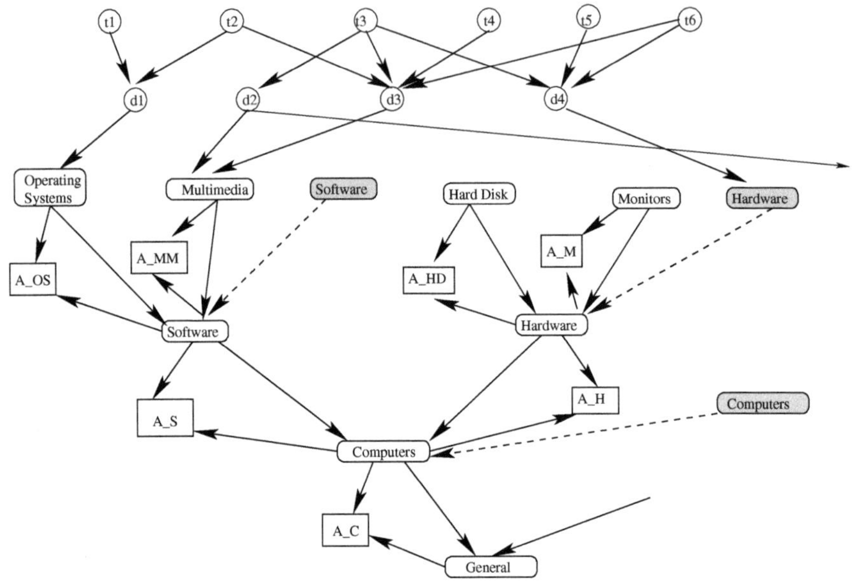

Fig. 4. The final hierarchical classifier.

Figure 4 presents a part of the final Bayesian network, obtained from the one displayed in Figure 3. The adequacy nodes have been represented with squares. For example, the node label with A_OS represents the adequacy of the "Operating Systems" category.

4.2 Assessment of the Probability Distributions

In this section, we shall present some guidelines about how to assess the quantitative information. For each node X_i, we need to estimate the conditional probabilities $p(x_i^+|pa(X_i))$ with $pa(X_i)$ being a given configuration of the values of the set of parents of X_i, $Pa(X_i)$, in the graph. In general, the estimation of these probability values is not an easy problem because the number of possible configurations $pa(X_i)$ (any assignment of values to all the variables in $Pa(X_i)$), and therefore the number of conditional probabilities that we need to estimate and store, grows exponentially with the number of parents in X_i. For example, if X_i has 20 parents (and this may be a common situation in our model), we need 2^{20} (around one million) probability distributions, hence we cannot use a standard approach. Since computing and storing the conditional probabilities becomes prohibitive, we propose an approach that has been successfully used by the BNR model in the field of Information Retrieval [2][5] which is based on the use of canonical models of multicausal interaction [11].

[5] The subgraph containing only terms and document nodes is the basis for the BNR model [2].

We shall therefore discuss the possible alternatives for each type of node:

- *Term nodes*: Because all the terms are root nodes, marginal distributions need to be stored. We propose that identical probability be used for all the terms, $p(t^+) = p_0$, $\forall T \in \mathcal{T}$:

$$p(t_i^+) = p_0 \text{ and } p(t_i^-) = 1 - p_0 \tag{2}$$

- *Document nodes*: In order to estimate the conditional probabilities of relevance of a document D_j, $p(d_j^+|pa(D_j))$, we shall consider the canonical model used in [2], i.e.

$$p(d_j^+|pa(D_j)) = \sum_{\substack{T_i \in D_j \\ t_i^+ \in pa(D_j)}} w(T_i, D_j) \tag{3}$$

where $w(T_i, D_j)$ represents the weight of term T_i in document D_j and the expression $t_i^+ \in pa(D_j)$ in eq. (3) means that only the weights w_{ij} are included in the sum such that the value assigned to the corresponding term T_i in the configuration $pa(D_j)$ is t_i^+. Therefore, the greater the number of relevant terms in $pa(D_j)$, the greater the probability of relevance of D_j. In addition, the weights $w(T_i, D_j)$ verify that $0 \leq w(T_i, D_j)\ \forall i, j$ and $\sum_{T_i \in D_j} w(T_i, D_j) \leq 1\ \forall j$ and are computed by means of the following expression:

$$w(T_i, D_j) = \frac{tf_{ij} idf_i^2}{\sum_{T_k \in D_j} tf_{kj} idf_k^2} \tag{4}$$

with tf_{ij} being the frequency of the term in the document and idf_i the inverse document frequency of the term T_i in the document training set. Different alternatives for computing these weights might be the *cosine formula*, probabilistic techniques [6], etc.

- *Basic Categories:* Similar to the previous case, the set of basic categories (including virtual nodes) would have a large number of parents, one for each document that has been pre-classified under the category. It would therefore be convenient to use a canonical model similar to the one used for document nodes, i.e.

$$p(b_j^+|pa(B_j)) = \sum_{\substack{D_i \in B_j \\ d_i^+ \in pa(B_j)}} w(D_i, B_j) \tag{5}$$

Different approaches might be considered to define the support that each document gives to the class, i.e. the link weights $w(D_i, B_j)$. For instance, we can consider that all the documents equally support the class, or that these weights depend on the set of terms indexing the document. A different approach which is suitable for web categorization is to explore the definition of these weights considering some measure of the quality of web pages that can be obtained using the links pointing to these pages, and this is similar to HITS [8] or PageRank [1] algorithms.

- *Complex Categories:* In the case of a complex category, S_i, the situation is different since we can distinguish between the information supported by its associated virtual node S', i.e. the information that comes from those documents that have been explicitly pre-classified under the category, and the information provided by those (sub)categories included in it, $Pa(S_j) \setminus S'_j$. Although an approach similar to the previous one can be used, it seems natural to consider that the relevance of the virtual node has a different strength, and therefore a different treatment, than the remaining parents in the category.

  We therefore propose that $p(s_j^+|pa(S_j))$ be computed using the following convex combination where the parameter $\alpha_j, 0 \leq \alpha_j \leq 1$, is used to discriminate between the strength of the relevance supported by its virtual node, S'_j, and the strength of the relevance degree supported by its proper subclasses. For instance, the value α_j might depend on the number of documents pre-classified under $Pa(S_j) \setminus S'_j$. If there is a large number of such documents, we can therefore consider that the (sub)classes are well-informed and α_j should be lower than if we do not have enough documents classified into $Pa(S_j) \setminus S'_j$.

$$p(s_j^+|pa(S_j)) = \begin{cases} \frac{1-\alpha_j}{Z_j-1} \sum_{\substack{C_i \in Pa(S_j) \setminus S'_j \\ c_i^+ \in pa(S_j)}} w(C_i, S_j) & \text{if } s_j'^+ \notin pa(S_j) \\ \\ \alpha_j + \frac{1-\alpha_j}{Z_j-1} \sum_{\substack{C_i \in Pa(S_j) \setminus S'_j \\ c_i^+ \in pa(S_j)}} w(C_i, S_j) & \text{if } s_j'^+ \in pa(S_j) \end{cases} \tag{6}$$

  where $Z_j = \sum_{C_k \in Pa(S_j)} w(C_k, S_j)$. In addition, $w(C_i, S_j)$ with $C_i \in Pa(S_j)$ represents the weight of the parent category C_i in S_j. We assume that $w(S'_j, S_j) = 1$. Different alternatives should be considered to estimate the weights $w(C_i, S_j)$ with $C_i \in Pa(S_j) \setminus S'_j$, for instance to consider that all the (sub)categories are equally probable (uniform distribution) or to consider that these weights depend on the ratio of the number of terms (final features) describing the (sub)class C_i with respect to the number of terms describing the class S_j.

- *Adequacy Nodes:*
  The probability distributions stored in these nodes will be used to measure the strength of the information supported by the categories related with a node C_i in that the more relevant $Ch(C_i)$ is, the less adequate C_i is. This estimation is simple since an adequacy node A_i has only two parents: category C_i and its child $Ch(C_i)$. For instance, if $C_j = Ch(C_i)$, these distributions should be computed using

$$\begin{aligned} p(a_i^+|c_i^+, c_j^+) &= \beta_i \quad \text{and} \quad p(a_i^+|c_i^+, c_j^-) = 1 \\ p(a_i^+|c_i^-, c_j^+) &= 0 \quad \text{and} \quad p(a_i^+|c_i^-, c_j^-) = 0 \end{aligned} \tag{7}$$

where β_i could be a value between 0 and 1 in such a way that the smaller β_i is, the greater importance we are giving to the nodes connected with C_i, i.e. $Ch(C_i)$.

5 Categorizing Web Pages: Inference

In this section, we shall examine how web page categorization is performed in the Bayesian network classifier introduced in the previous section. Our objective is therefore to find a category that best represents the content of the new document (in the case of single label categorization) or several categories (in the case of multi-classification). The best class or classes are selected according to the highest posterior probabilities of adequacy of each category given the content of the web page.

Classification is initially carried out by instantiating the terms belonging to the new document in the network to the relevant value, acting as evidences, i.e. $p(t^+) = 1$. From that point, a propagation process is performed on the entire network. On account of the size of the network (in particular, the high number of documents and term nodes) and its topology, the application of classic propagation algorithms could be a very time-consuming task.

Taking advantage of the canonical model to estimate the probability distributions in documents and basic and complex category nodes, the propagation process could be reduced to the evaluation of probability functions in documents and basic and complex category nodes. This inference method has been widely and successfully applied in the Information Retrieval framework with the Bayesian Network Retrieval and the Context-based Influence Diagram models [2, 3]. Moreover, this propagation has been proved to be exact, i.e. it computes the same values as a classic algorithm [2] but very efficiently. Finally, and in order to compute the posterior probability of the adequacy of a class given a document, a simple computation of Bayes' rule is only performed in some of the nodes belonging to the set $\mathcal{A}$.

Therefore, given the new unseen document $D = \{T_a, T_b, \ldots, T_k\}$, whose terms act as evidences, i.e. $E = D = \{T_a, T_b, \ldots, T_k\}$, the classification is carried out in the following stages:

1. Calculate the relevance probability of document nodes given the evidences, i.e. $p(d^+ \mid E)$:

$$\forall D_j \in \mathcal{D},\ p(d_j^+|E) = \sum_{T_i \in Pa(D_j)\setminus E} w(T_i, D_j)\, p(t_i^+) + \sum_{T_i \in Pa(D_j)\cap E} w(T_i, D_j). \tag{8}$$

2. Obtain the posterior probability of each basic category, B, or virtual node, S', i.e. $p(b^+ \mid E)$ or $p(s'^+ \mid E)$, respectively. In this case, we need to evaluate the following equation:

$$\forall C_j \in \mathcal{C}_b,\;\; p(c_j^+|E) = \sum_{D_i \in Pa(C_j)} w(D_i, C_j)\, p(d_i^+|E). \tag{9}$$

3. Compute the posterior probability of each complex category node, S, i.e. $p(s^+ \mid q)$ is performed evaluating the following formula:

$$\forall S_j \in \mathcal{C}_c, p(s_j^+|E) = \alpha_j p(s_j'^+ \mid E) + \frac{(1-\alpha_j)}{Z_j - 1} \sum_{C_i \in Pa(S_j)\setminus S_j'} w(C_i, S_j)\, p(c_i^+|E), \tag{10}$$

with S_j' being the associated virtual node corresponding to the S_j complex category, and $\alpha_j \in [0,1]$ a parameter that helps us to give more or less importance to the content of the internal category S_j.
4. Finally, in the penultimate stage, the probability of each adequacy node, A, must be computed, i.e. $p(a^+ \mid E)$. In this case, if $C_j = Ch(C_i)$, these values are obtained simply by solving this equation:

$$\forall A_i \in \mathcal{A},\;\; P(a_i^+ \mid E) = \sum_{\mathbf{c_i},\mathbf{c_j}} p(a_i^+ \mid \mathbf{c_i}, \mathbf{c_j}) p(\mathbf{c_i}, \mathbf{c_j} \mid E), \tag{11}$$

where $\mathbf{c_i}$ and $\mathbf{c_j}$ represent the two values that C_i and C_j can take, respectively. In order to put this model into practice, it is therefore necessary to assess the bi-dimensional posterior probabilities corresponding to each category C_i and the category C_j where it is contained, $p(C_i, C_j|E)$. Obtaining these values may be a time-consuming process due to the large amount of calculations required on classification time. This is why we propose the use of a first approximation assuming that both categories are independent given the evidence, i.e.

$$p(C_i, C_j|E) = p(C_i|E)\, p(C_j|E)\,. \tag{12}$$

We therefore only need to compute the adequacy values for each category given a new document E, using equations (9) and (10). The final expression that enables this value to be calculated is given by the following equation:

$$\begin{aligned}\forall A_i \in \mathcal{A},\;\; P(a_i^+ \mid E) &= \sum_{\mathbf{c_i},\mathbf{c_j}} p(a_i^+ \mid \mathbf{c_i}, \mathbf{c_j}) p(\mathbf{c_i} \mid E) p(\mathbf{c_j} \mid E) \\ &= \beta_i p(c_i^+ \mid E) p(c_j^+ \mid E) + p(c_i^+ \mid E)(1 - p(c_j^+ \mid E)) \\ &= p(c_i^+ \mid E)[1 - (1-\beta_i) p(c_j^+ \mid E)]. \end{aligned} \tag{13}$$

5. Select the classes with the highest posterior probability. As mentioned before, if only a single category is assigned, the one with the highest posterior probability is selected. In case of multi-label classification, we could opt to select the k first categories or those with a posterior probability greater than a threshold, ensuring that two selected classes do not belong to the same branch.

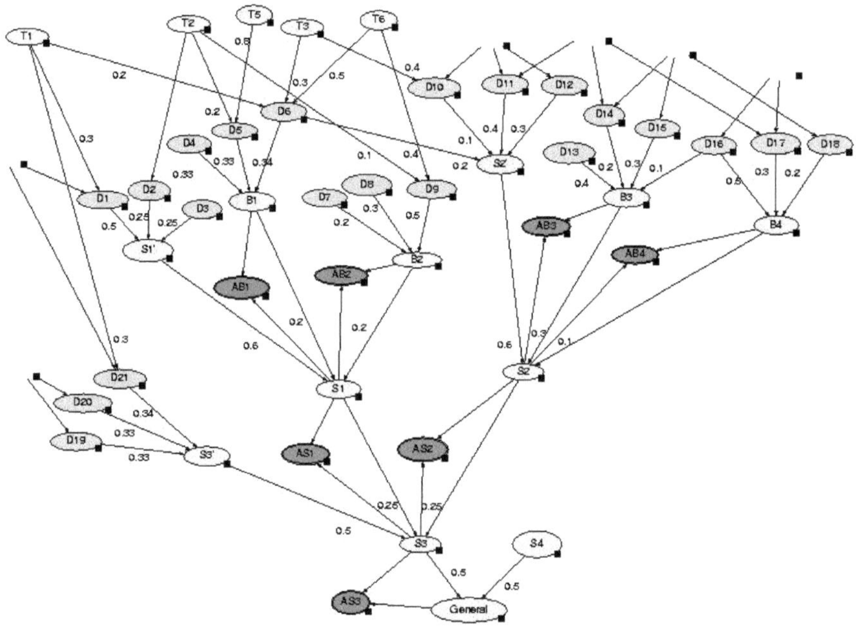

Fig. 5. Example of a Hierarchical Bayesian Network-based Categorization Model.

All the computations of the posterior probabilities in Steps 1 to 3 can be performed very efficiently using a single traversal of the graph, considering that all the nodes in $\mathcal{A}$ have been removed, as well as the arcs pointing to them, starting only from the instantiated terms in E, provided that the prior probabilities of relevance have been calculated and stored within the structure. An algorithm that computes all the posterior probabilities $p(d^+|E)$, $p(b^+|E)$, $p(s'^+|E)$ and $p(s^+|E)$, starts from the terms in E and carries out a width graph traversal until it reaches the documents that require updating, computing $p(d^+|E)$ using eq. (8). Starting from these nodes, it carries out a depth graph traversal to compute $p(c^+|E)$, only for those basic and complex units that require updating, using eq. (9) and (10).

Example:

In order to illustrate the behavior of the proposed model, let us consider the example in Figure 5. So as to clarify the figure, some term nodes have been omitted[6], although they have been used for inference tasks. In order to set

[6] We think that in order to show how the evidences are related to the different categories, it is enough to consider the a posteriori probability of relevance for the set of pre-classified documents. This is why most term nodes are omitted in Figure 5.

the quantitative values, we use the scheme proposed in Subsection 4.2, where the used weights are displayed near the arcs. The prior probabilities of all the terms have been set to 0.5, the values α in Equation 6 have been set as $\alpha_{S1} = \alpha_{S2} = 0.6$ and $\alpha_{S3} = 0.5$. Finally, the values β in Equation 7 have been set to: $\beta_{ABi} = 0.75, i \in \{1,2,3,4\}$; $\beta_{AS1} = \beta_{AS2} = 0.8$ and $\beta_{AS3} = 0.85$. It should be noted that with this criterion, the strength given to those categories relating to a general class is lower than the strength given to those categories relating to the more specific ones. This performance seems to be reasonable since, given a hierarchy, the relationships between categories at the bottom (general classes) might be more spurious than at the top (specific classes).

Firstly, it is worth remembering that when there is no evidence, the probability for all the nodes is 0.5, except for adequacy nodes taking the values $ABi = 0.44, i \in \{1,2,3,4\}$, $AS1 = AS2 = 0.45$ and $AS3 = 0.46$. These values might be used as reference in the next discussion. It should also be noted that, a priori, a document tends to be classified under the most general category.

In the example, we shall classify two new documents, illustrating different situations that should appear. The different rows of Table 1 show the relevance probability for those pre-classified documents, basic categories or virtual nodes, complex categories and adequacy nodes that require updating. These values have been obtained by propagating the evidence through the network using the above equations. It should be noted that by showing these probability values, we can also discuss the results obtained by a flat classifier (considering only nodes in $\mathcal{C}_b$) and the basic hierarchical model (without the improvements discussed in Section 4.1, i.e. without adequacy nodes and considering the probabilities of simple and complex categories). In order to perform the final classification, we shall consider the three most relevant categories.

Let us assume that the first unseen document to be classified, i.e. the evidence, is $E_1 = \{T3, T5, T6\}$. In this case, $E1$ is only related to documents $D4, D6, D9$ and $D10$ that have been pre-classified under $B1$, $B2$ and $S2$ categories. In this case, considering the adequacy, the new document will be classified (in order) under $B1$, $B2$ and $S2$. It should be noted that the class $S1$, with an adequacy greater than $S2$, has not been selected because some of its (sub)classes were previously assessed. The same results will also be obtained for flat and basic hierarchical classification.

Finally, we use a second document, E_2, with 16 terms being considered. Looking at row E_2 in Table 1, it can be seen that this new set of terms provide strong evidences to half the set of pre-classified documents, being related to all the categories. In this case, considering the adequacy values, the document will only be classified under $S1$ and $S2$. It should be noted that the document should not be classified under $S3$ because it has been previously classified under $S1$. Although the model selects two different categories on the same level of the hierarchy, we can distinguish two different situations. On one hand, $E2$ is closely related to the categories $B1$ and $B2$ ($P(B1|E2) = 0.81$ and $P(B2|E2) = 0.82$) but it is also similarly related to

Table 1. $p(\cdot|E_i)$ for the updated nodes

E_1	$D5 = D6 = 0.9; D9 = 0.95; D10 = 0.7$
	$B1 = 0.77; B2 = 0.73; S2' = 0.6$
	$S1 = 0.60; S2 = 0.56; S3 = 0.54$
	$A_{B1} = 0.68; A_{B2} = 0.64; A_{B3} = A_{B4} = 0.44; A_{S1} = 0.53; A_{S2} = 0.50; A_{S3} = 0.48$
E_2	$D1 = 1; D3 = 0.6; D4 = 0.6; D5 = 0.9; D6 = 0.93; D7 = 0.85;$
	$D9 = 1; D10 = 0.7; D11 = 0.8; D16 = 0.65; D21 = 0.79$
	$B1 = 0.81; B2 = 0.82; B3 = 0.52; B4 = 0.58; S1' = 0.78; S2' = 0.73; S3' = 0.60$
	$S1 = 0.79; S2 = 0.64; S3 = 0.66$
	$A_{B1} = 0.65; A_{B2} = 0.66; A_{B3} = 0.44; A_{B4} = 0.49; A_{S1} = 0.68; A_{S2} = 0.56; A_{S3} = 0.59$

the category $S1$ ($P(S1'|E2) = 0.78$). The model takes this fact into account and selects the category $S1$. On the other hand, $S2$ has been selected because $E2$ is strongly related to $S2$ ($P(S2'|E2) = 0.73$) but slightly related with the subcategories $B3$ and $B4$. It is interesting to note that although $S1'$ and $S2'$ are quite relevant, the lower adequacy of $S3$ can be explained by considering the strength of $S3'$ in $S3$ and that there is not enough evidence for those documents pre-classified under $S3$ ($P(S3'|E2) = 0.6$). Finally, and regarding the categorization obtained by the flat and the basic hierarchical classifiers, both will classify the document under $B1$, $B2$ and $S2$. These models can not classify the document under $S1$ unless $P(S1'|E2) > \max\{P(B1|E2), P(B2|E2)\}$.

6 Concluding Remarks and Future Works

In this paper, we have presented a theoretical framework for classifying web pages in an existing hierarchy of categories belonging to a directory. This classifier is based on Bayesian networks. Using a very efficient propagation algorithm in the network, given a new web document, the model determines the most appropriate class or category. One of the main advantages of this classifier compared to others facing the same problem is that there is no need to perform a training stage because the topology is given by the structure of the directory and the probability distributions could be computed directly from the data; only the parameters α and β should be tuned. Another important aspect is that the size of the directory could increase without this representing a problem for the classifier. In such a case, it will not be re-trained again and only new weights must be computed. The model can also assign multiple classes to a web document, only by selecting the highest k classes in the ranking.

By way of future work, we intend to evaluate the model in a real web environment (such as for example, the Yahoo directory) in order to test its performance and make any necessary modifications.

Acknowledgments

This work has been supported by the Spanish Fondo de Investigación Sanitaria, under Project PI021147.

References

1. S. Brin and L. Page, (1998). The anatomy of a large-scale hypertextual Web search engine. Computer Networks and ISDN Systems, 30:1–7, pp. 107–117.
2. L.M. de Campos, J.M. Fernández-Luna, J.F. Huete, (2003) The BNR model: Foundations and performance of a Bayesian network retrieval model, International Journal of Approximate Reasoning 34:265–285.
3. L.M. de Campos, J.M. Fernández-Luna, J.F. Huete, (2004) Using context information in structured document retrieval: an approach based on influence diagrams, Information Processing and Management 40(5):829–847.
4. S. Dumais, H. Chen (2000). Hierarchical Classification of Web Content. In Proceedings of the SIGIR Conference 256–263.
5. I. Frommholz (2001). Categorizing Web Documents in Hierarchical Catalogues. In Proceedings of the 23rd Conference on Information Retrieval
6. N. Govert, M. Lalmas and N. Furh (1999). A probabilistic description-oriented approach for categorising web documents. In Proc. of the ACM Intern. Conference on Information Knowledge and Management, 475–482.
7. F. V. Jensen (1996). An Introduction to Bayesian Networks. University College London Press, London.
8. J. Kleinberg. (1999) Authoritative sources in a hyperlinked environment. Journal of the ACM, 46:5, pp. 604–632.
9. D. Koller, M. Sahami (1997). Hierarchically classifying documents using very few words. In Proceedings of the 14th International Conference on Machine Learning 170–178.
10. D. Mladenić (1998). Turning Yahoo into an Automatic Web-page Classifier. In Proceedings of the 13th European Conference on Artificial Intelligence 473–474.
11. J. Pearl (1988). Probabilistic Reasoning in Intelligent Systems: Networks of Plausible Inference. Morgan and Kaufmann, California.
12. M. E. Ruiz, P. Srinivasan (2002). Hierarchical Text Categorization Using Neural Networks. Information Retrieval 5:87–118, 2002
13. F. Sebastiani (2002). Machine Learning in Automated Text Categorizarion. ACM Computing Surveys 34(1):1–47.
14. R. Schapier, E. Singer and A. Singhal (1998) Boosting and Rochio applied to text filtering. In Proc. of the SIGIR'98, 21st Intern. Conference on Research and Developement in Information Retrieval. 215–223.
15. A. Sun, E. Lim (2001). Hierarchical Text Classification and Evaluation. In Proceedings of the IEEE International Conference on Data Mining 521–528.
16. A. Sun, E. Lim, W. Ng, J. Srivastava (2004). Blocking Reduction Strategies in Hierarchical Text Classification. In IEEE Transactions on Knowledge and Data Engineering, 18 (10), 1305–1308.
17. `www.yahoo.com`

18. Y. Yang, J. Zhang and B. Kisiel. (2003). A scalability of classifiers in text categorization. In Proc. SIGIR'03, Intern. Conference on Research and Developement in Information Retrieval. 96–103.
19. Y. Yang and J. Pedersen (1997). A comparative study on feature selection in text categorization. In Proc. of International Conference on Machine Learning. 412–420.
20. A.S. Weigend, E.D. Weiener, J.O. Pedersen (1999). Exploiting Hierarchy in Text Categorization. Information Retrieval 1:193–216.

Personalized Knowledge Models Using RDF-Based Fuzzy Classification

Vincenzo Loia and Sabrina Senatore

Dipartimento Matematica e Informatica – Universita' di Salerno, via Ponte Don Melillo – 84084 Fisciano (SA), Italy

Summary. Due to the constant growth of the Web, the development of new techniques for improving the comprehensibility of the Web assumes a prominent role, in both research and commercial domains. Although emergent trends customize web content, it is a gruelling task to capture web structures, effective semantics and unambiguous results. Sometimes the user's inquiry is so excessively precise that no match occurs; often, returned information is too general to satisfy the user's expectation; finally, the same information can describe different web contexts, returning disappointing results. The exigency for global semantic coherence moves toward the Semantic Web which provides the hunted tools to obtain machine-readable data, transforming user-skilled information into semantic-oriented knowledge.

This work designs a system for gathering semantic-based information, using a user-driven classification, given a collection of RDF documents; this approach proposes a structural rather than content-based perspective, in order to reveal the semantics behind the data information. Furthermore, the system provides specific facilities, supplying customized categorization of collected information, in order to reflect personal cognition and viewpoints.

1 Introduction

By observing the web scenario, the first perception is the critical emergency of semantic reliability, due to the lack of strategies for the interoperability and the management of heterogeneous resources such as data, services and hypermedia. Technical and educational challenges presume methods for exploiting diverse descriptive metadata, involving facilities for enhanced web discovery, for collecting and triggering web services, through user profiles, ontologies, formal semantics and proof models. The strong need for semantics tends to model web resources assuring appreciable machine-oriented understanding, in order to represent the Web as an integrated environment where data can be shared and processed by automated tools as well as by people. Most of the web content is designed to be presented to humans, even though the results of search engines are often irrelevant, due to the difficulty of capturing the

V. Loia and S. Senatore: *Personalized Knowledge Models Using RDF-Based Fuzzy Classification*, StudFuzz **197**, 45–64 (2006)
www.springerlink.com

effective semantics related to user background which often does not meet the meaning of terms indexed by search engines. In addition search engines do not take into account the user preferences, returning the results according to own typical ranking [12]. Search engines index or categorize the Web resources according to relevant words, but they are not able to capture the intended context enclosed into user queries, often confusing, generic and not explicative.

The Semantic Web suggests some ways to tackle these hindrances: systems are designed to overcome the plain match of words in documents and queries, versus a more refined match based on topic, data type, relations between data. Furthermore, through approximation-based rules and uncertainty management of information space, Semantic Web accesses essential, pruned information. In this sense, the Semantic Web represents an extension of the Web where information assumes a well-defined meaning, providing a universally accessible infrastructure that helps to guarantee more effective discovery, automation, integration and reuse across heterogeneous applications [9].

The Resource Description Framework [1] defines a syntactical data model for representing knowledge in a more structured form (resources as URI), in order to provide more expressive layers of logic through agent-oriented language [3]. In particular, the RDF approach seems to overcome the restrictions of classical user-oriented HTML pages, where human expressiveness, meaning vagueness and complexity of data sources, make it more complicated to provide a right interpretation of user intentions. On the other hand, because the HTML standard blends content and presentation into a single representation, end users are often caught inside the idea and perception of the designer [22]. The well-built RDF formalism and, at the same time, the flexible approach to the declarative representation of resources is a key direction in providing more facilities to an agent-based framework in terms of more rigorous parsing activities for capturing semantically meaningful data.

At the same time, there is the need to support the user web navigation, through customized techniques of information retrieval and web mining. In the last years, several approaches have been moving in this direction. According to the traditional techniques of information retrieval, agent-based systems are been developed for enhanced approaches to document categorization and exploration [5]; other systems converge to a semantically advanced web search [20] or toward web filtering approaches based on fuzzy linguistic methods for recommending structured (SGML-based) documents [10].

Annotation methodologies [11] improve the task of finding out resources, through easy but manual interpretation of web pages. Other approaches [4], [19] and [14] overcome manual annotations and exploit methodologies based on the integration of heterogeneous information from distributed resources, through an ontology-based semantic layer, for acquiring, structuring and sharing information between humans and machines.

Although our approach is far from cited works, points out the user-oriented RDF classification: based on the well-defined RDF structure, it gives major relevance to ontological aspect of data, rather than traditional human-oriented, content-based information.

This paper describes a customized system for semantic-based information discovery through techniques of fuzzy clustering and fuzzy rules. The proposed system aims at classifying RDF-based documents, in terms of the semantics of their metadata. For metadata, we intend the tagged words (viz. enclosed in angular parenthesis) in an RDF-based document. In details, this approach focuses on the following activities:

- the gathering of RDF-based documents through an advanced web searching activity (for instance, exploiting the enhanced features of well-known search engines);
- the parsing of the documents and the elicitation of most representative metadata;
- the building of feature space, according to the selected metadata in the RDF context of documents;
- the clustering activities for grouping semantically closer RDF pages;
- the interpretation of the obtained clusters through fuzzy rules;
- the further classification of new, incoming RDF documents, exploiting the fuzzy rules.

The purpose of this approach is to classify and interpret the conceptual information enclosed in the metadata.

In the remainder of this paper, after an introductive section about the RDF-based vision of information, a sketch of usage scenarios provides a general idea of system performance and its functional objectives. Furthermore, focusing on the architecture, details are supplied on the main components, with the respective theoretical approach. Experimental evaluation completes the overall description of the proposed framework.

2 Preliminary Considerations: From a Human to Machine-Oriented Vision of Information

Most researches in Web searching or Web mining domains are based on information retrieval systems for HTML-based documents. Indeed, the absence of a standard in the construction of a HTML document complicates a markup-framed interpretation of a web page: our previous approaches exploited agent-based facilities [15–17] to discern the layer-based analysis of a web page, to extract information enclosed in the tags, to tailor accurate user profiles (by navigation tracking), through a parsing activity and a granularity-based matching of web documents (such as structure, layout, content, etc). With the stressing exigency of semantics, the development of RDF-based languages for describing machine-interpretable information highlights a clear separation of content from presentation. The information is semantically marked up, rather

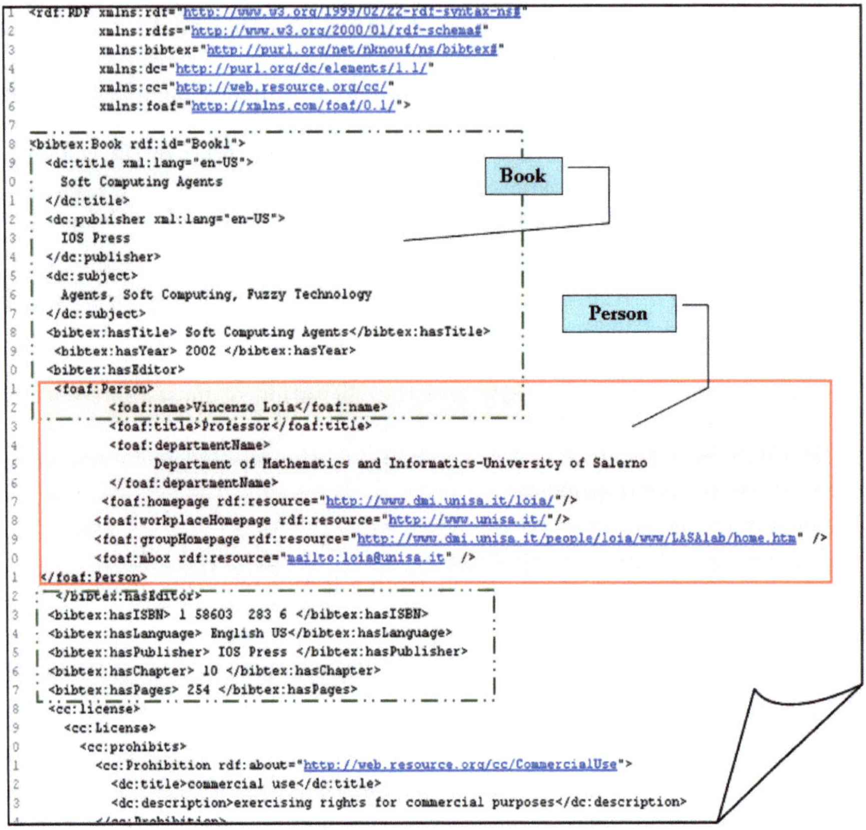

Fig. 1. RDF code.

than formatted for aesthetic visualization, increasing the possible alternatives about the information presentation.

This way, the same information can exhibit different views, emphasizing only aspects related to current human background or personal formation, by grouping properties in a unique representation or showing partial descriptions of instance resources. Therefore, for example, an undergraduate student could be interested in information on a specific course; another (PhD.) student could be interested in some papers, collected in a book. Figure 1 shows the RDF description about this information; although the goal resource is the same, the sought information is different, covering specific interests (highlighted in the figure). Both students would like to get only the required information: the former seeks the course lecture (lined square), while the other student looks for only the book data (dashed squares), but they do not care about information related to each other.

Compared with HTML standard, an RDF document can represent information with additional details; through triple-based statements, meaningful metadata provides conceptual information about the nature of an object while the enclosed values (surrounded by mark-ups) give a specific identity to object class. RDF guarantees machine-oriented interpretation of data, but it becomes an obstacle for the human understanding, considering people suddenly desire to capture the information (during the HTLM page visualization) related to their specific topics. On the light of these considerations, the next section anticipates the system performance, presenting the working scenario from a user's viewpoint, in order to outline the user-driven approach.

2.1 Usage Scenario

The proposed system is designed for users with some expertise in the information retrieval domain, because it proposes an advanced approach to get information through the classification of a given collection of resources. In order to illustrate the intended use of system, let us reconsider our PhD student; he is looking for resources in RDF format, because his topic of research is the Semantic Web. He often consults some RDF-oriented pages seeking specific information. He would like to capture which information enclosed in the RDF pages could be relevant to his researches, without inspecting the whole RDF collection or site. In particular, he would prefer to get flexible groupings (or classes) of RDF pages which reflect his specific topics (stressing particular words or concepts of interest), in order to group the RDF pages according to some similarities.

Figure 2 highlights these aspects, depicting the main sequence of actions attained by the user interface. In order to provide configuration parameters, initial snapshots show how the student selects (points 1) and loads (point 2) the collection of RDF pages. The system maintains a local knowledge database where RDF pages, collected in a previous spidering activity, are stored. The local database is periodically updated, every time the spidering (not shown here) is launched on a Web domain. Locally stored RDF pages opportunely are loaded for the clustering activity. This way, the user can either select to load stored RDF pages or insert the URL domain to analyze.

After the acquisition of RDF collection (point 2 of Figure 2), the student can insert representative terms (word-based hints) to characterize personalized classifications; the words selection is an optional phase (point 3 in Figure 3): if the user does not select the appropriate option (clicking on the relative combo-box), default setting is applied (no user-driven selection is considered). These added words are meaningful to describe his search topics. After the parameters setting, the system is ready to perform the classification on given RDF collection, on the basis of user preferences.

3 Architecture Overview

Figure 4 depicts the core modules of the overall architecture; the following three distinct phases describe the system performance.

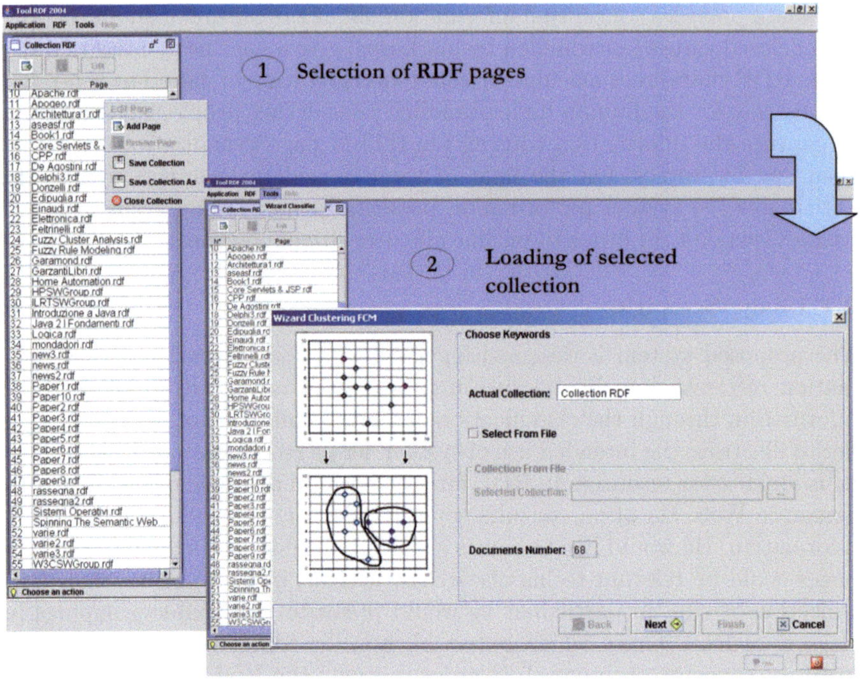

Fig. 2. Steps 1-2: Selection of RDF pages to define current collection.

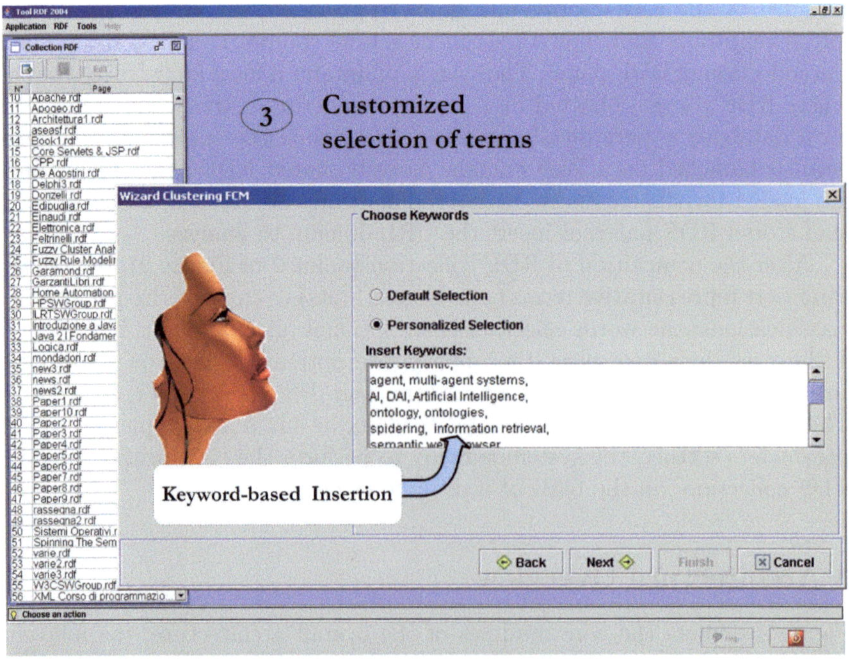

Fig. 3. Step 3: Selection of meaningful words.

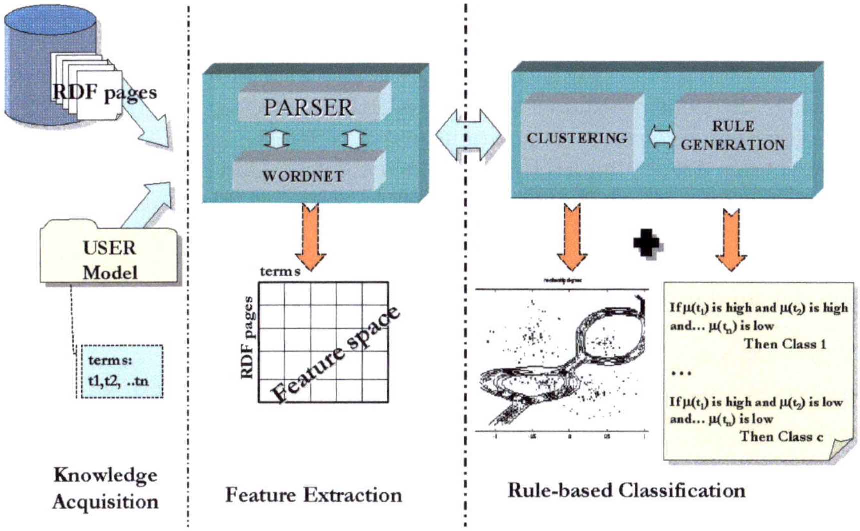

Fig. 4. General View.

- *Knowledge Acquisition* phase: it deals with the user interaction (presented in the previous depicted usage scenario), in order to catch input data. Collected *RDF pages* and defined *User Model* wholly describe the input data.
- *Feature Extraction* phase: this step achieves the parsing of RDF pages to extract more meaningful metadata (in general, terms) for defining the feature space and characterizing the term-document matrix.
- *Rule-based Classification* phase: according to the computed feature space, an RDF classification is returned. As evidenced in Figure 4, this step consists of two main activities: the *Clustering* and *Rule Generation*. The first one performs a clustering procedure on the incoming RDF pages, getting a membership-based partition of them; then, the *Rule Generation* activity translates these partitions into a more human-oriented description, based on fuzzy rules, that, through fuzzy labels, supplies a more interpretable groupings of the pre-clustered RDF pages. Furthermore, *Rule Generation* activity enables classification of additional RDF pages, captured during the spidering, avoiding to re-perform the clustering algorithm, on the basis of the produced labels.

In order to directly provide a global overview of interaction activities of the system, the next section is devoted to showing the whole system's working flow. In the sequel, some attention is dedicated to the phases just introduced.

3.1 Workflow Scenario

A sketch of the main working activities is provided in Figure 4. At the beginning, the system acquires the selected RDF collection and the (optional) set of words (*Knowledge Acquisition* phase). Then, a parsing activity is performed: the *Parser*, interacting with *WordNet* [7], filters the *terms* of the *User Model* (acquired through the user interface, see Section 2.1), focusing on the proper meaning and correlated sense (the polysemy is one of principal factors by which a search engine returns inappropriate search results). At the same time, the *Parser* validates the given RDF/XML documents and elicits the terms (evaluating the metadata of RDF pages with respect of the user words, given in input). The parsing activity is accomplished through SiRPAC parser [24], distributed by W3C, in order to validate RDF/XML documents, on the basis of triple-based statements.

As results, a term-document matrix is built, where each row describes an RDF page through a weight-based vector (*Feature Extraction* phase). The achieved matrix is processed by the *Clustering* module which returns the clusters describing the data matrix. Then the *Rule Generation* module translates the clusters in a rule-compliant form, easier to interpret by humans (*Rule-based Classification* phase). Furthermore, the system accomplishes a spidering activity (not shown in the figure), when a user request of complementary data occurs or when specific conditions are met (high frequency of user-suggested terms occur in the RDF pages); then, the spidering action is triggered, starting by some URLs of RDF collection. The spiders reach the RDF-based pages, parse them in order to evaluate their relation to users topic and then send back the page to the host destination, where it is stored.

Finally, gathered pages are returned to the host and stored in the database for further data extraction analysis (through *Rule Generation* module). At this stage, the system is able to return more exhaustive results, finalized to closely characterize the user's intentions. Details about the classification phase is given in the following, deepening the functions of the *Clustering* and *Rule-based Generation* modules.

4 Knowledge Acquisition

Collected *RDF pages* (with the referenced RDF Schema) and the terms describing the *User Model* represent the information came from from the outside environment. In particular, the *Knowledge Acquisition* phase handles the information captured through the user interface shown in Figure 2 for the preliminary setting of classification phase. Besides, it enables an additional preliminary spidering activity [17,18] for collecting RDF pages. Launched on a specific domain, it inspects a web site to download sought pages and locally stores them. The user can select and load (a selection of) RDF pages, saved in the database. Periodically the spidering activity checks the inspected sites, in order to discover changing or updating of the RDF pages. Considering the limited diffusion of a semantic, machine-oriented approach, the local storing of RDF resources is permissible again; indeed a planning of an extension through intelligent spiders is taken into account, in order to parse RDF pages at destination, avoiding to download and save much information.

The stored RDF collection represents the system awareness; although it consists of static RDF statements, it is presented to the user in a very dynamic way, tailored to his viewpoint, through the given descriptive *User Model*. In other words, the same knowledge can be framed by different view points, emphasizing different aspects of it, according to the user's expectation.

5 Features Extraction

The parsing activity of the RDF collection elicits the terms which represent the features in the term-document matrix, through the vector-based representation of each RDF page. In details, the *Parser* module interacts with the linguistic database *WordNet*, which deals with sets of terms, *synsets*, in order to select the appropriate "word senses". The *Parser* module accomplishes two subtasks: 1) by analysing multiple senses of each given term (of *User Module*), it tries to draw out a richer sequence of words that better represents the given user's intent; 2) produces a ranked list of more relevant metadata extracted in the RDF collection, taking into account the given terms. The input terms influence the rank of the list, on the basis of corresponding synonyms, hyponyms, correlated words, and so on. The top of resulting list includes the most relevant terms, candidate to represent the column headings of data matrix, whose values, estimated for each parsed RDF document, form the feature space (see Figure 4).

5.1 Relevance Measurement of the Features

The selection of relevant metadata represents, at semantic level, the collection of RDF documents. Recall that an RDF model describes each resource through triple-based statements: classes, properties and values. RDF pages describe data exploiting pre-constructed *definition* of classes and properties (using the data-structure described by RDF Schemas, ontologies, etc., whose references are declared inside the RDF documents). Bearing in mind that an RDF statement can be nested in another statement, in the calculus of relevant metadata, the level of nesting is considered.

In particular, the analysis of RDF documents emphasizes information based on different abstraction levels:

RDF dictionaries: all dictionaries (RDF Schema and ontologies) declared (by URI references) inside an RDF document enable representation of web information, through formal semantics specification.

RDF metadata: metadata is special data (described as RDF tags), surrounded by angle parenthesis, defined into dictionaries for describing the context of web resources. A parsing activity at this level enables individualization of the relevant concepts of examined RDF pages.

content of RDF tag: text or values associated with RDF tags, characterizing different (instances of) resources, in correspondence of the class specifications given by dictionaries. This level provides an interpretation of effective data associated to metadata.

Our analysis converges on the RDF metadata which represent classes and properties rather than the relative assumed values. In order to select metadata, suitable to represent the features in the term-document matrix, a measure of relevancy is computed, for each metadata in an RDF collection. Let us assume the following notation:

Collection of RDF pages: let P be the set of r RDF pages:

$$P = \{P_1, P_2, \ldots, P_r\}$$

Collection of schemas or dictionaries: Let D be the set of all schemas and – without loss of generality – dictionaries and/or ontologies used in the collection P:

$$D = \{D_1, D_2, \ldots, D_m\}$$

Dictionaries related to the current RDF page: let P_i (with $1 \leq i \leq r$) be generic pages of collection P; the set of dictionaries of the page P_i, declared inside it, is:

$$DP_i = \{D_{i_1}, D_{i_2}, \ldots, D_{i_m}\}$$

where $D_{i_h} \in D$ for $1 \leq h \leq m$.

So, each RDF page P_i depicts resources through triple-based statements; each statement describes a resource, represented by an instance of a class (defined into the dictionaries DP_i), a named property plus the value of that property for that instance (that can be a literal or another resource). Besides, more instances of the same class (or equivalently, more resources of the same type) can exist inside the same RDF document and a statement can be composed of some nested statements.

For each RDF page P_i, only instances of classes and properties are evaluated. Our approach, for each RDF instance, computes the degree of nesting of the instance in the statement and two measures (detailed in the sequel): the *accuracy* and *relevance* values associated with that instance. These measures, the relevance in particular, permits a selection of adequate metadata, in order to represent the features.

Accuracy: Fixed a class C of the dictionary DP_i, let us consider A_1, $A_2, \ldots, A_h$ (nested) statements in P_i, that describe (or, more specifically, have as subject) instances of C (for example, a number of properties associated to the same resource). Let us define a function $\pi(A_s, C)$ that represents the number of distinct properties relative to class C, within the statement A_s. Thus we define the accuracy by the following formula:

$$Accuracy(A_s, C) = \frac{\pi(A_s, C)}{\sum_{j=1}^{h} \pi(A_j, C)} \tag{1}$$

This value indicates the detail or granularity degree by which each instance of class is described on the RDF page.

Instance Relevance: It is a final value that represents the weight that the statement A_s assumed in the context of page P_i and it is so computed:

$$Inst\_Relevance(A_s, C) = \frac{Accuracy(A_s, C)}{nesting\ level} \tag{2}$$

This expression points out the influence of a *nesting level* of the statement A_s, describing the instance of class C, with respect to the accuracy by which the instance is described. This measure computes the relevance associated with all statements in an RDF page: for each page, information relative to all the instances of classes (or resources) is collected.

An analogous analysis is attained for the property names associated with these classes in the statements of an RDF page: each resource (instance) can have different properties or more occurrences of the same property. Let us recall that a property represents a predicate, which describes a relation in a statement. So, for each (univocally identified) property name, associated with an instance of a class (through a relative dictionary DP_i) the nesting level and *relevance* value are computed. Now, for each RDF page P_i, we define:

Property Relevance: Let p be a property associated to an instance of class C and let $Inst\_Relevance(A_s, C)$ be the previously computed parameter for a generic statement A_s; the relevance degree for the property p is so defined:

$$Prop\_Relevance(p, C) = Inst\_Relevance(A_s, C) \cdot \#p \tag{3}$$

where $\#p$ represents the number of occurrences of property p in the statement A_s. This expression highlights the strong dependence on the relevance of the instance to which the property is associated, specifically the value of $Prop\_Relevance$ results greater or at least equal to $Inst\_Relevance(A_s, C)$, given an instance of C, on which the property is defined. This way, if an instance of a class has more occurrence of the same property, the relative *property relevance* measure will assume greater values than the *instance relevance* measure.

In summary, two relevance measures have been previously built: the former characterizes the relevance of each resource described into statements, the other one evaluates the relevance of each property, associated with the given resources. In order to define the appropriate feature space in the term-document matrix, a digest (summarized) relevance value is computed for each metadata (class and property name) of the RDF page. So, given a page, for each class name C (defined by an RDF dictionary), the sum of all relevance measures $Inst\_Relevance(A_s, C)$ is computed for all instances A_s, $\forall s$. In the same way, the sum of relevance values $Prop\_Relevance(p, C)$ for each property name p is computed.

We call these summarized relevance measures *Metadata_Relevance(M)* where M represents a class C or a property p names. This way, a collection of relevance-based metadata is associated with each RDF page. Finally, in order to evaluate a global relevance of each RDF metadata M, with regard to the given RDF collection, the averages on all *Metadata_Relevance(M)* for each distinct M, are computed. So, a ranked list of relevance values is obtained, corresponding to a list of metadata belonging to a dictionary DP of the whole collection P of RDF pages. The metadata associated with a ranked list represent the default selection which the system proposes, when no user model is given as input.

Otherwise, a different construction of ranked list is computed. During the parsing of RDF pages, the text surrounded by metadata is analyzed to discover words (synonyms, hyponyms, general terms, etc.) related to the terms of the *User Model*, through the interaction with WordNet. Discovered related words increase the relevance of the metadata which encloses them. Additional information about the metadata (considering the occurrences of correlated words) is evaluated when a matching is found, guaranteeing an effective, user-driven ranked list of most relevant metadata. Figure 5 gives a snapshot of an

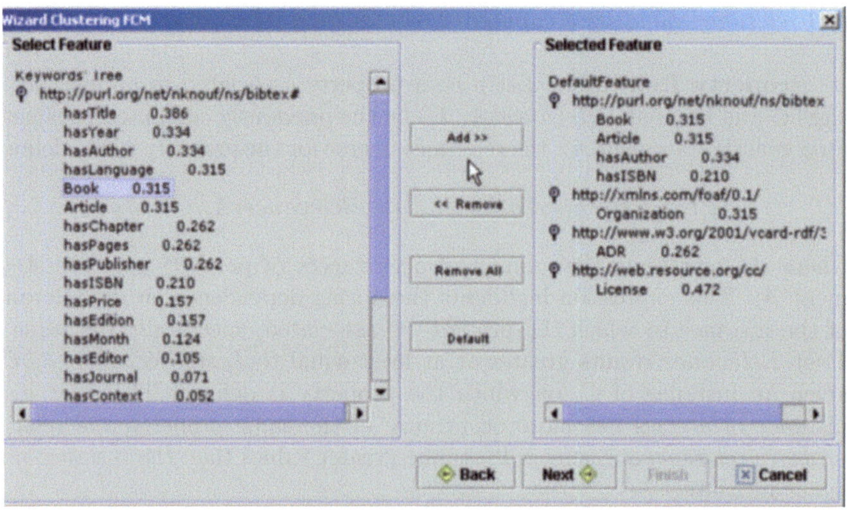

Fig. 5. Ranked list of computed terms for the features selection.

interactive interface that suggests a ranked list of computed RDF metadata, for representing the features. Although the system supplies the appropriate list of metadata, the user can directly modify the well-arranged list of features, according to his idea of relevance.

Once the features selection is accomplished, the term-document matrix can be computed: each row describes an RDF page, based on a vector representation, whereas each element of the vector is a value associated with a selected term (word or metadata) designed as feature. In particular, each value of the vector is the previously computed *Metadata_Relevance(M)* for the metadata M, normalized to 1.

6 Rule-based Classification

The term-document matrix represents the input for the last phase of the system performance, achieved by *Clustering* and *Rule Generation* modules (see Figure 4). The *Clustering* module generates K clusters about the patterns which characterize the term-document matrix, according to a prior defined setting. Through the *Rule Generation* module, the clusters are translated in human-understandable, rule-based expressions.

As introduced in Section 3.1, a spidering activity is accomplished in the *Rule-based Classification* phase, when additional RDF information is required. Starting from URLs of some relevant classified pages, the spider agents reach the pages, parse them and decide if they have to be retrieved or not (according to the user topics).

6.1 Clustering of RDF Pages

The clustering phase is accomplished by the well-known FCM algorithm [2], particularly useful for flexible data organization. Generally, FCM algorithm takes as input a data matrix where each row is a characteristic vector which represents an RDF document. As explained above, the features are metadata (in general terms) extracted by parsing of RDF pages on the basis of user suggestions, given in input. Specifically, each row of *term-document* matrix is a weight-based vector that represents an RDF page $P \longleftrightarrow \underline{x} = (x_1, x_2, \ldots, x_n)$, where each component of vector is a value in correspondence of a feature term. After the FCM execution, partitions of RDF documents are returned, in a prior fixed number K of clusters.

Considering that the complexity of the FCM algorithm is proportional to data size, it is not practical to handle frequent updating of RDF collection (through a re-performing the FCM-based clustering), every time that the spidering action returns new RDF pages. So, the ingoing web pages are classified through fuzzy-based rules [6] that are generated, analyzing the resulted clusters on the RDF collection. In literature, fuzzy if-then rules are often used for this kind of task, because they well adapt themselves to provide a human description of structures in data [13]. Major details are illustrated in the next section.

6.2 Rules Generation

This activity enables a further classification of additional RDF documents, on the basis of existing clusters. Through fuzzy rules obtained by the FCM clustering, incoming RDF pages (discovered during the spidering activity) can be assigned to a class of the already computed classification. This phase can be framed into a knowledge management approach (for instance, a bookmarking-like activity) where RDF-oriented semantics enrich the local information base and refine the local RDF cognizance.

Each page P is given in input to FCM as a vector $\underline{x} = (x_1, x_2, \ldots, x_n) \in [0,1]^n$, where each $x_j, j = 1, \ldots, n$ represents the weight value (or normalized relevance) of the jth feature (or metadata) evaluated on the page P. The FCM execution produces a partition of the data matrix in a fixed number K of clusters. Each page P has a membership value $u_{i,P} \in [0,1]$ in the cluster C_i, with $i = 1, \ldots, K$.

The goal is to describe each cluster through a fuzzy rule, where each argument is a fuzzy set with a membership function μ.

According to the cylindrical extension (a projection-based method of n-dimensional argument vector: see [6]), the generic i-th fuzzy cluster C_i can be described by the following membership functions:

$$\mu_{i_1}, \quad \mu_{i_2}, \ldots, \mu_{i_n} \tag{4}$$

obtained by the projection of the membership values $u_{i,P}$ for each page P to each axis (vector component) of the n-dimensional space and then, interpolating the projected values. Then, the gained membership functions are interpreted as linguistic labels. This way, the membership values of a n-dimensional vector $\underline{x} = (x_1, x_2, \ldots, x_n)$ can be evaluated in the fuzzy sets "x_1 *is* μ_1", "x_2 *is* μ_2", … "x_n *is* μ_n", where to each μ_j $(j = 1, \ldots, n)$ is associated a linguistic label.

Exploiting the conjunction (= *and*) of these fuzzy sets (achieved using the minimum), the evaluation of membership degree $\mu_i(\underline{x})$ of a vector $\underline{x}$ in the i-th cluster with $i = 1, \ldots, K$, can be defined:

$$\mu_{i_1}(x_1) \ \wedge \ \mu_{i_2}(x_2) \ \wedge \ \ldots \ \wedge \ \mu_{i_n}(x_n) \tag{5}$$

where $\wedge$ indicates the minimum. It simply follows that the fuzzy rule associated for the i-th class (thanks to an assignment of linguistic label to projected membership function, clusters can be interpreted as classes) is so described:

If $(x_1 \ is \ \mu_{i_1})$ *and* $(x_2 \ is \ \mu_{i_2})$ *and* … *and* $(x_n \ is \ \mu_{i_n})$ **then**
$\underline{x}$ *is in the class/cluster* C_i *with* $\mu_i(\underline{x})$

Fig. 6. Rules extraction by features analysis.

The rule expresses the relation which exists between the membership functions assumed by characteristic vector (evaluate in the antecedent rule) and the global membership value for the i-th class.

Then, the generic incoming page, depicted as vector $\underline{x}$ is assigned to j-th class such that:

$$\mu_j(\underline{x}) = \max_{1 \leq i \leq n} \mu_i(\underline{x}) \tag{6}$$

Figure 6 provides a screenshot that graphically shows the tendency of membership functions for each feature, in each cluster. In detail, fuzzy rules enable assignment of incoming RDF pages to the existing (FCM-based) classes; sometimes, some misclassifications can occur because new pages could not accurately be assigned to classes. There are two possible explanations: the information is not sufficient to classify them (low membership values for each class) or the classification may introduce an approximation error. These pages become "undecided" due to the fact that much information (with further features) occurs to correctly classify them, it follows a new classification phase is necessary.

7 Experimental Results

Although the heterogeneity of web resources evinces the exigency of machine-processable information, the diffusion of semantic, machine-readable applications is yet limited. This proposed system can be framed in Virtual Organizations or intranet-based activities, where, according to profiles of employees, different user-driven characterizations of information can be tailored. In our approach, a user interacts with the system, formulating the customized context of interest, through a query.

Some experiments have been carried out on a set of 120 RDF pages downloaded by FOAF site [8] or retrieved through advanced search exploiting well-known search engines. To evaluate the system performance, an accurate selection of RDF pages (dealing with some elicited topics) has been achieved for act on the specific categories of topic, according to the fixed a priori number of clusters. In the following, we sketch three experiments.

Table 1 shows the terms of *User Model* and the correspondent representative metadata (features), associated by the *Parser* module (see Figure 1); the prefix of each metadata is referred to the namespace describing it: for instance, the prefix *foaf:* is bound to the FOAF dictionary for describing people, social contacts and working groups, through the relative namespace http://xmlns.com/foaf/0.1/.

Table 2 presents the prototypes obtained after the clustering algorithm has been performed: for each experiment the coordinates of prototypes are given. The values of the prototypes is referred to metadata, with respect to the ordering given in Table 1. Finally, Table 3 sketches the experiments results: each experiment is composed of two phases, one evaluates only the returned information after the classification whereas, the other phase, through the spidering activity, achieves the classification of incoming pages, by fuzzy rules. Experiment 1 considers 7 features, some of them are related to the input

Table 1. Parameters configuration: keywords vs. metadata

Exp. No.	USER KeYWORDS	METADATA
1	books and papers related to computational web intelligence	foaf:Document, dc:title, foaf:Project, wn:Book-1 rdf:description, foaf:interest, foaf:Person
2	scientific publications related to the Web domain	foaf:Document, bibtex:Inproceedings, foaf: Person, bibtex:Conference, dc:title, bibtex:Book, bibtex:Article, foaf:interest bibtex:hasTitle, bibtex:hasAuthor
3	homepages of people working for FOAF community groups	foaf: Person, foaf:groupHomepage, foaf:interest, foaf:workplaceHomepage, dc:title, foaf:homepage, foaf:Project, foaf:projectHomepage, foaf:name, rdfs:seeAlso, foaf:Group, foaf:knows, contact:nearestAirport

Table 2. Prototypes analysis

Exp. No.	Cluster #	Prototype Values
1	3	0.65, 0.12, 0.86, 0.12, 0.26, 0.34, 0.66
		0.77, 0.62, 0.05, 0.57, 0.56, 0.69, 0.74
		0.91, 0.53, 0.13, 0.21, 0.61, 0.19, 0.38
2	3	0.71, 0.10, 0.79, 0.14, 0.77, 0.34, 0.16, 0.43, 0.11, 0.01
		0.21, 0.52, 0.26, 0.72, 0.63, 0.51, 0.59, 0.03, 0.62, 0.51
		0.45, 0.02, 0.68, 0.57, 0.50, 0.24, 0.56, 0.33, 0.02, 0.16
3	4	0.76, 0.27, 0.75, 0.32, 0.36, 0.76, 0.56, 0.19, 0.32, 0.41, 0.74, 0.56, 0.87
		0.82, 0.12, 0.21, 0.22, 0.46, 0.34, 0.76, 0.73, 0.68, 0.61, 0.18, 0.80, 0.17
		0.91, 0.56, 0.75, 0.72, 0.66, 0.68, 0.15, 0.53, 0.87, 0.41, 0.32, 0.86, 0.37
		0.87, 0.72, 0.67, 0.67, 0.53, 0.79, 0.19, 0.45, 0.80, 0.28, 0.84, 0.74, 0.44

Table 3. System performance in terms of recall and relative error (percentage)

Experiment No.	No. features	waited recall %	actual recall %	error %
1	7	70%	60%	11%
1 (with Fuzzy Rules)	7	70%	64%	11%
2	10	87%	71%	6%
2 (with Fuzzy Rules)	10	88%	70%	5%
3	13	87%	77%	4%
3 (with Fuzzy Rules)	13	90%	82%	4%

topic (books and papers related to computational web intelligence, see Table 1). Specific information like *computational web intelligence* are not relevant for the classification, because this approach considers the semantic aspect of data, represented by the metadata. In this experiment, the computed features only partially reflect the user intention.

Note that an advanced parsing of textual words enclosed in the metadata is considered, in order to support the interpretation of the terms of the *User Model* and get an appropriate selection of metadata, as candidate features.

Anyway, providing an interpretation to the relative three prototypes (Experiment 1), the first one represents RDF pages that deals with documents and projects related to peoples; indeed, the values associated to the respective metadata *foaf:Document, foaf:Project and foaf:Person* are the highest. The second prototype describes pages of people through their personal information (foaf:interests) and related documents (foaf:Document, wn:Book-1). The last one mostly represents documents. In Table 3, evaluating the two phases of classification, this experiment presents similar (actual) recall values and a disputable error of evaluation that represents the percentage of unwanted (returned, but not relevant) results. The results based on fuzzy rules (see Table 3, Experiment 1 with FR), classify well all incoming pages, producing the same error percentage of the clustering approach.

Similar considerations can be made about the whole Experiment 2 where the features set is bigger (the topic is related to books too, but it considers all the scientific publication on the web area: web intelligence, web systems, web information retrieval technology, etc.). The same way, the interpretation of the prototypes in Table 2 is easy to get.

Looking at Table 3, the error is smaller in the rule-based approach, although the expected recall is far from the actual one (only some new pages are classified). Better results have been reached for the Experiment 3 (with 13 features): it shows a discrete performance in the rule-based approach (returned pages do not influence the number of unwanted pages). Anyway, the search target of the Experiment 3 is to seek all the web pages of people which work for FOAF community groups. After the clustering activity (the number of cluster $K = 4$), the prototypes assemble working groups, evidencing different characteristics: the first prototype marks all the documents emphasizing information as homepages, interests and nearest airport to the place of working groups (higher values for the metadata *foaf:homepage, foaf:interest, foaf:Group, conctact:nearesrAirport*). The second one describes group of people that work to the same project (high values for *foaf:Person, foaf:Project, foaf:projectHomepage, foaf:knows* metadata). The last two are similar, they provide for each person all the information related to their working activities (interests, relationships and references to several web pages). In particular, the last one better characterizes the group of work (higher value for the metadata *foaf:Group*). Suitable results are evidenced in Table 3 for the rule-based actual recall where most of RDF documents, returned by spidering are well-classified.

Additional experimental results are in progress; anyway, further improvements are taken into account, in order to enrich the feature space of term-document matrix which represents the focal point for increasing the system performance.

8 Conclusions

The heterogeneity and the lack of standards for the Web communities represent the prominent problem of the web discovery activities. Although RDF language represents machine-readable interpretation of data, at the same time, it constitutes an obstacle for the human understanding, considering people may in a moment wish to retrieve information related to their own specific topics. In order to tackle this problem, different solutions are exploited: hybrid approaches combine classical search techniques with a semantic model of an ontology-based domain [23]; annotation technologies [11, 21] allow the association of the metadata to web resources supporting shared interpretation, although their quality depends on the human expertise employed in manual annotations.

This paper represents a novel contribution to the retrieval of information according to the users interests. A combined framework is designed in order to couple the expressive power of a semantic-oriented approach with a customized information retrieval. Our approach, rather than returning all pages related to a requested topic, identifies an arranged collection of knowledge, organized in accordance with concept-based features instead of simple keywords. Our RDF classification provides a proficient perspective of the conceptual information enclosed into RDF pages; the user-driven selection of features, enables to classify semantically related pages according to the user viewpoints. This flexible representation of a domain is useful in many information retrieval issues: automatic meta-indexing, RDF-based querying and concept-based visualization are some of the possible applications derivable from our proposal.

References

1. Dave Beckett and Brian McBride. RDF/XML Syntax Specification (Revised) W3C Recommendation 10 February 2004. Available http://www.w3.org/TR/rdf-syntax-grammar/.
2. Bezdek, J.C. *Pattern Recognition and Fuzzy Objective Function Algorithms.* Plenum Press, N. York, 1981.
3. DARPA-DAML. DARPA Agent Markup Language, 2003. http://www.daml.org/.
4. J.B. Domingue, M. Dzbor, and E. Motta. Magpie: supporting browsing and navigation on the semantic web. In *Proceedings of the 9th international conference on Intelligent user interface*, pages 191–197. ACM Press, 2004.
5. H. Eui-Hong, D. Boley, M. Gini, R. Gross, K. Hastings, G. Karypis, V. Kumar, B. Mobasher, and J. Moore. WebACE: A Web Agent for Document Categorization and Exploration. In Katia P. Sycara and Michael Wooldridge, editors, *Proceedings of the 2nd International Conference on Autonomous Agents (Agents'98)*, pages 408–415, New York, 1998. ACM Press.
6. F. Hoppner and F. Klawonn and R. Kruse and T. Runkler. *Fuzzy Cluster Analysis – Methods for Image Recognition.* J. Wiley, N. York, 1999.
7. Fellbaum C. WordNet An Electronic Lexical Database. 1998.
8. FOAF. The friend of a friend (foaf) project, 2003. http://www.foaf-project.org/.
9. J. Hendler, T. Berners-Lee, and E. Miller. Integrating applications on the semantic web. *Institute of Electrical Engineers of Japan*, 122(10):676–680, October 2002.
10. Enrique Herrera-Viedma and Eduardo Peis. Evaluating the informative quality of documents in SGML format from judgements by means of fuzzy linguistic techniques based on computing with words. *Inf. Process. Manage.*, 39(2):233–249, 2003.
11. J. Kahan, M. Koivunen, E. Prud'Hommeaux, and R.R. Swick. Annotea: an open RDF infrastructure for shared web annotations. In *Proceedings of the 10th International World Wide Web Conference*, pages 623–632, 2001.
12. L. Kershberg, W. Kim, and A. Scime. A presonalized agent for Semantic Taxonomy-Based Web Search. *Electronic Commerce Research and Applications (ECRA)*, 1(2), 2003.
13. A. Klose. Extracting fuzzy classification rules from partially labeled data. *Soft Computing Springer-Verlag*, 2003.
14. H. Knublauch. An AI tool for the real world - knowledge modeling with protege. *Java World*, June 2003. Walkthrough of Protege.

15. V. Loia, W. Pedrycz, S. Senatore, and M.I. Sessa. Proactive utilization of proximity-oriented information inside an Agent-based Framework. In *17th International FLAIRS Conference*, Miami Beach, Florida, May 17-19 2004.
16. V. Loia, W. Pedrycz, S. Senatore, and M.I. Sessa. Support web navigation by means of cognitive proximity-driven assistant agents. *Journal of the American Society for Information Science and Technology*, 2004. accepted.
17. V. Loia, S. Senatore, and M.I. Sessa. Discovering Related Web Pages through Fuzzy-Context Reasoning. In *Proceedings of 11th IEEE International Conference on Fuzzy Systems*, volume 1, pages 150–155, Hawaii, 12-17 May 2002. IEEE PRESS.
18. V. Loia, S. Senatore, and M.I. Sessa. Similarity-based SLD Resolution its Role for Web Knowledge Discovery. In *Special Issue on "Advances in Possibilistic Logic and Related Issues"*, volume 144, pages 151–171. Fuzzy Sets & Systems, 2004.
19. Alexander Maedche, Steffen Staab, Nenad Stojanovic, Rudi Studer, and York Sure. SEAL – A framework for developing SEmantic Web PortALs. *Lecture Notes in Computer Science*, 2097:1–22, 2001.
20. N. Guarino and C. Masolo and G. Vetere. OntoSeek: Using Large Linguistic Ontologies for Accessing On-Line Yellow Pages and product Catalogs. *IEEE Intelligent System*, 14(3):70–80, 1999.
21. I.A. Ovsiannikov, M.A. Arbib, and T.H. McNeill. Annotation Tecnology. *International Journal of Human-Computer Studies*, 50(4):329–362, 1999.
22. D. Quan and D.R. Karger. How to Make a Semantic Web browser. In *Proceedings of WWW 2004*. ACM Press, 2004.
23. C. Rocha, D. Schwabe, and M. Poggi de Arag ão. A hybrid approach for Searching in the Semantic Web. In *Proceedings of WWW 2004*, pages 374–383. ACM Press, 2004.
24. W3C SiRPAC. A simple RDF parser and compiler. 1999. http://www.w3.org/RDF/Implementa-tions/SiRPAC/.

A Genetic Programming Approach for Combining Structural and Citation-Based Evidence for Text Classification in Web Digital Libraries

Baoping Zhang[1], Weiguo Fan[1], Yuxin Chen[1], Edward A. Fox[1], Marcos André Gonçalves[2], Marco Cristo[2], and Pável Calado[3]

[1] Department of Computer Science, Virginia Polytechnic Institute and State University, Blacksburg, VA 24061 USA
{bzhang, wfan, yuchen, fox}@vt.edu
[2] Department of Computer Science, Federal University of Minas Gerais, Belo Horizonte, MG, Brazil
{mgoncalv, marco}@dcc.ufmg.br
[3] Pável Calado, IST/INESC-ID, Lisbon, Portugal
pavel@tagus.ist.utl.pt

Summary. This paper investigates how citation-based information and structural content (e.g., title, abstract) can be combined to improve classification of text documents into predefined categories. We evaluate different measures of similarity, five derived from the citation structure of the collection, and three measures derived from the structural content, and determine how they can be fused to improve classification effectiveness. To discover the best fusion framework, we apply Genetic Programming (GP) techniques. Our empirical experiments using documents from the ACM digital library and the ACM classification scheme show that we can discover similarity functions that work better than any evidence in isolation and whose combined performance through a simple majority voting is comparable to that of Support Vector Machine classifiers.

1 Introduction

In the last few years, automated classification of text into predefined categories has attracted considerable interest, due to the increasing volume of documents in digital form and the ensuing need to organize them. However, traditional content-based classifiers are known to perform poorly when documents are noisy and contain little text [3, 23].

Particularly, Web Digital Libraries (DL) (i.e., digital libraries accessible through the Web) offer a number of opportunities and challenges for classification. In digital libraries, information is explicitly organized, described, and

B. Zhang et al.: *A Genetic Programming Approach for Combining Structural and Citation-Based Evidence for Text Classification in Web Digital Libraries*, StudFuzz **197**, 65–83 (2006)
www.springerlink.com

managed and community-oriented services are built to attend specific information needs and tasks. DLs can be thought as being in the middle of a large spectrum between databases and the Web. In the Web, we have a very unstructured environment and very little is assumed about users. On the other hand, databases assume very rigid structures/standards and very specialized users.

Accordingly, DLs offer the opportunity to explore the rich and complex internal (semi-)structured nature of documents and metadata records in the classification task as well as the social networks implied by specific communities as expressed for example by citation patterns in the research community. On the other hand, many DLs which are created by aggregation of other sub-collections/catalogs, suffer from problems of quality of information. One such problem is incompleteness (e.g., missing information). This makes it very hard to classify documents using traditional content-based classifiers like SVM, *kNN* or Naive Bayes. Another quality problem is imprecision. For example, citation-based information is often obtained with OCR, a process which produces a significant number of errors. In this work we try to overcome these problems by applying automatically discovered fusion techniques of the available evidence to the classification problem. Particularly, we investigate an inductive learning method – Genetic Programming (GP) – for the discovery of better fused similarity functions to be used in the classifiers and explore how this combination can be used to improve classification effectiveness.

Experiments were performed on the ACM Digital Library using the ACM classification scheme. Three different content-based similarity measures applied to the abstract and title fields were used in the combination: bag-of-words, Cosine, and Okapi. Five different citation-based similarity measures also were used: bibliographic coupling, co-citation, Amsler, and Companion (authority and hub). The new similarity functions, discovered through GP, where applied to *kNN* classifiers showed a significant improvement in macro-F1 over the best similarity functions in isolation. Furthermore, the performance of a simple majority voting of the *kNN* classifiers with the GP functions produced performance comparable to that of content-based SVM classifiers using the same training and test data.

This paper is organized as follows. In Section 2 we introduce background on Genetic Programming. We present our approach to solve the classification problem in Section 3. Section 4 describes how the GP framework and the discovered similarity functions are applied to the classification problem. We conduct two sets of experiments to evaluate this framework and summarize the experimental findings in Section 5. Section 6 discusses the related works to this study and Section 7 concludes the paper and points out future research directions.

2 Background

Genetic Programming (GP), an extension of Genetic Algorithms (GAs), is an inductive learning technique designed following the principles of biological inheritance and evolution [30]. Genetic Programming has been widely used and proved to be effective in solving optimization problems, such as financial forecasting, engineering design, data mining, and operations management. GP makes it possible to solve complex problems for which conventional methods can not find an answer easily.

In GP, a large number of individuals, called a population, are maintained at each generation. An individual is a potential solution formula for the target problem. All these solutions form a space, say, Σ. An individual can be stored using complex data structures like a tree, a linked list, or a stack. A fitness function ($f(\cdot) : \Sigma \rightarrow R$) is also needed in Genetic Programming. A fitness function takes the solution space, Σ, as its domain and returns a real number for each individual in the space. Hence tentative solutions, represented by individuals, can be evaluated and ordered according to their return values. The return value of a fitness function must appropriately measure how well an individual, which represents a solution, can solve the target problem.

GP searches for an "optimal" solution by evolving the population generation after generation. It works by iteratively applying genetic transformations, such as reproduction, crossover, and mutation, to a population of individuals to create more diverse and better performing individuals in subsequent generations. The reproduction operator directly copies or, in a more appropriate term, clones some individuals into the next generation. The probability for an individual to be selected for Reproduction should be proportional to its fitness. Therefore the better a solution solves the problem, the higher probability it has to enter the next generation. While Reproduction keeps the best individuals in the population, Crossover and Mutation introduce transformation and so provide variations to enter into the new generation. The crossover operator randomly picks two groups of individuals, selects the best (according to the fitness) individual in each of the two groups as parent, exchanges a randomly selected gene fragment of each parent and produces two "children". Thus, a "child" may obtain the best fragments of its excellent parents and so may surpass them, providing a better solution to the problem. Since parents are selected from a "competition", good individuals are more likely to be used to generate offspring. The mutation operator randomly changes a gene code of an individual. Using these genetic operators, subsequent generation keeps individuals with the best fitness in the last generation and takes in "fresher air", providing creative solutions to the target problem. Better solutions are obtained either by inheriting and reorganizing old ones or by lucky mutation, simulating Darwinian Evolution.

In order to apply GP to the problem of classification, several required key components of a GP system need to be defined. Table 1 lists these essential components along with their descriptions.

Table 1. Essential GP Components

Components	Meaning
Terminals	Leaf nodes in the tree structure. i.e. x, y as in Figure 1.
Functions	Non-leaf nodes used to combine the leaf nodes. Commonly numerical operations: +, -, *, /, log.
Fitness Function	The objective function GP aims to optimize.
Reproduction	A genetic operator that copies the individuals with the best fitness values directly into the population of the next generation without going through the crossover operation.
Crossover	A genetic operator that exchanges subtrees from two parents to form two new children. Its aim is to improve the diversity as well as the genetic fitness of the population. This process is shown in Figure 1.
Mutation	A genetic operator that replaces a selected individual's subtree whose root is a picked mutation point with a randomly generated subtree.

3 Our Approach

3.1 GP System Configurations

We set up the configurations of the GP system used for similarity function discovery as shown in Table 2.

Table 2. Modeling setup for classification function discovery by GP. Refer to Table 1 for explanations of the various components

Terminals	We use features discussed in Section 3.2 as terminals.
Functions	+, *, /, sqrt
Fitness Function	Algorithm 1 (see below)
Genetic Operators	Reproduction, Crossover, Mutation

Algorithm 1 (below) details our fitness evaluation function which GP intends to optimize within a particular class.

Algorithm 1:

Let L_p, L_r be empty lists
For each document D in class C
  Let $L_p = L_p$ union (the set of $|C|$ documents most similar to D)
  Let $L_r = L_r$ union (the set of $|C|$ documents most similar to D and also not already in L_r)
end for
Let $P =$ (no. of documents in L_p that are of class $C)/|L_p|$
Let $R =$ (no. of documents in L_r that are of class $C)/|C|$
$F = 2PR/(P+R)$

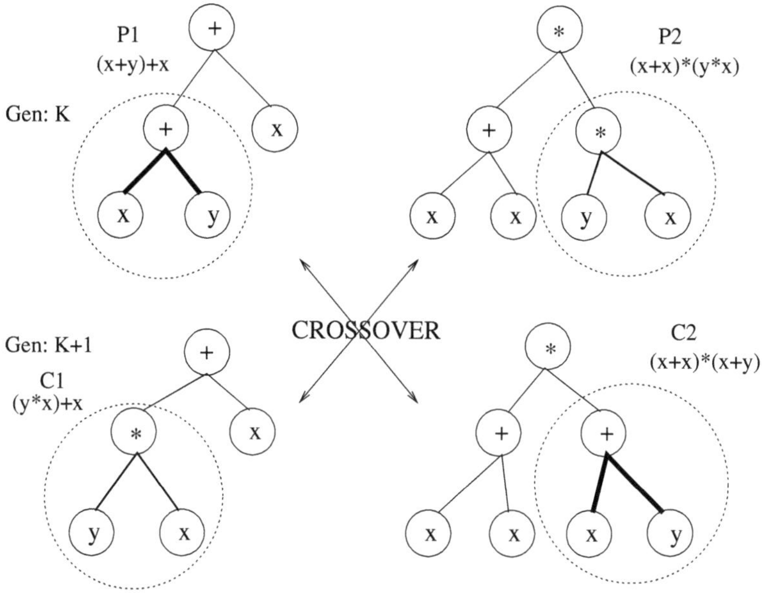

Fig. 1. A graphical illustration of the crossover operation.

A good similarity function, i.e., a similarity function with a high fitness value, is one that, when applied to a document d_i of class C, ranks many documents from class C as similar to d_i. The higher the value of F, the better the function. It is worth to notice that the choice of fitness function can have a huge impact in the final classification performance [11]. Experiments with different fitness functions are currently being performed.

3.2 Used Terminals

We combined features regarding content-based structural information and features regarding citation-based information together to serve as the terminals in our GP system.

Structural Similarity Measures

To determine the similarity between two documents we used three different similarity measures applied to the content of abstract and title of documents separately: Bag-of-Words, Cosine, and Okapi [39]. This gave us six similarity measures, represented as document × document matrices: AbstractBagOfWords, AbstractCosine, AbstractOkapi, TitleBagOfWords, TitleCosine, and TitleOkapi. More specifically, the documents are represented as vectors in the Vector Space Model [41]. Suppose we have a collection with N distinct index terms t_j. A document d_i can be represented as follows: $d_i = (w_{i1}, w_{i2}, \ldots, w_{iN})$, where w_{ij} represents the weight assigned to term t_j in document d_i. For the bag-of-words measure, the similarity between two documents d_1 and d_2 can be calculated as the following:

$$\textit{bag-of-words}(d_1, d_2) = \frac{|\{d_1\} \cap \{d_2\}|}{|d_1|} \tag{1}$$

where $\{d_i\}$ corresponds to the set of terms occuring in document d_i. For the Cosine measure, the similarity between two documents can be calculated as the following [42]:

$$cosine(d_1, d_2) = \frac{\sum_{i=1}^{t} w_{1i} * w_{2i}}{\sqrt{\sum_{i=1}^{t} w_{1i}^2 * \sum_{i=1}^{t} w_{2i}^2}} \tag{2}$$

For the Okapi measure, the similarity between two documents can be calculated as the following:

$$Okapi(d_1, d_2) = \sum_{t \in d_1 \cap d_2} \frac{3 + tf_{d_2}}{0.5 + 1.5 * \frac{len_{d2}}{len_{avg}} + tf_{d_2}} * \log \frac{N - df + 0.5}{df + 0.5} * tf_{d_1} \tag{3}$$

Here, tf is the term frequency in a document and df is the document frequency of the term in the whole collection. N is the number of documents in the whole collection, len is the length of a document, and len_{avg} is the average length of all documents in the collection.

From Eqs. (1), (2), and (3), we can see that the cosine similarity matrix is symmetric while the bag-of-words and okapi similarity matrices are not.

Citation-based Similarity Measures

To determine the similarity of subject between two documents we used five different similarity measures derived from link structure: co-citation, bibliographic coupling, Amsler, and Companion (authority and hub).

Co-citation was first proposed by Small [43], as a similarity measure between scientific papers. Two papers are co-cited if a third paper has citations to both of them. This reflects the assumption that the author of a scientific paper will cite only papers related to his own work. To further refine this idea, let d be a document and let P_d be the set of documents that cite d, called the *parents* of d. The co-citation similarity between two documents d_1 and d_2 is defined as:

$$cocitation(d_1, d_2) = \frac{P_{d_1} \cap P_{d_2}}{|P_{d_1} \cup P_{d_2}|} \tag{4}$$

Equation (4) tells us that, the more parents d_1 and d_2 have in common, the more related they are. This value is normalized by the total set of parents, so that the co-citation similarity varies between 0 and 1.

Also with the goal of determining the similarity between papers, Kessler [26] introduced the measure of bibliographic coupling. Two documents share one unit of bibliographic coupling if both cite a same paper. The idea is based on the notion that authors who work on the same subject tend to cite the same papers. More formally, let d be a document. We define C_d as the set of documents that d cites, also called the *children* of d. Bibliographic coupling between two documents d_1 and d_2 is defined as:

$$bibcoupling(d_1, d_2) = \frac{C_{d_1} \cap C_{d_2}}{|C_{d_1} \cup C_{d_2}|} \quad (5)$$

Thus, according to Eq. (5), the more children in common document d_1 has with document d_2, the more related they are. This value is normalized by the total set of children, to fit between 0 and 1.

In an attempt to take the most advantage of the information available in citations between papers, Amsler [1] proposed a measure of similarity that combines both co-citation and bibliographic coupling. According to Amsler, two papers A and B are related if (1) A and B are cited by the same paper, (2) A and B cite the same paper, or (3) A cites a third paper C that cites B. Thus, let d be a document, let P_d be the set of parents of d, and let C_d be the set of children of d. The Amsler similarity between two documents d_1 and d_2 is defined as:

$$amsler(d_1, d_2) = \frac{(P_{d_1} \cup C_{d_1}) \cap (P_{d_2} \cup C_{d_2})}{|(P_{d_1} \cup C_{d_1}) \cup (P_{d_2} \cup C_{d_2})|} \quad (6)$$

Equation (6) tell us that, the more links (either parents or children) d_1 and d_2 have in common, the more they are related.

Finally, taking a different approach, Dean and Henzinger proposed the Companion algorithm [8] for Web pages. Given a Web page d, the algorithm finds a set of pages related to d by examining its links. Companion is able to return a degree of how related the topic of each page in this set is to the topic of page d. This degree can be used as a similarity measure between d and other pages. We use a similar approach where web pages correspond to documents and links to citations.

To find a set of documents related to a document d, the Companion algorithm has two main steps. In step 1, we build the set $\mathcal{V}$, the vicinity of d, that contains the parents of d, the children of the parents of d, the children of d, and the parents of the children of d. This is the set of documents related to d. In step 2 we compute the degree to which the documents in $\mathcal{V}$ are related to d. To do this, we consider the documents in $\mathcal{V}$ and the citations among them as a graph. This graph is then processed by the HITS algorithm [29], which returns a degree of *authority* and *hubness* for each document in $\mathcal{V}$. Intuitively, a good authority is a document with important information on a given subject. A good hub is a document that cites many good authorities. Companion uses the degree of authority as a measure of similarity between d and each document in $\mathcal{V}$. For a more detailed description of the Companion and HITS algorithms, the user is referred to [8] and [29], respectively.

4 The Framework for Classification

All of the similarity measures discussed in Section 3.2 were represented as document × document matrices and served as the terminals in the GP system. With the above settings, the overall classification framework is as follows:

1. For each class, generate an initial population of random "similarity trees".
2. For each class, perform the following sub-steps on training documents for N_{gen} generations.
   a) Calculate the fitness of each similarity tree.
   b) Record the top N_{top} similarity trees.
   c) Create new population by:
      i. Reproduction, which copies the top (in terms of fitness) trees in the population into the next population.
      ii. Crossover. We use the tournament selection to select, with replacement, six random trees from the population. The top two among the six trees (in terms of fitness) are selected for crossover and they exchange subtrees to form trees for the next generation.
      iii. Mutation, which creates new trees for the next population by randomly mutating a randomly chosen part of selected trees.
3. Apply the "best similarity tree" of each class (i.e., the first tree of the last generation) on a set of testing documents to a *kNN* algorithm (see below).
4. Combine the output of each classifier through a simple majority voting.

Steps 1 and 2 concern the training process within GP which intends to discover better similarity functions for each class. However, the discovered functions can only be used to calculate the similarity between any two documents. In order to evaluate the performance of those functions in the classification task, we used a strategy based on a nearest neighbor classifier. This classifier assigns a category label to a test document, based on the categories attributed to the k most similar documents in the training set. The most widely used such algorithm was introduced by Yang [46] and is referred to, in this work, as *kNN*. The *kNN* algorithm was chosen since it is simple and makes a direct use of similarity information.

In the *kNN* algorithm, to a given test document d is assigned a relevance score $s_{c_i,d}$ associating d to each candidate category c_i. This score is defined as:

$$s_{c_i,d} = \sum_{d' \in \mathcal{N}_k(d)} similarity(d, d') f(c_i, d') \tag{7}$$

where $\mathcal{N}_k(d)$ are the k nearest neighbors (the most similar documents) of d in the training set and $f(c_i, d')$ is a function that returns 1 if document d' belongs to category c_i and 0 otherwise. In Step 3 of our framework the generic similarity function of *kNN* is substituted by the functions discovered by GP for each class.

In multi-classification problems with n classes, we effectively end up with n *kNN* classifiers using the described framework. In order to produce a final classification result, we combine the output of all n classifiers using a simple *majority voting* scheme, whereby the class of a document d_i is decided by the most common class assigned by all the n classifiers. In case of ties, we assign d_i to the larger class. Besides its simplicity we chose to use majority voting in our framework (Step 4) to: 1) help alleviate the common problem of overfitting found in GP training [13] and; 2) help boost performance by allowing *kNN* classifiers to apply different similarity functions which explore and optimize the characteristics of each particular class in different ways. A reasonable alternative here would be to generate only one "global" similarity function instead of n "local" ones. However, discovery of such a globally unique similarity function, besides the potential of suffering overfitting, was too demanding in terms of training time and necessary computational resources while the applied "per class" training allowed easy distribution of the training task. Nonetheless, we are working on parallel strategies to allow "global vs. local" experiments.

5 Experiments

To test the hypotheses that GP is able to adapt itself to find the best similarity functions we run two sets of experiments following the framework of the previous section. For these experiments, we used only the first level of the ACM classification scheme (11 categories, A to K) (http://www.acm.org/class/1998/) and a subset of the ACM collection with 30K metadata records corresponding to those classified under only one category in the first level. Experiments on other levels of the ACM classification scheme are currently underway. The ACM digital library suffers from most of the problems we mentioned before. For example, only 42% of the records have abstracts, which makes it very hard to classify them using traditional content-based classifiers. For these records, the only available textual content is title. But titles contain normally only 5 to 10 words. Citation information was created with OCR and had a significant number of errors. A very imprecise process of matching between the citation text and the documents, using adaptations of techniques described in [19,31,32], had to be performed. This introduced noise and incompleteness in the citation-based similarity matrices computed with measures such as co-citation or bibliographic coupling. The other impact factors were the large searching space and skewed distributions of some categories.

Stratified random sampling (cf. Section 5.1) was used to create training and test collections. The combination of these experiments should provide us with insights about the capability of the GP-based discovery framework.

5.1 Sampling

The collection used in our experiments has 30,022 documents. Each terminal or feature described is a similarity matrix which contains the similarity between each pair of documents. Using half or more of the whole collection as our training data, the required resources, as CPU time, and amount of memory, would be enormous. The time required to discover a proper classification framework also would be significant. To reduce the high cost of resources and at the same improve efficiency, sampling was used. A sample is a finite part of a statistical population whose properties are studied to gain information about the whole [35]. Sampling is the act, process, or technique of selecting a suitable sample, or a representative part of a population for the purpose of determining parameters or characteristics of the whole population. A random sample is a sample selected based on a known probability that each elementary unit will be chosen. For this reason, it is sometimes referred to as a probability sample. A stratified sample is one type of random sample. A stratified sample is obtained by independently selecting a separate simple random sample from each population stratum. A population can be divided into different groups based on some characteristic. In our case, the documents belonging to each category of the ACM classification scheme (first level) corresponded to different population strata. We can then randomly select from each stratum a given number of units which may be based on a proportion, like 15%, for each category. However special attention needs to be paid to skewed categories. For example, category E only has 94 documents while the average size of the other categories is in the range of thousands. 15% of 94 only gives us 14 documents and this would be too small to serve as the sample to discover the whole category's characteristic. In this case, we might want to have larger samples. Classification statistics for the whole collection would be used to control the sampling procedure. That is, baselines based on the samples would be compared with baselines for the whole collection to ensure that the samples mirror the whole collection as well as possible. We generate two sets of training samples using stratified sample strategy. The first set used a random 15% sample for large classes and 50% for skewed classes (A, E, and J). And the second set used a random 30% sample for large classes and 50% for skewed classes (A, E, and J)[1]. The rest of the samples will be used for testing and performance comparison. All results reported in later sections are based on test data sets only.

[1] In the remainder of the paper, we use 15% to refer to the first sample set and 30% to refer to the second sample set.

5.2 Baselines

In order to demonstrate that the combination of different features by GP is able to provide better classification results, we need to compare it with the classification statistics of each feature in isolation (baselines). We used F1 as our comparison criteria. F1 is a combination of precision and recall. Precision is defined as the proportion of correctly classified records in the set of all records assigned to the target class. Recall is defined as the proportion of correctly classified records out of all the records having the target class. F1 is defined as 2*Precision*Recall / (Precision + Recall). It is worth to notice that F1 is an even combination of precision and recall. It reduces the risk that you can get perfect precision by always assigning zero categories or a perfect recall by always assigning every category. The result we want is to assign the correct categories and only the correct categories, maximizing precision and recall at the same time, and therefore maximizing F1. Table 3 shows the evidence that performs the best when applied to a *kNN* algorithm for a specific category in isolation in the test collections, among all similarity evidence, based on macro F1. Table 4 shows the average macro F1 over all categories for each similarity evidence, also in isolation.

Table 3. Best baseline for each category

Class	Macro F1/ class (15%)	Best Evidence (15%)	Macro F1/ class(30%)	Best Evidence (30%)
A	40.00	Title_BagOfWords	43.56	Title_BagOfWords
B	63.89	Amsler	70.58	Amsler
C	60.32	Title_Okapi	63.01	Title_Cosine
D	67.97	Title_Okapi	69.03	Title_Okapi
E	20.69	Title_Cosine	15.38	Title_Cosine
F	45.83	Amsler	53.15	Amsler
G	63.37	Title_Okapi	66.27	Title_Okapi
H	65.58	Title_Okapi	69.27	Title_Okapi
I	58.90	Title_Okapi	61.84	Title_Okapi
J	22.63	Title_Cosine	18.58	Title_Cosine
K	66.42	Title_Cosine	68.38	Title_Cosine
Avg F1	52.33	–	54.43	–

From Table 3 it can be seen that title-based evidence is the most important for the majority of the classes. For those classes whose best performer was a citation-based evidence, Amsler was the best measure. From Table 4, it can be seen that the best types of evidence are the title-based ones, followed by citation-based and abstract-based evidence, respectively. This should be expected since title is the only evidence which appears in all the documents while the information provided by the citation structure is very incomplete and imprecise.

Table 4. Macro F1 on individual evidence

Evidence	Macro F1 (15%)	Macro F1(30%)
Abstract_BagOfWords	17.64	19.50
Abstract_Cosine	32.59	34.29
Abstract_Okapi	32.60	33.86
Bib_Coup	31.27	34.73
Amsler	37.44	41.23
Co-citation	21.88	27.31
Comp_Authority	26.53	32.09
Comp_Hub	29.93	33.95
Title_BagOfWords	45.68	49.20
Title_Cosine	50.41	52.53
Title_Okapi	50.06	52.53

5.3 Experimental Set Up

We run several experiments on the two training samples using different parameters. Particularly, we noticed that a larger population size and different random seeds[2] produce better results. On the other hand, they have a huge effect on the training time. The settings for our GP system are shown in Table 5. In the next section, we only report performance of the best tree in the training sets applied to the test sets.

Table 5. GP system experimental settings

Population size	400 (only for 15% sample), 300
Crossover rate	0.7
Mutation rate	0.25
Reproduction rate	0.05
Generations	30 (only for 15% sample), 20
No. of seeds	4 (maximum)

5.4 Experimental Results

We demonstrate the effectiveness of our classification framework in three ways: 1) by comparing its performance against the best baselines per class in isolation; 2) by comparing it against a majority voting of classifiers using those best baseline similarity functions; and 3) by comparing our experimental results with the results achieved through a content-based SVM classifier[3]. While

[2] Random seed impacts population initialization, which will accordingly affect the final learning results.

[3] For content we used a concatenation of title + abstract.

the third comparison may seem inappropriate since the classifiers are trained and applied to different types of content, it does provide a good idea of the core performance of our method, clearly showing it as a valid alternative in classification tasks similar to the ones used in this paper. The SVM classifier has been extensively evaluated for text classification on reference collections, thus offering a strong baseline for comparison. A SVM classifier was first used in text classification by Joachims [24]. It works over a vector space, where the problem is to find a hyperplane with the maximal margin of separation between two classes. This hyperplane can be uniquely constructed by solving a constrained quadratic optimization problem, by means of quadratic programming techniques.

In a comparison class by class between the majority GP (Table 6 columns 6 and 7) and the best evidence (Table 3) in isolation, the majority GP outperforms the best evidence in 10 out of 11 classes in both samples (only the performance for class A is worse).

Table 6. Comparison between Majority Voting using the best evidence per class in isolation, SVM and Majority GP

Class	Majority Best Evidence		SVM		Majority GP	
	15%	30%	15%	30%	15%	30%
A	32.61	39.77	44.65	52.19	32.97	41.42
B	63.35	68.33	68.41	72.94	75.43	78.19
C	62.22	64.27	65.06	68.49	66.96	69.81
D	68.28	69.32	72.13	74.67	75.22	76.44
E	17.54	12.00	11.76	4.13	26.32	20.25
F	41.83	47.36	51.01	56.03	55.76	61.04
G	64.90	68.22	64.65	70.3	70.41	74.30
H	67.94	71.43	71.45	74.03	74.68	78.15
I	59.80	63.05	64.21	68.88	69.03	72.48
J	20.10	16.39	18.38	19.93	24.46	24.15
K	67.09	69.45	70.08	73.58	70.01	72.67
Avg F1	51.42	53.60	54.71	57.74	58.30	60.81

When comparing the majority GP (Table 6 columns 6 and 7) against the majority using the best evidence (Table 6 columns 2 and 3) it is clear that the former presents better performance: we obtain a gain of 13.38% in the 15% sample and 13.35% in the 30% sample.

Finally, it can be seen from Table 6 that the performance of SVM is slightly worse than that of the majority GP, which suggests that we have a comparable classification method.

6 Related Work

In the World Wide Web environment, several works have successfully used link information to improve classification performance. Different information about links, such as anchor text describing the links, text from the paragraphs surrounding the links, and terms extracted from linked documents, has been used to classify documents. For example, Furnkranz et al. [18], Glover et al. [20] and Sun et al. [45] show that anchor text and the paragraphs and headlines that surround the links helped improve the classification result. Similarly, Yang et al. [47] claimed that the use of terms from linked documents works better when neighboring documents are all in the same class.

Other researchers applied learning algorithms to handle both the text components of the Web pages and the links between them. For example, Joachims et al. [25] studied the combination of support vector machine kernel functions representing co-citation and content information. Cohn et al. [6] show that a combination of link-based and content-based probabilistic methods improved classification performance. Fisher and Everson [17] extended this work by showing that link information is useful when the document collection has a sufficiently high density in the linkage matrix and the links are of high quality.

Chakrabarti et al. [3] estimate the category of test documents by studying the known classes of neighboring training documents. Oh et al. [36] improved on this work by using a filtering process to further refine the set of linked documents to be used. Calado et al. [2] proposed a Bayesian network model to combine the output of a content-based classifier and the information provided by the document's link structure.

Both GP and GA have been applied to the information retrieval field [7,11,12,14–16,21,22,33,37,38] as well as for data classification. In [4], Cheang et al. proposed a Genetic Parallel Programming (GPP) classifier to evolve parallel programs for Data Classification problems. Eggermont et al. [10] proposed several methods using techniques from the field of machine learning to refine and reduce the search space sizes for evolving decision trees for data classification. They showed how the classification performance got improved by reducing the search space in which a tree-based Genetic Programming (gp) algorithm searches. Kishore et al. [28] proposed a methodology for GP-based n-class pattern classification. They modeled the given n-class problem as n two-class problems. And a genetic programming classifier expression (GPCE) was evolved as a discriminant function for each class. In [27], they integrated the GP classifier with feature space partitioning (FSP) for localized learning to improve pattern classification. Genetic Programming has also been used to text classification through the use of a parse-tree. In [5], Clack et al. used Genetic Programming to route in-bound documents to a central classifier which autonomously routed documents to interested research groups within a large organization. The central classifier used a parse tree to match the aspects of a document to nodes of the tree. The tree ultimately reduces to a single numberical value – the classification or "confidence value" during evaluation.

Castillo et al. [9] developed a multistrategy classifier system for document classification. They applied different types of classifiers (e.g., Naive Bayes, Decisions Trees) to different parts of the document (e.g., titles, references). Genetic algorithms are applied for feature selection as well as for combining the output of the different classifiers.

Differently, our proposed work tries to combine several types of content-based and citation-based evidence through Genetic Programming to improve text classification. Each evidence is represented as a document-document similarity matrix. GP is used to discover new and better similarity functions by combining the available evidence, therefore reducing the aforementioned problems of scarce and imprecise information. There is little prior work on applying GP to large scale text classification. We seek to fill the near-void that exists in this area; our approach appears to have promise for broad impact.

7 Conclusion

In this paper, we considered the problem of classification in the context of a document collections where textual content is scarce and imprecise citation information exists. A framework for tackling this problem based on Genetic Programming has been proposed and tested. Our experimental results on two different sets of documents have demonstrated that the GP framework can be used to discover better similarity functions that, when applied to a *kNN* algorithm, can produce better classifiers than ones using individual evidence in isolation. Our experiments also showed that the framework achieved results as good as traditional content-based SVM classifiers.

Our future work includes an extensive and comprehensive analysis why, when and how GP-based classification framework works and how the obtained different similarity trees are related. In addition, we want to improve scalability through parallel computation. Other sampling strategies like active sampling [34, 40, 44] will also be used to within our classification framework. We also want to test this framework in different document collections to see its viability (e.g., the Web). We also will use the GP framework to combine the current results with content-based classifiers as the SVM classifiers used in the comparisons. Besides that, we want to improve our current evidence, for example, using better methods for citation matching, by trying to fix some OCR errors and using different matching strategies. Finally, new terminals (features) representing additional evidence may be explored. For example, matrices representing relations from other data spaces like author information and patterns of authorship in certain categories can be explored.

Acknowledgements

This research work was funded in part by the NSF through grants DUE-0136690, DUE-0121679 and IIS-0086227 to Edward A. Fox(Baoping Zhang and Yuxin Chen). Marcos André Gonçalves is supported by CAPES, process 1702-98 and a fellowship by American Online (AOL). Pável Calado is supported by MCT/FCT scholarship grant SFRH/BD/4662/2001. Marco Cristo is supported by Fucapi, Technology Foundation, Manaus, AM, Brazil.

References

1. Robert Amsler. Application of citation-based automatic classification. Technical report, The University of Texas at Austin, Linguistics Research Center, Austin, TX, December 1972.
2. Pável Calado, Marco Cristo, Edleno Silva de Moura, Nivio Ziviani, Berthier A. Ribeiro-Neto, and Marcos André Gonçalves. Combining link-based and content-based methods for Web document classification. In *Proceedings of CIKM-03, 12th ACM International Conference on Information and Knowledge Management*, pages 394–401, New Orleans, US, 2003. ACM Press, New York, US.
3. Soumen Chakrabarti, Byron Dom, and Piotr Indyk. Enhanced hypertext categorization using hyperlinks. In *Proceedings of the ACM SIGMOD International Conference on Management of Data*, pages 307–318, Seattle, Washington, June 1998.
4. Sin Man Cheang, Kin Hong Lee, and Kwong Sak Leung. Data classification using genetic parallel programming. In E. Cantú-Paz, J. A. Foster, K. Deb, D. Davis, R. Roy, U.-M. O'Reilly, H.-G. Beyer, R. Standish, G. Kendall, S. Wilson, M. Harman, J. Wegener, D. Dasgupta, M. A. Potter, A. C. Schultz, K. Dowsland, N. Jonoska, and J. Miller, editors, *Genetic and Evolutionary Computation – GECCO-2003*, volume 2724 of *LNCS*, pages 1918–1919, Chicago, 12–16 July 2003. Springer-Verlag.
5. Chris Clack, Johnny Farringdon, Peter Lidwell, and Tina Yu. Autonomous document classification for business. In *AGENTS '97: Proceedings of the first international conference on Autonomous agents*, pages 201–208. ACM Press, 1997.
6. David Cohn and Thomas Hofmann. The missing link - a probabilistic model of document content and hypertext connectivity. In Todd K. Leen, Thomas G. Dietterich, and Volker Tresp, editors, *Advances in Neural Information Processing Systems 13*, pages 430–436. MIT Press, 2001.
7. I. De Falco, A. Della Cioppa, and E. Tarantino. Discovering interesting classification rules with genetic programming. *Applied Soft Computing*, 1(4F):257–269, May 2001.
8. Jeffrey Dean and Monika Rauch Henzinger. Finding related pages in the World Wide Web. *Computer Networks*, 31(11–16):1467–1479, May 1999. Also in Proceedings of the 8th International World Wide Web Conference.
9. M. Dolores del Castillo and José Ignacio Serrano. A multistrategy approach for digital text categorization from imbalanced documents. *SIGKDD Explor. Newsl.*, 6(1):70–79, 2004.
10. J. Eggermont, J. N. Kok, and W. A. Kosters. Genetic programming for data classification: Refining the search space. In T. Heskes, P. Lucas, L. Vuurpijl, and W. Wiegerinck, editors, *Proceedings of the Fivteenth Belgium/Netherlands Conference on Artificial Intelligence (BNAIC'03)*, pages 123–130, Nijmegen, The Netherlands, 23-24 October 2003.

11. Weiguo Fan, Edward A. Fox, Praveen Pathak, and Harris Wu. The effects of fitness functions on genetic programming-based ranking discovery for web search. *Journal of the American Society for Information Science and Technology*, 55(7):628–636, 2004.
12. Weiguo Fan, Michael D. Gordon, and Praveen Pathak. Personalization of search engine services for effective retrieval and knowledge management. In *The Proceedings of the International Conference on Information Systems 2000*, pages 20–34, 2000.
13. Weiguo Fan, Michael D. Gordon, and Praveen Pathak. Discovery of context-specific ranking functions for effective information retrieval using genetic programming. *IEEE Transactions on Knowledge and Data Engineering*, 16(4):523–527, 2004.
14. Weiguo Fan, Michael D. Gordon, and Praveen Pathak. A generic ranking function discovery framework by genetic programming for information retrieval. *Information Processing and Management*, 40(4):587–602, 2004.
15. Weiguo Fan, Michael D. Gordon, Praveen Pathak, Wensi Xi, and Edward A. Fox. Ranking function optimization for effective web search by genetic programming: An empirical study. In *Proceedings of 37th Hawaii International Conference on System Sciences*, Hawaii, 2004. IEEE.
16. Weiguo Fan, Ming Luo, Li Wang, Wensi Xi, and Edward A. Fox. Tuning before feedback: combining ranking function discovery and blind feedback for robust retrieval. In *Proceedings of the 27th Annual International ACM SIGIR Conference*, U.K., 2004. ACM.
17. Michelle Fisher and Richard Everson. When are links useful? Experiments in text classification. In F. Sebastianini, editor, *Proceedings of the 25th annual European conference on Information Retrieval Research, ECIR 2003*, pages 41–56. Springer-Verlag, Berlin, Heidelberg, DE, 2003.
18. Johannes Furnkranz. Exploiting structural information for text classification on the WWW. In *Intelligent Data Analysis*, pages 487–498, 1999.
19. Lee Giles. Citeseer: An automatic citation indexing system. December 16 1998.
20. Eric J. Glover, Kostas Tsioutsiouliklis, Steve Lawrence, David M. Pennock, and Gary W. Flake. Using Web structure for classifying and describing Web pages. In *Proceedings of WWW-02, International Conference on the World Wide Web*, 2002.
21. M. D. Gordon. User-based document clustering by redescribing subject descriptions with a genetic algorithm. *Journal of the American Society for Information Science*, 42(5):311–322, June 1991.
22. Michael Gordon. Probabilistic and genetic algorithms for document retrieval. *Communications of the ACM*, 31(10):1208–1218, October 1988.
23. Norbert Gövert, Mounia Lalmas, and Norbert Fuhr. A probabilistic description-oriented approach for categorizing web documents. In *Proceedings of the 8th International Conference on Information and Knowledge Management CIKM 99*, pages 475–482, Kansas City, Missouri, USA, November 1999.
24. Thorsten Joachims. Text categorization with support vector machines: learning with many relevant features. In *Proceedings of ECML-98, 10th European Conference on Machine Learning*, pages 137–142, Chemnitz, Germany, April 1998.
25. Thorsten Joachims, Nello Cristianini, and John Shawe-Taylor. Composite kernels for hypertext categorisation. In Carla Brodley and Andrea Danyluk, editors,

Proceedings of ICML-01, 18th International Conference on Machine Learning, pages 250–257, Williams College, US, 2001. Morgan Kaufmann Publishers, San Francisco, US.
26. M. M. Kessler. Bibliographic coupling between scientific papers. *American Documentation*, 14(1):10–25, January 1963.
27. J. K. Kishore, L. M. Patnaik, V. Mani, and V. K. Agrawal. Genetic programming based pattern classification with feature space partitioning. *Information Sciences*, 131(1-4):65–86, January 2001.
28. J. K. Kishore, Lalit M. Patnaik, V. Mani, and V. K. Agrawal. Application of genetic programming for multicategory pattern classification. *IEEE Trans. Evolutionary Computation*, 4(3):242–258, 2000.
29. Jon M. Kleinberg. Authoritative sources in a hyperlinked environment. *Journal of the ACM (JACM)*, 46(5):604–632, 1999.
30. John R. Koza. *Genetic programming: On the programming of computers by natural selection.* MIT Press, Cambridge, Mass., 1992.
31. S. Lawrence, C. L. Giles, and K. Bollacker. "Digital Libraries and Autonomous Citation Indexing". *IEEE Computer*, 32(6):67–71, 1999.
32. Steve Lawrence, C. Lee Giles, and Kurt D. Bollacker. Autonomous citation matching. In Oren Etzioni, Jörg P. Müller, and Jeffrey M. Bradshaw, editors, *Proceedings of the Third International Conference on Autonomous Agents (Agents'99)*, pages 392–393, Seattle, WA, USA, 1999. ACM Press.
33. M. J. Martin-Bautista, M. Vila, and H. L. Larsen. A fuzzy genetic algorithm approach to an adaptive information retrieval agent. *American Society for Information Science*, 50:760–771, 1999.
34. Andrew Kachites McCallum and Kamal Nigam. Employing EM and pool-based active learning for text classification. In *Proc. 15th International Conf. on Machine Learning*, pages 350–358. Morgan Kaufmann, San Francisco, CA, 1998.
35. Frederic C. Misch, editor. *Webster's Ninth New Collegiate Dictionary.* Merriam-Webster Inc., Springfield, Massachusetts, 1988.
36. Hyo-Jung Oh, Sung Hyon Myaeng, and Mann-Ho Lee. A practical hypertext catergorization method using links and incrementally available class information. In *Proceedings of the 23rd annual international ACM SIGIR conference on research and development in information retrieval*, pages 264–271. ACM Press, 2000.
37. P. Pathak, M. Gordon, and W. Fan. Effective information retrieval using genetic algorithms based matching function adaptation. In *Proceedings of the 33rd Hawaii International Conference on System Science (HICSS)*, Hawaii, USA, 2000.
38. Vijay V. Raghavan and Brijesh Agarwal. Optimal determination of user-oriented clusters: an application for the reproductive plan. In John J. Grefenstette, editor, *Proceedings of the 2nd International Conference on Genetic Algorithms and their Applications*, pages 241–246, Cambridge, MA, July 1987. Lawrence Erlbaum Associates.
39. S. E. Robertson, S. Walker, and M. M. Beaulieu. Okapi at TREC-4. In *NIST Special Publication 500-236: The Fourth Text REtrieval Conference (TREC-4)*, pages 73–96, 1995.
40. Maytal Saar-Tsechansky and Foster Provost. Active learning for class probability estimation and ranking. In Bernhard Nebel, editor, *Proceedings of the Seventeenth International Conference on Artificial Intelligence (IJCAI-01)*, pages

911–920, San Francisco, CA, August 4–10 2001. Morgan Kaufmann Publishers, Inc.
41. Gerard Salton. *Automatic Text Processing.* Addison-Wesley, Boston, Massachusetts, USA, 1989.
42. Gerard Salton and Chris Buckley. Term-weighting approaches in automatic text retrieval. *Information Processing and Management*, 24(5):513–523, 1988.
43. Henry G. Small. Co-citation in the scientific literature: A new measure of relationship between two documents. *Journal of the American Society for Information Science*, 24(4):265–269, July 1973.
44. A. Srinivasan. A study of two sampling methods for analysing large datasets with ILP. *Data Mining and Knowledge Discovery*, 3(1):95–123, 1999.
45. Aixin Sun, Ee-Peng Lim, and Wee-Keong Ng. Web classification using support vector machine. In *Proceedings of the fourth international workshop on Web information and data management*, pages 96–99. ACM Press, 2002.
46. Yiming Yang. Expert network: effective and efficient learning from human decisions in text categorisation and retrieval. In W. Bruce Croft and Cornelis J. van Rijsbergen, editors, *Proceedings of SIGIR-94, 17th ACM International Conference on Research and Development in Information Retrieval*, pages 13–22, Dublin, IE, 1994. Springer Verlag, Heidelberg, DE.
47. Yiming Yang, Seán Slattery, and Rayid Ghani. A study of approaches to hypertext categorization. *Journal of Intelligent Information Systems*, 18(2-3):219–241, 2002.

Part II

Semantic Web

Adding a Trust Layer to Semantic Web Metadata

Paolo Ceravolo, Ernesto Damiani, and Marco Viviani

Università degli Studi di Milano, Dipartimento di Tecnologie dell'Informazione, Via Bramante, 65 – 26013 Crema (CR) – Italia
{ceravolo, damiani, viviani}@dti.unimi.it
http://ra.crema.unimi.it

Summary. We outline the architecture of a modular Trust Layer that can be superimposed to generic semantic Web-style metadata generation facilities. Also, we propose an experimental setting to generate and validate trust assertions on classification metadata generated by different tools (including our *ClassBuilder*) after a process of metadata standardization. Our experimentation is aimed at validating the role of our Trust Layer as a non-intrusive, user-centered quality improver for automatically generated metadata.

1 Introduction

The growing interest in studying and developing systems for generating and managing metadata is strictly connected to the increasing need of sharing and managing new knowledge about heterogeneous data available within organizations. Typically, metadata provide annotations specifying content, quality, type, creation, and even location of a data item. Although a number of specialized formats based on *Resource Description Framework* (RDF) are available, in principle metadata can be stored in any format such as a free text sentences, *Extensible Markup Language* (XML) fragments, or database records.

There are a number of advantages in using information extracted from data instead of data themselves. First of all, because of their small size compared to the data they describe, metadata are much more easily shareable than data. Thanks to metadata shareability, information about data is usually readily available to anyone seeking it. Thus, metadata make information discovery easier and reduce data duplication. On the other hand, metadata can be generated by a number of sources (the data owner, other users, automatic tools) and may or may not be digitally signed by their author. Therefore, in general metadata generated by different sources have non-uniform *trustworthiness*. In order to take full advantage of them, it is therefore fundamental that users are aware of each metadata level of trustworthiness and that metadata trustworthiness is continuously updated, e.g. based on the reported view

P. Ceravolo et al.: *Adding a Trust Layer to Semantic Web Metadata*, StudFuzz **197**, 87–104 (2006)
www.springerlink.com

of the user community. Trustworthiness assessment is even more important when the original source of metadata is an automatic metadata generator like our *ClassBuilder*[1], whose error rate, though hopefully low, is in general not negligible. Our aim is to develop a *Trust Layer*, based on a *Trust Manager* module capable of collecting human behavior in the navigation of data (or metadata) and to compute variations to trust values on metadata. In our approach, continuous non-intrusive monitoring of user behavior is complemented with explicit collection of votes on metadata based on user roles. Our trust layer will be validated by:

1. programmable agents playing the roles of users navigating data and metadata;
2. a trust visualization module, called *Publication Center*, capable of showing evolving *trust landscapes* on metadata and of computing *trust-based* views on them.

2 The Architecture

Before describing in detail our proposed Trust Layer, let us add some preliminary remarks on related work. Traditionally [3, 11] research approaches distinguish between two main types of trust management systems, namely *Centralized Reputation Systems* and *Distributed Reputation Systems*.

There is a clear distinction between *trust* and *reputation*. In general the trust T can be computed based on its reputation R, that is $T = \phi(R, t)$, where t is the time elapsed since when the reputation was last modified.

In centralized reputation systems, trust information is collected from members of the community in the form of their *ratings* on resources. The central authority collects all the ratings and computes a global score for each resource.

In a distributed reputation system there is no central location for submitting ratings and computing resources' reputation scores; instead, there are a number of distributed stores (*trust repositories*) where ratings can be submitted. In a "pure" peer-to-peer (P2P) setting, each user has its own repository of trust values.

In our approach, trust is attached to metadata describing resources rather than to resources themselves. We start from an initial "metadata base" storing assertions describing resources. This first layer of metadata is produced through an untrusted production process, such as an automatic data classifier. For this reason, we set up an assessment mechanism to upgrade it. While interacting with resources, users can provide important feedback on metadata trustworthiness. These information is then captured and transformed in

[1] ClassBuilder is a knowledge extraction tool developed by the Knowledge Management Group of the Department of Information Technologies of the University of Milan, based on ideas explained in [6, 8]. It can be downloaded from `http://ra.crema.unimi.it/kiwi6.html#ClassBuilder`

a second metadata layer composed of assertions expressing the level of trust of the assertions of the first layer[2]. This second layer can be computed by a central server or by distributed clients; in both cases, the trust degree associated to each assertion must be aggregated and the result provided to all interested clients. On the basis of this aggregated trust degree and on their attitude, clients can produce their *personal views* on the metadata base, using a movable threshold to discard untrusted assertions.

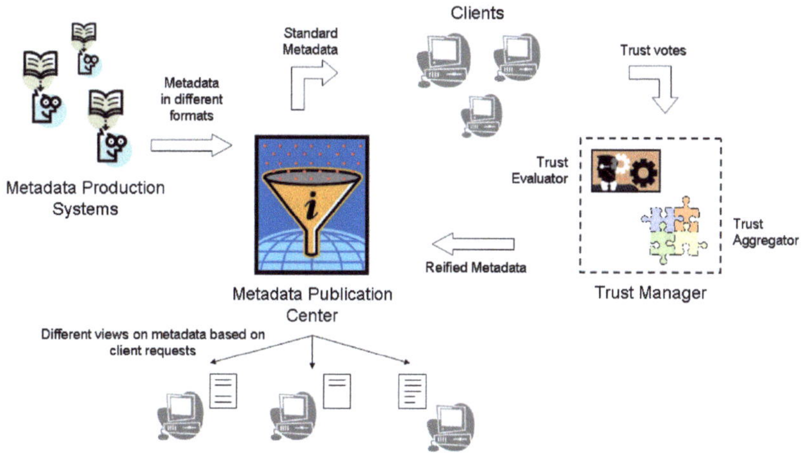

Fig. 1. The architecture of the Trust Layer.

Figure 1 shows a detailed view of this scenario. The architecture of our Trust Layer is composed of a centralized *Metadata Publication Center* that collects and displays metadata assertions, possibly in different formats and coming from different sources. Our Publication Center can be regarded as a semantic search engine, containing metadata provided by automatic generators crawling the web or by other interested parties. Of course, our Center will assign different trust values to assertions depending on their origin: assertions manually provided by a domain expert are much more reliable than automatically generated ones submitted by a crawler.

All metadata in the Publication Center are indexed and a group of *Clients* interacts with them by navigating them and providing implicitly (with their behavior) or explicitly (by means of an explicit vote) an evaluation about metadata trustworthiness. This trust-related information is passed by the Publication Center on to the Trust Manager in the form of new assertions. Trust assertions, which we call *Trust Metadata*, are built using the well-known technique of *reification* [19,20]. This choice allows our system to interact with

[2] Although we shall not deal with digital signatures in this chapter, it is important to remark that meta-assertions could be enriched with a *digital signature*, providing a way of indicating clearly and unambiguously the author of the assertions.

heterogeneous sources of metadata because our Trust Metadata syntax is *not dependent* on the format of the original assertions.

Let us now examine the process in more detail. In a first phase, clients navigate the collection of metadata by selecting a single assertion (or an *assertion pattern*). The result is (a list of links to) the unfiltered set of resources indexed by the selected assertion(s). According to their evaluation of results, clients may explicitly (e.g. by clicking on a "Confirm Association" button) assign a rating to the quality of the link, i.e. to the underlying assertion. More frequently, a non-intrusive technique will be used: when users click on data they are interested in in order to display them, the system takes this action as an implicit vote confirming the quality of the association between data and metadata (after all, it was the metadata that allowed the user to reach the data). Of course, explicit votes count more than implicit ones; also, aggregation should be performed by taking into account voters' profiles (when available) and a wealth of other context metadata. We shall elaborate further on these issues in the remainder of the paper. For now, it suffices to say that once enough votes have been collected and aggregated, our Trust Manager is able to associate *community trust values* to Trust Metadata assertions. After this phase, trust values are made available to all interested parties, and clients can compute different views on the original metadata based on our Trust Metadata (e.g. all metadata whose trust value $\geq x$); we will call these *Trust Constraints.* A suitable *aggregation function* is used to collect votes and update trust values during the whole system life; hence, trust constraints are continually updated as well.

Summarizing our architecture, our Trust Manager is composed of two modules:

- *Trust evaluator*: it examines metadata and evaluates their reliability;
- *Trust aggregator*: it aggregates all inputs coming from the trust evaluator clients according to a suitable aggregation function.

The overall Trust Manager is the computing engine behind the Publication Center module that provides interested parties with a visual overview on the metadata reliability distribution.

All these modules can be deployed and evolve separately; taken together, they compose a complete Trust Layer, whose components communicate by means of standard *web services* interfaces specified in `WSDL`. The same standard interfaces are used by distributed clients to access the Trust Layer services. This will give us the opportunity to test the whole system despite the fact that single modules can evolve with different speeds.

2.1 The Metadata Format

As mentioned in Section 2 our system is aimed at improving the quality of a metadata base automatically updated by metadata generators and manually added to by human users. In general, metadata generation systems produce metadata in a standard format. In other words, resources are associated to assertions expressing a statement in a form allowed by a shared vocabulary. Our Trust Layer generates a second metadata layer, composed of special-purpose metadata expressing a degree of trust on the first layer's metadata. The availability of this second layer improves the quality of the output of all metadata production procedures where the error rate is not negligible. We tested our architecture on a corpus of metadata automatically produced by a clustering algorithm running on a set of semi-structured documents [6] and producing assertions in RDF format. Our ClassBuilder tool implements a set of functions for detecting typical instances from a document corpus. The detection process allows, as a final result, to associate a resource (a document of the corpus) to a class (a typical instance of the domain).

Figure 2 shows some sample assertions produced by ClassBuilder.

```
<?xml version="1.0" encoding="UTF-8"?>

<rdf:RDF xmlns:rdf="http://www.w3.org/1999/02/22-rdf-syntax-ns#"
         xmlns:rdfs="http://www.w3.org/2000/01/rdf-schema#"
         xmlns:ont="http://OntoExtractor.crema.unimi.it/onto/Music.rdfs#"
         xmlns:indi="http://kiwi.crema.unimi.it/kiwi-schema#">

  <rdf:Description rdf:about="http://MV.crema.unimi.it/file/Album2.xml">
    <indi:assert rdf:resource=
      "http://OntoExtractor.crema.unimi.it/onto/Music.rdfs#SingleAlbum"/>
  </rdf:Description>

  <rdf:Description rdf:about="http://MV.crema.unimi.it/file/Album4.xml">
    <indi:assert rdf:resource=
      "http://OntoExtractor.crema.unimi.it/onto/Music.rdfs#SingleAlbum"/>
  </rdf:Description>

  <rdf:Description rdf:about="http://MV.crema.unimi.it/file/Album1.xml">
    <indi:assert rdf:resource=
      "http://OntoExtractor.crema.unimi.it/onto/Music.rdfs#SingleArtistAlbum"/>
  </rdf:Description>

  <rdf:Description rdf:about="http://MV.crema.unimi.it/file/Album5.xml">
    <indi:assert rdf:resource=
      "http://OntoExtractor.crema.unimi.it/onto/Music.rdfs#GreatesHits"/>
  </rdf:Description>

</rdf:RDF>
```

Fig. 2. ClassBuilder's RDF-based metadata.

In our application the assertions are essentially links between resources and concepts. Following the RDF standard, the assertions are triples composed of a `subject`, a `predicate` and an `object`. The `subject` and the `object` are general resources or classes from a shared schema, and the `predicate` is a relation between classes and resources following the constraints of the shared schema.

In the specific case of our example the `rdf:Description` element refers to a resource by means of the `rdf:about` property. Besides, `rdf:Description`

```
<?xml version="1.0" encoding="UTF-8"?>
<rdf:RDF xmlns:rdf="http://www.w3.org/1999/02/22-rdf-syntax-ns#"
         xmlns:rdfs="http://www.w3.org/2000/01/rdf-schema#"
         xmlns:ont="http://OntoExtractor.crema.unimi.it/onto/Music.rdfs#"
         xmlns:indi="http://kiwi.crema.unimi.it/kiwi-schema#">

  <rdfs:Class rdf:about=
    "http://OntoExtractor.crema.unimi.it/onto/Music.rdfs#SingleAlbum"
    rdfs:comment="An album containing a single musical track">
    <rdfs:subClassOf rdf:resource="http://www.w3.org/2000/01/rdf-schema#Resource"/>
  </rdfs:Class>

  <rdfs:Class rdf:about=
    "http://OntoExtractor.crema.unimi.it/onto/Music.rdfs#SingleArtistAlbum"
    rdfs:comment="An album containing multiple tracks belonging to the same artist">
    <rdfs:subClassOf rdf:resource="http://www.w3.org/2000/01/rdf-schema#Resource"/>
  </rdfs:Class>

  <rdfs:Class rdf:about=
    "http://OntoExtractor.crema.unimi.it/onto/Music.rdfs#GreatestHits"
    rdfs:comment="An album containing tracks belonging to albums of the same artist">
    <rdfs:subClassOf rdf:resource="http://www.w3.org/2000/01/rdf-schema#Resource"/>
  </rdfs:Class>

</rdf:RDF>
```

Fig. 3. The ClassBuilder Music RDF Schema.

refers to a class present in the domain vocabulary (the RDF Schema referred in Fig. 3).

Metadata assertions binding a resource to a concept are produced as output by most metadata generators. A system supporting a manual annotation process for metadata production should produce more complex assertions such as those in Fig. 4.

In this setting, a metadata navigation tool like our Semantic Navigator[3] (see Fig. 5) is enriched with a "Confirm Association" button assigning a rating to the quality of the selected metadata assertion.

During the experimentation, a fraction of the client simulators express explicit votes with a pre-set (low) fixed error rate; both the probability of expressing explicit votes and the error rate will be experimental parameters. Other simulators produce implicit votes, simulating users that use metadata to reach a data item and click on it to display it. Modelling implicit vote rate of correctness is more difficult, as it depends on several interacting causes. We rely on a noisy channel probability distribution [5] taking into accounts probability of errors due to distraction, etc. Noisy channel models have been applied to a wide range of simple tasks involving human users, including spelling. Comparatively less research has gone into applying channel models to more complex tasks, such as form filling. Our channel model for mistakes is based on the distribution of human errors in generic text editing. The model

[3] The Semantic Navigator is our open-source tool used to navigate metadata and to obtain custom views on domain ontologies.
It can be downloaded from `http://ra.crema.unimi.it/kiwi6.html#Semantic_Navigator`

```
<?xml version="1.0" encoding="UTF-8"?>
<rdf:RDF xmlns:rdf="http://www.w3.org/1999/02/22-rdf-syntax-ns#"
         xmlns:rdfs="http://www.w3.org/2000/01/rdf-schema#"
         xmlns:ont="http://OntoExtractor.crema.unimi.it/onto/Music.rdfs#"
         xmlns:indi="http://kiwi.crema.unimi.it/kiwi-schema#">

  <rdf:Description rdf:about="http://MV.crema.unimi.it/file/Album2.xml">
    <indi:assert rdf:resource=
      "http://OntoExtractor.crema.unimi.it/fileRdF/Statement.rdf#ST_0"/>
    <indi:assert rdf:resource=
      "http://OntoExtractor.crema.unimi.it/fileRdF/Statement.rdf#ST_8"/>
  </rdf:Description>

  <rdf:Statement rdf:about=
    "http://OntoExtractor.crema.unimi.it/fileRdF/Statement.rdf#ST_1">
    <rdf:subject rdf:resource=
      "http://OntoExtractor.crema.unimi.it/onto/Music.rdfs#Album"/>
    <rdf:predicate rdf:resource=
      "http://OntoExtractor.crema.unimi.it/onto/Music.rdfs#recorded_in"/>
    <rdf:object rdf:resource=
      "http://OntoExtractor.crema.unimi.it/onto/Music.rdfs#Milan"/>
  </rdf:Statement>

  <rdf:Statement rdf:about=
    "http://OntoExtractor.crema.unimi.it/fileRdF/Statement.rdf#ST_8">
    <rdf:subject rdf:resource=
      "http://OntoExtractor.crema.unimi.it/onto/Music.rdfs#Artist"/>
    <rdf:predicate rdf:resource=
      "http://OntoExtractor.crema.unimi.it/onto/Music.rdfs#has_name"/>
    <rdf:object rdf:resource=
      "http://OntoExtractor.crema.unimi.it/onto/Music.rdfs#Abbado"/>
  </rdf:Statement>

</rdf:RDF>
```

Fig. 4. Sample RDF complex assertions.

assumes that a noisy flow of click actions is generated as follows: Given a vocabulary D of concepts, a person chooses (based on a concept $d \in D$) a link l to follow, according to a probability distribution $P(d)$. Then the person attempts to click on l, but cognitive noise (i.e., imperfect knowledge or understanding of the vocabulary) induces the person to click on a link l' associated to term $d' \in D$, according to the distribution $P(d'|d)$. For instance, when a person uses our Semantic Navigator we would expect $P(archaeology|archaeology)$ (expressing the probability that the person clicking on a resource indexed under the archaeology concept actually considers the resource to be related to archaeology) to be very high, $P(archaeology|history)$ to be relatively high and $P(archaelogy|chemistry)$ to be extremely low. In the remainder of this chapter, we will refer to the channel model as our *error model.*

2.2 Modelling User Behavior in Implicit Voting

In the case of simple assertions, we assume the vocabulary D to be organized according to the RDF Schema standard, i.e. as a graph of related concepts (semantic network). Our error model defines the *level k confusion set* C_k of

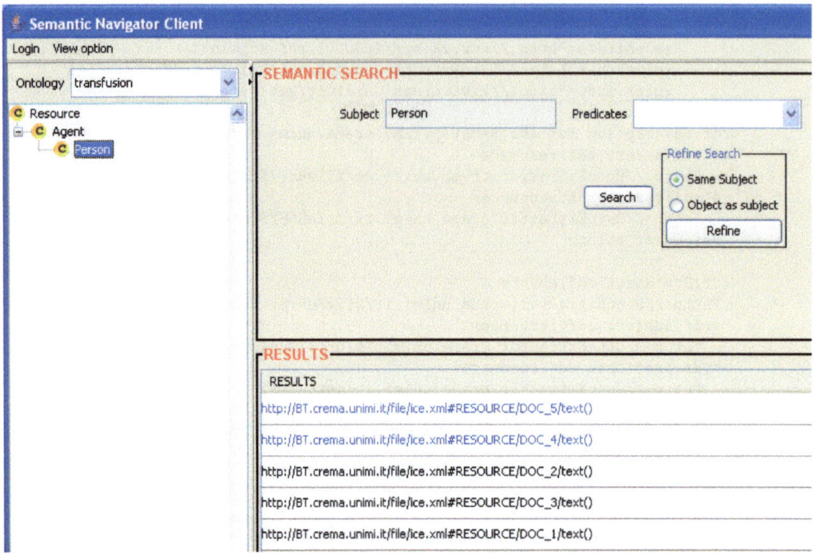

Fig. 5. The interface of our Semantic Navigator (in this case showing a custom view on an ontology).

a vocabulary term d to include d itself, as well as all other concepts in the semantic network D such that they are reachable from d in less than k steps. Let now C be the number of terms in the confusion set C_k of d for a suitable value of k. Then, we define

$$P(d'|d) = \frac{1-\alpha}{C-1} \tag{1}$$

Equation (1) holds unless $d' = d$, in which case $P(d|d) = \alpha$. Value α is a tunable experimental parameter which should be more or less close to 1 depending on the user's level of competence on the specific domain modelled by the semantic network D. Supposing that to each user u is associated a vocabulary D_u expressing her competence, the simple model of Eq. (1) can be enriched so that the probability of expressing a mistaken vote will also depend on the user profile. Other context information can also be taken into account.

2.3 Trust Assertions' Format

As we already mentioned, we are not forced to restrict the applicability of our Trust Layer to systems using metadata in RDF format. For instance, metadata are often provided in XML format. In those cases we can design a suitable XSLT transformation sheet and convert our metadata base in a format suitable to being annotated as shown in Figs. 2 and 4.

When a user expresses his implicit or explicit vote on a first layer's assertion, he is evaluating how much the association between the resource and the statement is trustworthy. In the specific case this is bound by the `rdf:Description` tag. Using the reification mechanism it is easy to produce an assertion expressing a degree of trust on an association. We call this type of metadata *Trust Metadata*; Fig. 6 shows a typical Trust Metadata assertion, whose associated RDF Schema (RDFS) is shown in Fig. 7. Our `Description` element uses a property called `has-trust` for associating a statement in any form to a `fuzzyval`.

```
<?xml version="1.0" encoding="UTF-8"?>

<rdf:RDF xmlns:rdf="http://www.w3.org/1999/02/22-rdf-syntax-ns#"
         xmlns:rdfs="http://www.w3.org/2000/01/rdf-schema#"
         xmlns:ont="http://OntoExtractor.crema.unimi.it/onto/Music.rdfs#"
         xmlns:indi="http://kiwi.crema.unimi.it/kiwi-schema#">

  <rdf:Statement rdf:about=
    "http://OntoExtractor.crema.unimi.it/fileRdF/Statement.rdf#ST_65">

    <rdf:subject>
      <rdf:Description rdf:about=
         "http://MV.crema.unimi.it/file/Album4.xml">
         <indi:assert rdf:resource=
           "http://OntoExtractor.crema.unimi.it/onto/Music.rdfs#SingleAlbum"/>
      </rdf:Description>
    </rdf:subject>

    <rdf:predicate rdf:resource=
      "http://kiwi.crema.unimi.it/kiwi-schema#has-trust"/>
      <rdf:object>0.7</rdf:object>

  </rdf:Statement>

</rdf:RDF>
```

Fig. 6. A sample of a Trust Metadata assertion.

In Fig. 8 the XML Schema defining the complex type `fuzzyval` as a double in the range from 0 to 1 is provided.

As we have seen, Trust Metadata assertions represent data exchanged between clients and the Trust Manager. Clients send Trust Metadata to the Trust Manager, communicating their personal degree of trust on a first layer's assertion. In a more sophisticated scenario, the Trust Manager itself can describe by means of metadata the different aggregation functions it uses and the allowed functions parameters. Clients can communicate to the Trust Manager the aggregation function they intend to use and the parameters they are requiring. The Trust Manager will then use this information for aggregating all Trust Metadata regarding the same first layer assertion. Periodically, the aggregation algorithm runs and new version of Trust Metadata is made available to clients. This new version can be use by clients to compute their custom view on the metadata base. This way, it is possible to display *trust*

```
<?xml version="1.0" encoding="UTF-8"?>

<rdf:RDF xmlns:rdf="http://www.w3.org/1999/02/22-rdf-syntax-ns#"
         xmlns:rdfs="http://www.w3.org/2000/01/rdf-schema#"
         xmlns:ont="http://OntoExtractor.crema.unimi.it/onto/Music.rdfs#"
         xmlns:indi="http://kiwi.crema.unimi.it/kiwi-schema#">

  <rdf:Property rdf:about="http://kiwi.crema.unimi.it/kiwi-schema#assert">
    <rdfs:label>assert</rdfs:label>
    <rdfs:comment>
      This property enable to relate a Statement to a Resource
      of our document base
    </rdfs:comment>
    <rdfs:domain rdf:resource="http://www.w3.org/2000/01/rdf-schema#Resource"/>
    <rdfs:range rdf:resource="http://www.w3.org/1999/02/22-rdf-syntax-ns#Statement"/>
  </rdf:Property>

  <rdf:Property rdf:about="http://kiwi.crema.unimi.it/kiwi-schema#has-trust">
    <rdfs:label>has-trust</rdfs:label>
    <rdfs:comment>
      This property enable to relate a Statement to a trust degree
    </rdfs:comment>
    <rdfs:domain rdf:resource="http://www.w3.org/1999/02/22-rdf-syntax-ns#Statement"/>
    <rdfs:range rdf:resource="http://kiwi.crema.unimi.it/Fuzzy.xsd#fuzzyval"/>
  </rdf:Property>

</rdf:RDF>
```

Fig. 7. RDF Schema for Trust Metadata assertion.

```
<?xml version="1.0" encoding="UTF-8"?>

<xs:schema xmlns:xs="http://www.w3.org/2001/XMLSchema"
    xmlns="http://example.com/crema/Fuzzy.xsd/">

  <xs:simpleType name="fuzzyval">
    <xs:restriction base="xs:double">
      <xs:minInclusive value="0"/>
      <xs:maxInclusive value="1"/>
    </xs:restriction>
  </xs:simpleType>
</xs:schema>
```

Fig. 8. The definition of the `fuzzyval` XML complex type.

landscapes, obtaining different data visualizations depending on different trust profiles and on the properties of the aggregation function.

In Section 3 we will see how the WOWA operator can be used to aggregate trust values.

3 The Reputation Computation Problem

There are two main problems in managing trust information about metadata: how to collect individual trust values and how to aggregate them. As we discussed in the previous sections, our system collects ratings based on implicit and/or explicit behavior of users that approach the Publication Center. Ratings deriving from implicit user behavior can be computed for example by the time spent by the user working on a resource. On the other hand, explicit votes are usually collected from a subset of the users, depending on their role or their knowledge about the domain (or both of them).

Anyway, the main problem to solve, once trust values have been obtained, is their aggregation. First of all it is necessary to consider if the system works with anonymous users or not. In the first case, every user behavior contributes in the same way to calculate a unique trust value on a resource. In the second case ratings have to be aggregated initially at user level, and subsequently at global level. Several principles for computing reputation and trust measures have been proposed [12]:

- *Summation or Average of ratings.* It is simply the sum of positive and negative ratings separately, keeping a total score as the positive score minus the negative score.
- *Bayesian Systems.* This kind of systems take binary ratings as input and are based on computing reputation scores by statistical updating of beta probability density functions (PDF).
- *Discrete Trust Models.* They are based on human discrete verbal statements to rate performances (ex. *Very Trustworthy, Trustworthy, Untrustworthy, Very Untrustworthy*).
- *Belief Models.* Belief theory is a framework related to probability theory, but where the sum of probabilities over all possible outcomes non necessarily add up to 1, and the remaining probability is interpreted as uncertainty.
- *Fuzzy Models.* Linguistically fuzzy concepts can represent trust, where membership functions describe to what degree an agent can be describes as, for example, trustworthy or not trustworthy.
- *Flow Models.* They compute trust by transitive iteration through looped or arbitrarily log chains.

In our system, we assume that the level of trust for an assertion is the result of the aggregation of *fuzzy values* summarizing the outcome of past human interactions with the indexed metadata. For this purpose the choice of the correct operator to use is fundamental.

3.1 Choice of the Aggregation Operator

The *Aggregation Operators* are mathematical objects that have the function of reducing a set of numbers into a unique representative (or meaningful) number. To do that it is possible to use one of the many operators that satisfies the definition of aggregation operator and its properties as illustrated in [14].

Definition 1. $A : \cup_{n \in \mathbb{N}} [0,1]^n \rightarrow [0,1]$ *is an aggregation operator on the unit interval if the following conditions hold:*

A1 $A(x) = x$ *when unary,*
A2 $A(0, \ldots, 0) = 0$ *and* $A(1, \ldots, 1) = 1$,
A3 $A(x_1, \ldots, x_n) \leq A(y_1, \ldots, y_n)$ *if* $(x_i \leq y_i)\ \forall i = 1, \ldots, n$.

A1 is the property of *Identity*, **A2** are called *Boundary Conditions* and **A3** resembles the *Monotonicity* of the operator A.

Many aggregation operators [17] satisfy these properties; the most widespread are those based on the *Weighted Mean* (WM) [1,15] and those based on the *Ordered Weighted Averaging operator* (OWA) [16], whose behavior is analyzed in [9,10] (a simple arithmetic average would perform a rough compensation between high and low values). The difference between the two functions is in the meaning of the weights combined with the input values. The OWA operator allows weighting the input values according to their size-based ordering[4]. Unfortunately, in the OWA operator, weights measure the importance of each input value (in relation to other values) according to its size, without relation to the information source that has generated it. In our system it is important to weight votes according to each client's reliability (computed for example based on its role). Moreover, votes can be aggregated differently depending on a number of other criteria (e.g. client location, connection medium, etc.). For this reason, we use a *Weighted OWA operator* (WOWA) [15] combining the advantages of the OWA operator and those of the weighted mean. We use two sets of weights: **p** corresponding to the *relevance of sources*, and **w** corresponding to the *relevance of values.*

Our solution represents an alternative to statistical or probabilistic approaches (e.g. Bayesian Systems or Belief Methods): in [2] an OWA-based solution is compared with analogous experiments run on a probabilistic approach [13] where reputation is defined as a probability and can be computed by an event-driven method using a Bayesian interpretation. A major advantage of fuzzy aggregation techniques is their speed. In [2] we showed that, even if the fuzzy solution has a slower start, the global convergence speed is faster than the EigenTrust algorithm described in [13].

3.2 The WOWA Operator

Definition 2. *Let* $\mathbf{p}$ *and* $\mathbf{w}$ *be weight vectors of dimension* n*:* $\mathbf{p} = [p_1\ p_2 \cdots p_n]$, $\mathbf{w} = [w_1\ w_2\ \cdots\ w_n]$*, such that (i)* $p_i \in [0,1]$ *and* $\sum_i p_i = 1$*; (ii)* $w_i \in [0,1]$ *and* $\sum_i w_i = 1$.
In this case a mapping $f_{\mathrm{WOWA}} : \mathbb{R}^n \rightarrow \mathbb{R}$ *is a Weighted Ordered Weighted Averaging (WOWA) operator of dimension* n *if*

$$f_{\mathrm{WOWA}}(a_1, a_2, \ldots, a_n) = \sum_i \omega_i a_{\sigma(i)} \tag{2}$$

[4] For example, as described in [18], it is possible to obtain an average from the scores assigned by judges in skating or fencing competitions at the Olympic Games, excluding extreme values.

where $\{\sigma(1), \sigma(2), \ldots, \sigma(n)\}$ *is a permutation of* $\{1, 2, \ldots, n\}$ *such that* $a_{\sigma(i-1)} \geq a_{\sigma(i)}$ *for all* $i = 2, \ldots, n$.
With $a_{\sigma(i)}$ *we indicate the* i^{th} *largest element in the collection* $\{a_1, a_2, \ldots, a_n\}$ *and the weight* ω_i *is defined as*

$$\omega_i = w^* \left(\sum_{j \leq i} p_{\sigma(j)} \right) - w^* \left(\sum_{j < i} p_{\sigma(j)} \right) \tag{3}$$

with w^* *a monotone increasing function (e.g. a polynomial) that interpolates the points* $(i/n, \sum_{j \leq i} w_j)$ *together with the point* $(0, 0)$. ω *is used to represent the set of weights* $\{\omega_i\}$: $\omega = \{\omega_1, \omega_2, \ldots, \omega_n\}$.

Compared with simple OWA, the WOWA operator allows to model more complex situations, e.g. ones in which not all voters are equal, regardless of the vote they express. Also, in our setting it will allow modelling the Trust Manager "attitude" (i.e., more or less confident) with respect to users expressing votes[5].

Following Def. 2, it is necessary, given a set or weights $\mathbf{p}$ and $\mathbf{w}$, and the data vector $\mathbf{a}$, to define $S = \{(i/n, \sum_{j \leq i} w_j) | i = 1, \ldots, n\} \cup \{0, 0\}$ and the function w^* interpolating S. There are two possible ways to proceed:

1. establishing a weighting vector $\mathbf{w}$ and then interpolating w^*;
2. starting with the definition of w^*.

In the first case, interpolation can be carried out by any method that from a set of ordered points in the unit interval defines a monotone and bounded function. With the second approach it is possible to derive the set of weights ω from w^*, where w^* is any monotone increasing function within the $[0, 1]$ interval with $w^*(0) = 0$ and $w^*(1) = 1$.

Referring to the first method, having for example a set of input values $\mathbf{a} = [a_1\ a_2\ a_3\ a_4]$ already ordered from the maximum to the minimum (avoiding in this way the ordering process), it is possible to choose the weighting vector $\mathbf{w}$ depending on the importance we want to assign to trust values. Choosing for example a weights vector $\mathbf{w}' = [.1\ .4\ .4\ .1]$ we enhance the impact of central values with respect to extreme ones. On the contrary, a choice like $\mathbf{w}'' = [.4\ .1\ .1\ .4]$ indicates that extreme values are the most important. Both interpolation functions w'^* and w''^* are shown in Fig. 9(a) and (b)[6]. Array $\mathbf{p}$ allows us to model reliability of trust values depending on their source. Considering $\mathbf{w}'$ (or w'^*) and $\mathbf{p}' = [.15\ .35\ .35\ .15]$ (where central values are

[5] Note that this attitude might well be expressed by the additional layer of service metadata describing the Trust Manager operation (Sect. 2.3). This would allow for searching and selecting a Trust Manager service based on a summary of its aggregation technique.

[6] Again, we remark that following the second approach one starts start by defining the interpolation function and then compute the weighting vector.

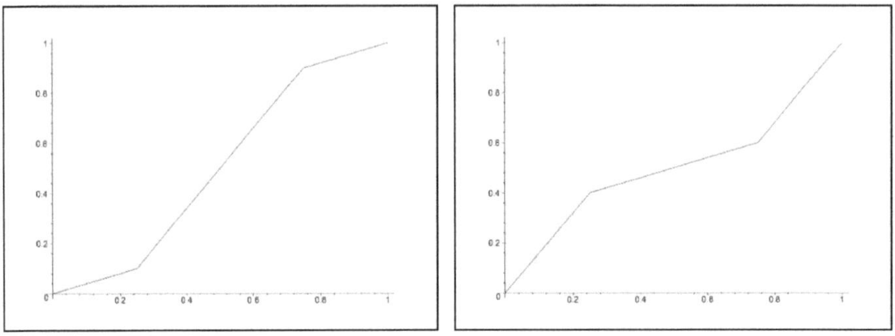

(*a*) (*b*)

Fig. 9. The interpolation functions w'^* (*a*) and w''^* (*b*).

considered the most reliable ones), it is easy to understand that weights ω_1 and ω_2 (corresponding to a_1 and a_2) are increased by applying weights w_i because the weights vector $\mathbf{w}$ gives more importance to central values. On the other hand, it is clear that $\mathbf{p}'' = [.35\ .15\ .15\ .35]$ leads to a lesser impact of the central ω_i.

3.3 An Example

In this section, we give an example illustrating two different approaches in assigning relevance to the data values. For the sake of simplicity, here we assume the array $\mathbf{w}$ is composed of values in the form $w_i = \frac{k}{n}$; in the first case, that we call *diffident approach*, $k = 1, 2, \ldots, n$ (Fig. 10(*a*)); in the second one, the *confident approach*, $k = n, n-1, \ldots, 1$ (Fig. 10(*b*)).

The "Diffident" Approach

Looking at Fig. 10(*a*) and considering the fact that the WOWA operator is applied to a permutation of data values a_i where $a_{\sigma(i-1)} \geq a_{\sigma(i)}, \forall i = 2, \ldots, n$ it is easy to understand that the first approach is "diffident", inasmuch it lessen the impact of high trust values. In this case, weight values are $w_i = k/n$, for $k = 1, \ldots, n$. These values have to be normalized by dividing them by their sum, as follows:

$$w_i = \frac{k/n}{\sum_{k=1}^{n} k/n} \,. \tag{4}$$

Let us now suppose that $\mathbf{a} = [.9\ .7\ .5\ .3\ .1]$ (values are already ordered for the sake of simplicity) and $\mathbf{w} = [\frac{1}{5}\ \frac{2}{5}\ \frac{3}{5}\ \frac{4}{5}\ 1]$. We obtain the normalized vector $\mathbf{w_n} = [\frac{1}{5}\ \frac{2}{5}\ \frac{3}{5}\ \frac{4}{5}\ 1]/\frac{15}{5} = [\frac{1}{15}\ \frac{2}{15}\ \frac{3}{15}\ \frac{4}{15}\ \frac{5}{15}]$. The next step consists in finding a function w^* interpolating the points $(i/n, \sum_{j \leq i} w_j)$:

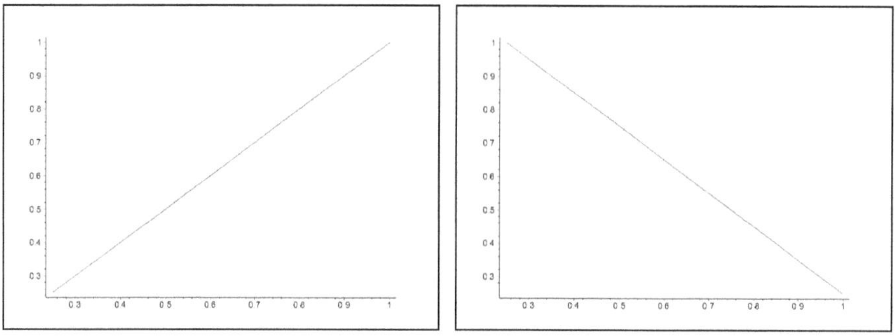

(*a*) (*b*)

Fig. 10. The weights $w_i = \frac{k}{n}$; in (a) $k = 1, 2, \ldots, n$; in (b) $k = n, n-1, \ldots, 1$.

$i = 1 \quad \left(\frac{1}{5}, w_1\right) = \left(\frac{1}{5}, \frac{1}{15}\right)$

$i = 2 \quad \left(\frac{2}{5}, w_1 + w_2\right) = \left(\frac{2}{5}, \frac{3}{15}\right) = \left(\frac{2}{5}, \frac{1}{5}\right)$

$i = 3 \quad \left(\frac{3}{5}, w_1 + w_2 + w_3\right) = \left(\frac{3}{5}, \frac{6}{15}\right) = \left(\frac{3}{5}, \frac{2}{5}\right)$

$i = 4 \quad \left(\frac{4}{5}, w_1 + w_2 + w_3 + w_4\right) = \left(\frac{4}{5}, \frac{10}{15}\right) = \left(\frac{4}{5}, \frac{2}{3}\right)$

$i = 5 \quad (1, w_1 + w_2 + w_3 + w_4 + w_5) = (1, 1)$

The equation of the interpolation function (Fig. 11) is therefore

$$w^*(x) = \frac{5}{6}x^2 + \frac{1}{6}x \ .$$

We use it to calculate final weights ω_i as follows, supposing $\mathbf{p} = [\frac{1}{15}\ \frac{4}{15}\ \frac{2}{15}\ \frac{5}{15}\ \frac{3}{15}]$:

$i = 1 \quad \omega_1 = w^*(p_1) = w^*\left(\frac{1}{15}\right) = \frac{2}{135}$

$i = 2 \quad \omega_2 = w^*\left(\sum_{i=1}^{2} p_i\right) - w^*(p_1) = w^*\left(\frac{1}{3}\right) - w^*\left(\frac{1}{15}\right) = \frac{2}{15}$

$i = 3 \quad \omega_3 = w^*\left(\sum_{i=1}^{3} p_i\right) - w^*\left(\sum_{i=1}^{2} p_i\right) = w^*\left(\frac{7}{15}\right) - w^*\left(\frac{1}{3}\right) = \frac{1}{9}$

$i = 4 \quad \omega_4 = w^*\left(\sum_{i=1}^{4} p_i\right) - w^*\left(\sum_{i=1}^{3} p_i\right) = w^*\left(\frac{4}{5}\right) - w^*\left(\frac{7}{15}\right) = \frac{11}{27}$

$i = 5 \quad \omega_5 = w^*\left(\sum_{i=1}^{5} p_i\right) - w^*\left(\sum_{i=1}^{4} p_i\right) = w^*(1) - w^*\left(\frac{4}{5}\right) = \frac{1}{3}$

We are now ready to compute the final value of f_{WOWA}[7]:

[7] In this case $a_{\sigma(i)} = a_i$ because vector **a** was already ordered.

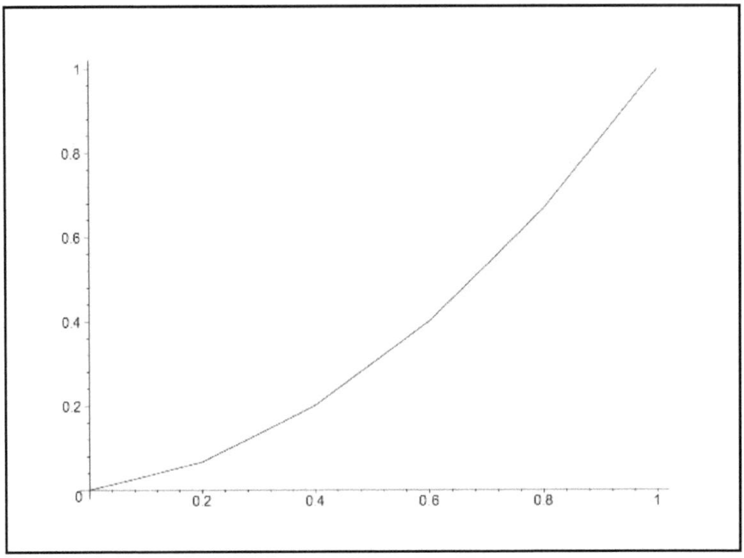

Fig. 11. The interpolation function w^* referred to the *diffident approach.*

$$f_{\mathrm{WOWA}}(.9,.7,.5,.3,.1) = \sum_{i=1}^{5} \omega_i a_i \;= .31\overline{7}\;.$$

The "Confident" Approach

In this case, we enhance the impact of high trust values in array **a** by choosing a function interpolating the points $(i/n, \sum_{j\leq i} w_j)$ where $w_i = \frac{k}{n}$, for $k = n, n-1, \ldots, 1$. Operating in the same way as before on the weighted normalized vector $\mathbf{w_n} = [\frac{5}{15}\ \frac{4}{15}\ \frac{3}{15}\ \frac{2}{15}\ \frac{1}{15}]$, we obtain the interpolation function expressed by the equation

$$w^*(x) = -\frac{5}{6}x^2 + \frac{11}{6}x\;.$$

As expected, the final trust value is higher respect to the "diffident" approach:

$$f_{\mathrm{WOWA}}(.9,.7,.5,.3,.1) = \sum_{i=1}^{5} \omega_i a_i \;= .54\overline{8}\;.$$

4 Conclusions

In this chapter, we have presented a complete approach for developing a Trust Layer service, aimed at improving the quality of automatically generated metadata. Our technique is based on non-intrusive collection of user feedback. A WOWA-based aggregation of individual trust values allows to take several effects into account, including users' attitude and sources reliability. When assertions are associated to identities of annotators, our schema can readily be extended to deal with trust on annotators as well as an assertion themselves [2]. Also, it is possible to envision many other usage scenarios, such as the one involving non-functional descriptors of web services published by service providers in a service-oriented software architecture. In this case, our approach could deal with the non-uniform reliability of service self-descriptions, which are liable to imprecisions and may lead to unexpected results [4].

Acknowledgments

This work was partially supported by contributions from Italian Ministry of Research Fund for Basic Research (FIRB) under project `RBAU01CLNB_001` "Knowledge Innovation for the Web Infrastructure" (KIWI).

References

1. Aczél, J.: On Weighted synthesis of judgements. Aequationes Math. **27** (1984) 288–307
2. Aringhieri, R., Damiani, E., De Capitani di Vimercati, S., Paraboschi, S., Samarati, P.: Fuzzy Techniques for Trust and Reputation Management in Anonymous Peer-to-Peer Systems. Journal of the American Society for Information Science and Technology (JASIST) (to appear) (2006)
3. Blaze, M., Feigenbaum, J., Ioannidis, J., Keromytis, A.: The role of trust management in distributed systems security. Secure Internet Programming (1999) 79–97
4. Bosc, P., Damiani, E., Fugini, M.: Fuzzy service selection in a distributed object-oriented environment. IEEE Trans. on Fuzzy Systems **9(5)** (2001) 682–698
5. Brill, E., Moore, R. C.: An improved error model for noisy channel spelling correction. Proc. of the 38th Annual Meeting of the ACL (2000) 286–293
6. Ceravolo, P., Nocerino, M. C., Viviani, M.: Knowledge extraction from semi-structured data based on fuzzy techniques. Knowledge-Based and Emergent Technologies Relied Intelligent Information and Engineering Systems KES'04 **2** (2004) 1257–1263
7. Damiani, E., De Capitani di Vimercati, S., Paraboschi, S., Samarati, P.: Managing and Sharing Servents' Reputations in P2P Systems. IEEE Trans. Knowl. Data Eng. **15(4)** (2003) 840–854
8. Damiani, E., Nocerino, M. C., Viviani, M.: Knowledge Extraction from an XML Data Flow: Building a Taxonomy based on Clustering Technique. Current Issues in Data and Knowledge Engineering, Proceedings of EUROFUSE 2004: 8th Meeting of the EURO Working Group on Fuzzy Sets (2004) 133–142
9. Fodor, J., Marichal, J. L., Roubens, M.: Characterization of the Ordered Weighted Averaging Operators. IEEE Trans. on Fuzzy Systems **3(2)** (1995) 236–240

10. Grabisch, M.: Fuzzy integral in multicriteria decision making. Fuzzy Sets and Systems **69** (1995) 279–298
11. Jøsang A.: The right type of trust for distributed systems. Proceedings of the 1996 Workshop on New Security Paradigms (1996)
12. Jøsang A., Roslan, I., Boyd, C.: A Survey of Trust and Reputation Systems for Online Service Provision. Decision Support Systems (to appear) (2005)
13. Kamvar, S., Schlosser, M., Garcia-Molina, H.: The EigenTrust Algorithm for Reputation Management in P2P Networks. Proceedings of the Twelfth International World Wide Web Conference (2003)
14. Mesiar, R., Komorníková, M.: Aggregation Operators. Proceedings of the XI Conference on applied Mathematics PRIM'96, Herceg D., Surla K. (eds.), Institute of Mathematics (1997) 193–211
15. Torra, V.: The weighted owa operator. International Journal of Intelligent Systems **12(2)** (1997) 153–166
16. Yager, R.: On ordered weighted averaging aggregation operators in multicriteria decision making. IEEE Transactions on Systems, Man and Cybernetics **18(1)** (1988) 183–190
17. Xu, Z.S., Da, Q.L.: An Overview of Operators for Aggregating Information. International Journal of Intelligent Systems **18** (2003) 953–969
18. Yager, R.: Quantifier Guided Aggregation Using OWA operators. International Journal of Intelligent Systems **11** (1996) 49–73
19. `http://msdn.microsoft.com/library/default.asp?url=/library/en-us/dnuddi/html/impmetadata.asp`
20. `http://www.w3.org/TR/rdf-primer/#reification`

A Fuzzy Linguistic Multi-agent Model Based on Semantic Web Technologies and User Profiles

Enrique Herrera-Viedma[1], Eduardo Peis[2], and José M. Morales-del-Castillo[2]

[1] Dpt. Computer Science and AI. University of Granada
viedma@decsai.ugr.es
[2] Dpt. Library and Information Science. University of Granada
epeis@ugr.es, josemdc@ugr.es

Summary. In this paper we define a model of a web multi-agent system that combines the use of Semantic Web technologies (like RDF and OWL languages to describe resources and define ontologies respectively) together with the application of user profiles to carry out its activity. It is designed using fuzzy linguistic techniques which allow dealing with information in a user-friendly way. The system activity is developed in two phases: retrieval phase to gather the documents from the Web, and feedback phase to update the elements used to filter the documents, i.e., user profiles and recommendation values. We focus on the analysis of the feedback phase. With this multi-agent system model the filtering possibilities are increased and, consequently, also the retrieval capabilities on the Web.

1 Introduction

The Internet harbours a vast amount of information that does not cease to grow exponentially. To avoid arriving to a situation of collapse, the actual model of Web needs to incorporate tools capable to handle and manage efficiently this huge quantity of resources. The solutions that exist nowadays are shown little efficient since they oblige users to lose great part of their time in deciding which documents are relevant and which not. Therefore it is becoming necessary to develop systems for searching and mining the Web that permit to improve the access to the information in an efficient way. At this moment, some of the more recurrent technologies to face this problem deal with the development of intelligent software agents [14, 27], the application of techniques of information filtering [37], and the development of the Semantic Web project [3, 4, 5].

Software agents applied to a Web-based environment are organized in distributed architectures [7, 13, 14, 26, 29] to mainly perform tasks of intermediation between users and the Web, e.g. assisting users in the information

E. Herrera-Viedma et al.: *A Fuzzy Linguistic Multi-agent Model Based on Semantic Web Technologies and User Profiles*, StudFuzz **197**, 105–120 (2006)
www.springerlink.com

retrieval process [7, 14, 26, 39]. These agents are entities capable to act in an autonomous way, processing and exchanging results with other agents [20], but need to have certain amount of knowledge about the system and the user to be able to develop these tasks. Having an extensive knowledge base about the user implies that the agent has access to an exhaustive and "first hand" information, but the user has to pay an expensive cost, being exposed to continuous consultations to maintain his/her information updated. An alternative solution is to provide the agent with a smaller knowledge base but supported on web ontologies [19, 20], forcing this way the agent to adopt an active stance and seek information about the user and the tasks that carries out in his/her interaction with the system. Agents are able to use different methods to get this information, e.g. by direct observation and imitation of users' behaviour (search strategies used, links visited during a session, etc.) or taking advantage of users' feedback [27].

Nevertheless the main problem of using agents is to find a flexible and agile communication protocol for exchanging information among agents, and between users and agents because of the great variety of forms the information is represented in the Web. A possible option that permits to reduce these agent-agent and user-agent communication problems is to apply fuzzy linguistic techniques that allow operating with the information by means of the use of linguistic labels [40]. The application of this flexible system of representation enables us to handle information with several degrees of truth, solving the problem of quantifying qualitative concepts. Some examples of the use of fuzzy linguistic tools in the design of multi-agent systems can be found in [10, 11, 22, 25].

On the other hand, information filtering techniques ease users the task of sorting out relevant documents from those that are not, thanks to the prior selection the system carries out of the resources that better fit users' needs, requirements and preferences. These needs, requirements and preferences are mostly defined in a normalized way in the form of user profiles [6]. The efficacy of these user profiles depends on the up-to-dateness of the information contained, so there should exist a mechanism capable to dynamically bring up to date this information and to reflect in "real time" the variations on users' behaviour in their interaction with the system. In [25] we proposed a fuzzy linguistic multi-agent system based on collaborative filtering tools but no characterization of user's preferences was used. As it is known, the problem is that the performance of these tools tend to fail when little is known about users. Therefore, the use of the user profiles could contribute to improve its activity.

Another possibility to improve the activity of a multi-agent system could be the use of some of the technologies of the Semantic Web project [3, 4, 5]. These technologies can be exploited to develop ontology-based infrastructures where agents can operate at semantic level with resources described using languages as XML (eXtensible Mark-up Language) or RDF (Resource Description Framework) [12, 30, 34, 35, 38] in a manner both interpretable by humans and machines. This common syntactic framework allows us to define an unique communication vocabulary among agents that could also be used to characterize the knowledge base of the system and even the semantics of resources or user profiles. In [41], for example, the use of the ontological language OWL (Ontology Web Language) as content language to exchange messages allows to enlarge the interoperability among agents (as a result of using a common convention for representing knowledge) and also to endow them with greater capacities for inference reasoning over resources.

The aim of this paper is to present a new model of fuzzy linguistic multi-agent system that involves the use of the Semantic Web technologies and user profiles dynamically updated to overcome the problems of the system presented in [25]. The Semantic Web technologies are used to endow the agents with a common communication language, to develop the ontologies of the system and to characterize resources and user profiles in a standardized way using RDF. We propose a method to automatically update the user profiles in function of the appraisals users provide about system's output. This updating method is designed using linguistic matching functions [23]. As in [25], the system activity presents two phases, retrieval and feedback. In this paper we focus on the latter.

The paper is structured as follows. Section 2 reviews the methodological tools employed in this research: the fuzzy linguistic tools, the filtering tools, and the Semantic Web technologies. Section 3 presents the new multi-agent model. Finally, some concluding remarks are pointed out in Section 4.

2 Methodological Tools

In this section, we present the tools that we apply to design our fuzzy linguistic multi-agent model.

2.1 Fuzzy Linguistic Tools

In the situations in which the information cannot be evaluated of quantitative form, but qualitative, it is necessary to apply a linguistic method. For example, when we try to qualify a phenomenon related to human perception, we use words in natural language instead of numerical values.

The *fuzzy linguistic approach* [40] and in particular, the *ordinal fuzzy linguistic approach* [21, 23, 24] are approximate techniques appropriate to deal with qualitative aspects of problems. An ordinal fuzzy linguistic approach is defined by considering a finite and totally ordered label set in the usual sense

$$S = \{s_i, i \in H = \{0, \ldots, T\}\}$$

and with odd cardinality (7 or 9 labels). The mid term representing an assessment of "approximately" 0.5 and the rest of the terms being placed symmetrically around it. The semantics of the linguistic term set is established from the ordered structure of the term set by considering that each linguistic term for the pair (s_i, s_{T-i}) is equally informative. For each label s_i is given a fuzzy number defined on the [0,1] interval, which is described by a linear trapezoidal membership function represented by the 4-tuple $(a_i, b_i, \alpha_i, \beta_i)$ (the first two parameters indicate the interval in which the membership value is 1.0; the third and fourth parameters indicate the left and right widths of the distribution). Furthermore, we require the following properties:

1. – *The set is ordered* : $s_i \geqslant s_j$ *if* $i \geqslant j$.
2. – *There is the negation operator* : $Neg(s_i) = s_j$, *with* $j = T - i$.
3. – *Maximization operator* : $MAX(s_i, s_j) = s_i$ *if* $s_i \geqslant s_j$.
4. – *Minimization operator* : $MIN(s_i, s_j) = s_i$ *if* $s_i \leqslant s_j$.

Additionally, we need aggregation operators to combine the linguistic information. In [22] is defined a useful operator, the Linguistic Ordered Weighted Averaging (LOWA) operator, which has been satisfactorily applied in different fields [21, 24, 25].

2.2 Filtering Techniques

Information filtering techniques deal with a variety of processes involving the delivery of information to people who need it. Operating in textual domains, *filtering systems* or *recommender systems* evaluate and filter the resources available on the Web (usually, stored in HTML or XML documents) to assist people in their search processes [32], in most cases through the use of filtering agents [33]. Traditionally, these systems have fallen into two main categories [31]. *Content-based filtering systems* filter and recommend the information by matching user query terms with the index terms used in the representation of documents, ignoring data from other users. These recommender systems tend to fail when little is known about user information needs, e.g. as happens when the query language has short expressive capabilities. A possible solution to this problem is using user profiles as a tool for characterising users through explicit and implicit inputs that define both personal and professional information. In fact the integration of user profiles in the filtering process renders a powerful tool that allows fast and efficient filtering [37]. On the other hand, *collaborative filtering systems* use the information provided by many users to filter and recommend documents to a given user, ignoring the representation of documents. It's very usual to group the users in specific categories or stereotypes that are characterized by a set of default preferences and rules that represent the information needs and common search habits of a group of related users. These recommender systems tend to fail when little is known about a user, or when he/she has uncommon interests [31]. Several researchers are exploring hybrid content-based and collaborative recommender systems to smooth out the disadvantages of each one of them [1, 9, 15, 31, 37].

2.3 Semantic Web Technologies

The Semantic Web is an extension of the present Web, in which the information is gifted of a well defined meaning, permitting a better cooperation between humans and machines [3, 4, 5]. It is based on two main ideas: the "semantic" mark up of resources and the development of "intelligent" software agents capable to understand and to operate with these resources at semantic level [3,19].

The semantic backbone of the model is RDF/XML [2], a language that provides the necessary infrastructure to codify, exchange and reuse structured metadata in order to make them directly interpretable by machines. RDF/XML structures the information in assertions (resource-property-value triples), and uniquely identifies resources by means of URI's (Universal Resource Identifier), allowing intelligent software agents the knowledge extraction from and inference reasoning over resources (such as documents, user profiles or even queries) using ontologies.

Ontologies are defined in the Semantic Web context as a collection of relevant concepts of the knowledge shared by the members of a specific domain, the relations these concepts establish between them, and the axioms defined upon these concepts and relations [8, 16, 17]. Several ontology languages can be used for the design of web-based ontologies like OWL [28], a language with a great expressive capacity that allow us, for example, to define ontologies maintaining the RDF/XML syntactic convention or even to use it as common vocabulary for exchanging messages among agents.

3 The Fuzzy Linguistic Multi-agent Model Based on Semantic Web and User Profiles

In [25] we defined a model of a fuzzy linguistic multi-agent system to gather information on the Web that presents a hierarchical architecture composed of seven action levels: *internet users, interface agent, collaborative filtering agent, task agent, content-based filtering agent, information agents* and *information sources.* The main novelty of this model is the introduction of collaborative filtering agents in its architecture in order to increase its information filtering capabilities on the Web. Then, it develops its activity in two phases:

- *Retrieval phase:* This first phase coincides with the information gathering process developed by the multi-agent model itself, i.e., this phase begins when a user specifies his/her query and finishes when he/she chooses his/her desired documents among the relevant documents retrieved and provided by the system.
- *Feedback phase:* This second phase coincides with the updating process of collaborative recommendations on desired documents existing in a collaborative recommender system, i.e., this phase begins when the interface agent

informs the documents chosen by the user to the collaborative filtering agent and finishes when the recommender system recalculates and updates the recommendations of the desired documents.

The main drawback of this model is that it does not utilize user profiles to characterize the user preferences and this limits its performance possibilities.

To overcome the limitations of the model presented in [25] we define a new and enhanced model of fuzzy linguistic multi-agent system that improves information retrieval by means of the application Semantic Web technologies to set a base for the operation of software agents and user profiles to enrich the filtering activity.

This model presents a hierarchical structure with six action levels (*internet users, interface agent, filtering agent, task agent, information access,* and *information bases*), also two main activity phases, and a set of agents (*interface agent, filtering agent, task agent, profile agent, recommendation agent, information agents*) that work depending on the phase activity (see Fig. 1):

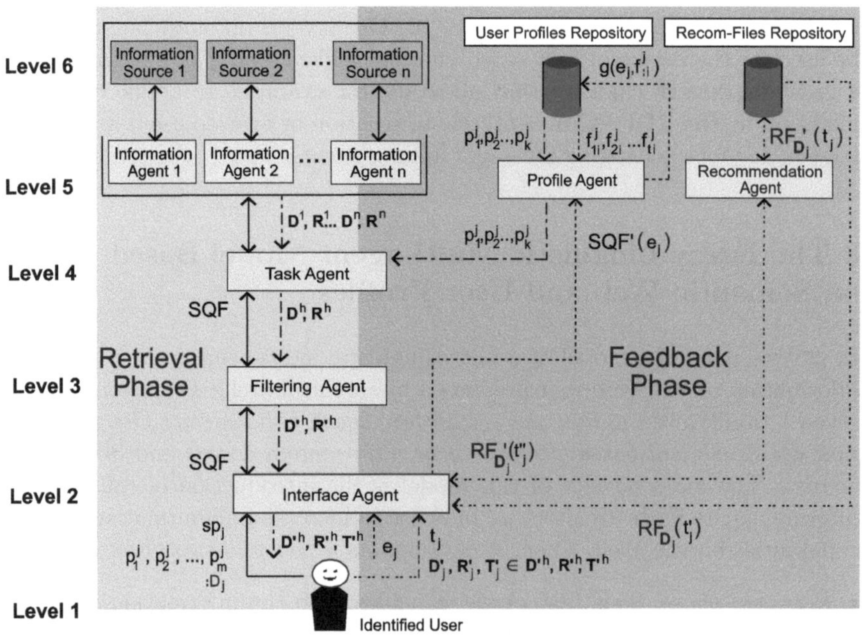

Fig. 1. Model architecture: Levels of action and processes.

- **Semantic Retrieval Phase**: This phase is similar to that developed by the model presented in [25], therefore, it coincides with the information gathering process developed by the multi-agent model itself. The main novelty of this phase is that the query language used is not a Boolean one, as in [25], but a semantic query language. The semantic information retrieval differs from the traditional information retrieval in the use of semantic query languages [18, 36], capable of comparing both literal and semantics structures. In such a way it is possible to obtain more accurate and contextualized answers to queries. The entities that participate in this phase are the following: *internet users (level 1), interface agent (level 2), filtering agent (level 3), task agent (level 4), information agent* as entity of information access *(level 5),* and *information sources* as the information bases that store the documents *(level 6).*

- **Feedback Phase**: This phase coincides with the updating process of both user profiles and collaborative recommendations on desired documents. Therefore, it involves two processes: "profile updating" process and "recommendation" process. Both processes need users' appraisals. In the first, they're given to assess the quality of the global answer provided by the system and, in the second one, to assess the retrieved documents. The "profile updating process" consists on the dynamic updating of the user profiles on the basis of the satisfaction degree the user expresses regard to the global results provided by the multi-agent system. In this process the entities that participate are the following: *internet users (level 1), interface agent (level 2), profile agent* as *entity of information access (level 5),* and the *user profiles repository* as *information bases* that store users' preferences *(level 6).* The "recommendation" process is similar to that defined in [25], therefore, it allows the user to express his/her opinions about any retrieved documents and using them the system can recalculate and update the stored recommendations about the desired documents. In this process the entities that participate are the following: *internet users (level 1), interface agent (level 2), recommendation agent* as *entity of information access (level 5),* and the *recommendation files repository* as *information bases (level 6).*

In the following subsection we analyze in detail the two action processes of the feedback phase.

3.1 Feedback Phase: User Profile Updating Process

Before to explain this process we have to clarify the inputs. This process works on three inputs:

1. *Semantic Query File (SQF):* This file is generated when the user formulates its requirements through the interface agent. This agent stores the user's ID_j, the search parameters (sp_j) that define his/her "semantic" query and a set of preferences $\{p_1^j, p_2^j, \ldots, p_m^j\}$ on a file in RDF format we call *Semantic Query File* (SQF). These preferences for the user j are $\{p_1^j, p_2^j, \ldots, p_k^j\}$, where $0 \leqslant k \leqslant m$, being m the number of properties used to define a user profile and $p_i^j \in F_i$, being F_i the expression domain associated to the property i. For example, the user could provide his/her preferences about any of these four categories ($m = 4$) of basic preferences:

- *Document Type*: This preference defines the type of document the user prefers to retrieve among the ones the system provides. For example, in this case $F_1 = \{SciArticle, Proceedings, BookChapter, all\}$.
- *Search Context:* It consists of general topic categories that represent the main areas of the system domain. For example, if the domain of work of our system is "information systems" we could define F_2 = {*decision support systems, information retrieval systems, geographical information systems, data mining, knowledge representation. all*}.
- *Search aim:* It defines those tasks the user is going to carry out with the information to be retrieved. Depending on the nature of these tasks the system is able to decide which resources should be retrieved and which should not. In this case we could define different task categories such as F_3 = {*research, teaching_bachelor, teaching_master, teaching_doctorate, studies_bachelor, studies_master, studies_doctorate, all*}.
- *Date:* It refers to the updating or publication date of the resources to be retrieved. A set of different time intervals are defined to cover a wide range of values that vary from few months to several years (for exhaustive searches). For example, F_4 = {*3 months, 6 months, 1 year, 3 years, 5 years, 10 years*}.

Then, in this case, an user could provide from none ($k = 0$) up to four ($k = 4$) possible values of preference. Usually he/her will give us just one or two, being the other values taken from his/her stored profile.

An example of a SQF is shown in Fig. 2.

2. *User profiles*: We assume a repository that stores the user profiles in the system. Each user profile is determined by the user's identity and characterized by the particular values that each user has assigned in the categories of basic preferences. Each possible preference of the user profile has associated a linguistic frequency property (tagged as <*freq*>) representing how often a specific value has been used in satisfactory search processes. Thus, if $F_i = \{d_{1i}, d_{2i}, \ldots, d_{ti}\}$ then we define $f_i^j = \{f_{1i}^j, f_{2i}^j, \ldots, f_{ti}^j\}$ as the set of frequency values associated with each possible value $l \in \{1, \ldots, t\}$ of the property i in the profile of the user j. The range of possible values for the frequencies is defined in a set of seven linguistic labels, S = {*always, almostAlways, mostTimes, sometimes, aFewTimes, almostNever, never*}, i.e., $f_{li}^j \in S$. The following example (Fig. 3) shows how can be represented the frequency of use of a specific preference in a user profile.
3. *Global satisfaction degree:* This input is a key element in the profiles updating process, allowing the elicitation of users' tacit preferences. Once the user has checked out the resources retrieved by the system, he/she is required to assess the results providing a linguistic satisfaction degree e_j, that give us an approximate idea of the "kindness" of the preference values selected to define a query. In such a way, the system will be able to select (when needed) those preference values that have provided more satisfying answers to a user, instead of those values that have been more often selected. This global satisfaction degree can be defined using a set of seven

```
<?xml version="1.0" encoding="iso-8859-1"?>
<rdf:RDF
  xmlns:rdf="http://www.w3.org/1999/02/22-rdf-syntax-ns#"
  xmlns="http://www.ugr.es/~kishimaru/sqf#"
  xml:base="http://www.ugr.es/~kishimaru/sqf">
  <Query rdf:ID="query384">
    <user_ID>user-022005</user_ID>
    <appraisal></appraisal>
    <preferences_e>
      <preferences rdf:ID=”pref_384”>
        <docType>SciArticle</docType>
        <context>User_Modeling</context>
        <aim>Research_Article</aim>
        <date>3months</date>
        <value>NULL</value>
      </preferences>
    </preferences_e>
    <search_parameters_e>
     ....
    </search_parameters_e>
  </Query>
</rdf:RDF>
```

Fig. 2. Semantic Query File (SQF).

```
  ...
  <DocType rdf:ID="docType-pr001">
     <type1_e>
        <Type1 rdf:ID="type1-pr001">
           <type>SciArticle</type>
           <freq>AlmostAlways</freq>
        </Type1>
     </type1_e>
     <type2_e>
        <Type2 rdf:ID="type2-pr001">
           <type>Proceedings</type>
           <freq>AlmostNever</freq>
        </Type2>
     </type2_e>
  ...
  </DocType>
  ...
```

Fig. 3. Representation of preferences in a user profile.

linguistic labels $S' = \{Total, veryHigh, high, medium, low, veryLow, Null\}$, i.e., $e_j \in S'$.

Assuming the above inputs the profile updating process is developed in the following steps:

- **Step 1:** Once the user has checked the set of resources retrieved (D'^h, R'^h), the interface agent asks the user to express his/her satisfaction degree ($e_j \in S'$) with respect to the overall performance of the system, and it is stored in the semantic query file SQF.
- **Step 2:** The file SQF is transferred from the interface agent to the profile agent.
- **Step 3:** The profile agent carries out the updating of the linguistic frequency of use $\{f^j_{1i}, f^j_{2i}, \ldots, f^j_{ti}\}$ of each preference i defined in the user profile. To do that, we propose to use some type of matching function similar to the matching functions defined in the information retrieval systems to model the weighted queries. In particular, we use linguistic matching functions to model threshold weights in weighted user queries [23]. Assuming the linguistic satisfaction degree $e_j \in S'$ expressed by a user, if the property $i$is used by the user with a value l, then its respective associated frequency $f^j_{li} \in S$ is updated by means of the following linguistic matching function $g : S' \times S \to S$:

$$g\left(e_j, f^j_{li}\right) = \begin{cases} s_{Min\{a+\beta,T\}} & if \;\; s_a \leqslant s_b \\ s_{Max\{0,a-\beta\}} & if \;\; s_b < s_a \end{cases}$$

$$s_a, s_b \in S \; |a, \; b \; \in \; H = \{0, \ldots, T\}$$

such that, (i) $s_a = f^j_{li}$; (ii) $s_b = e_j$; and iii) β is a bonus value that rewards/penalizes the frequencies of the preferences of the user profile, which can be defined depending on the closeness between f^j_{li} and e_j for example, $\beta = round(2|b - a|/T)$. We should point out that this function is a non-decreasing matching function as the traditional threshold matching functions.

3.2 Feedback Phase: Recommendation Process

This process needs two basic inputs:

1. *Recommendation Files:* We assume the system has access to a repository of recommendation files (RF). On each document accessed in some moment of its history in the system there exists an associated recommendation file in RDF format (see Fig. 4) where is contained information about all the appraisals made by users that have read it formerly. This document-RF relation must be defined in the proper ontology to allow agents operating with both resources (this will allow, for example, showing users a list of titles of documents together with their respective recommendation values

```
<?xml version="1.0" encoding="ISO-8859-1"?>
<rdf:RDF
xmlns:rdf="http://www.w3.org/1999/02/22-rdf-syntax-ns#"
xmlns="http://perso.wanadoo.es/kishimaru/recomFiles#"
xml:base="http://perso.wanadoo.es/kishimaru/recomFiles"
>
   <RecomFile rdf:ID="recomf001">
     <doc_ref>doc-pr008</doc_ref>
     <recom_value>High</recom_value>
     <recom_history>
        <R_history rdf:ID="histf001">
          <item>
            <RecomItem rdf:ID="form01-pr001">
              <appraissal>VeryHigh</appraissal>
              <topic>Data mining</topic>
              <user_ID>user-pr022005</user_ID>
            </RecomItem>
          </item>
          ....
        </R_history>
     </recom_history>
   </RecomFile>
</rdf:RDF>
```


Fig. 4. Representation of a Recommendation File (RF).

even though these data are located in different files). In a RF appears the ID of the document it belongs to, the current recommendation value and a set of log items containing previous appraisals about that document. Each log item is defined by an user's ID, his/her corresponding appraisal and the search context used in the query formulated by the user to retrieve that document (see "Search Context preference" in Section 3.1). This representation enables the adoption of different recommendation policies, allowing us, for example, to recalculate recommendation values based on the opinion of all the users, or just of some of them (e.g., using the appraisals given by those users who looked for information by the same topic). The decision to make a separate file defining the recommendations is justified by the dynamic nature of the recommendations. In such a way, we can modify the RF as many times as we like leaving untouched the representation of its associated document.

- *Recommendation value:* When a user checks out a document the systems asks him/her to recommend it to the rest of users of the system according to his/her opinion about the resource. In this way, each single person can help other users in the system to decide which resources are worthy to be read. In other words, the system enables its users to take advantage of the experience and knowledge of each others. The recommendation value (t_j)

provided by user j can be defined using a set of five linguistic labels $S''=$ $\{veryHigh, high, medium, low, veryLow\}$, i.e. $t_j \in S''$.

Then, the recommendation process is carried out in three steps:

- **Step 1:** Any document $D'_j \in (D'^h, R'^h, T'^h)$ has an associated relevance degree R'_j, and an associated set of "historical" recommendations $T'_j = \{t'_{1j}, t'_{2j}, \ldots, t'_{mj}\}$, being m any positive integer. When the user checks out the document D'_j is asked by the interface agent to appraise its global quality using a linguistic label $t_j \in S''$.
- **Step 2:** This appraisal is added to the RF associated to $D'_j(RF_{D'j})$ together with the "Search Context" preference value and the user's ID (stored in the SQF generated in the current search session), thus generating a new log item in the set of "historical" recommendation values.

In such a way, in a future search process the recommendation agent proceeds to dynamically recalculate the recommendation value of the document by means of an aggregation function, such as the LOWA operator [21, 22], that combines the whole set of historical recommendation values to calculate a new recommendation value (t''_j). This new value is stored in the *<recom_value>* tag of the $RF_{D'j}$, replacing the old t'_j, and can be used to be displayed in the list of retrieved resources, coupled with its respective document link.

4 Example

Suppose the following framework. John Quest, a researcher member of RECSEM (an academic institution specialized in the study of "information systems"), has to write a paper about "*semantic information systems*" for a prestigious scientific journal. John decides to search in RECSEM's documents repositories for recent resources about this topic to build a solid knowledge background for his work. He proceeds to log into RECSEM's site and reaches the search interface page. Then, he writes a pair of keywords ("*information systems*" and "*ontologies*") in the search box, and explicitly specifies that these keywords must be searched in the abstract of each document. After that, to scope the search, John selects *"knowledge representation"* as preferred search context, *"research"* as search aim and *"3months"* as preferred date of publication, but doesn't specify any value for the *"Document type"* preference. The system checks his profile and works out that the most satisfying value for the *"Document type"* preference is *"Scientific Article"*. With this information the different agents retrieve and filter those documents that better fit to John's requirements.

The resulting set of ranked documents is displayed on John's laptop screen and looks like this:

Each document appears with an associated relevance degree and a recommendation value. This extra information eases John's task of deciding which resources can be more useful for him. In this case he decides to choose the first

Table 1. Query answer sample

D_j^h	R_j^h	T_j^h
http://www.ugr.es/~Arep/N0021	0.95	high
http://www.ugr.es/~Arep/L57641	0.93	NULL
http://www.ugr.es/~Erep/P70435	0.87	medium

document of the list (D_1) because its relevance degree and recommendation are high. After casting an evaluative glance over the resource, and before John can check another document, the system asks him if he wants to appraise the resource (thus helping other users in future search processes). John decides to cooperate and assess this document selecting the linguistic label t' = *"very high"*. This appraisal, together with his user ID and the *"Search context"* preference value (*"knowledge representation"*) are added as a new "historical" recommendation item in the RF associated to D_1, ready to be used in future search processes.

Before John can leave the system, the interface agent asks him to express his opinion with respect to the global answer provided to his query. John, evaluating the quality of the resources he has checked out, considers that, in general, the results have been highly satisfactory, so he decides to select the linguistic label e_j = *"high"* (where $e_j \in S'$ = {*Total, veryHigh, high, medium, low, veryLow, Null*}). Then, the system proceeds to update John's profile rewarding the frequencies of the preference values used in this query. For example, let's focus on the *"Search context"* preference. The expression domain for this preference is F_2 = {*decision support systems, information retrieval systems, geographical information systems, data mining, knowledge representation, all*}, and its associated frequency in John's profile is f_{52}^j = *"aFewTimes"* (where $f_{li}^j \in S$ = {*always, almostAlways, mostTimes, sometimes, aFewTimes, almostNever, never*}) . Taking into account that $s_a \leqslant s_b$, the values of the indexes a and b are 2 and 4 respectively, and $T = 6$, we have that $\beta = 1$. So, the frequency for this preference value is increased in a factor of one, and therefore $(f_{52}^j)' = g$ (*high, aFewTimes*) = *"sometimes"*.

After the search process, John finishes his session with the feeling that he has saved a lot of time, and with the certainty that he has obtained relevant and useful resources for his purposes.

5 Concluding Remarks

In this paper, we have described the architecture and elements of a fuzzy linguistic multi-agent system specially designed to perform information retrieval and filtering tasks in domain dependant environments.

The key aspect of the system is the joint application of Semantic Web technologies and the use of user profiles. Semantic Web technologies provide the system with the necessary semantic infrastructure to improve inference and communication capacities of agents and with means to represent the information (resources and user profiles) in a common vocabulary both human and machine interpretable. The use of user profiles allows to characterize better the user and in such a way the performance of the system can be increased. Furthermore, we have proposed a method to dynamically update user profiles that allows adapting them to the changes observed in users' preferences by mean of fuzzy linguistic matching functions.

Our plans for future work include the development of an enhanced user profile updating process based on web usage analysis and rule discovery techniques, and the adaptation of this system to different work contexts such as e-commerce or e-learning. The application of stereotypes to classify users' profiles will also be contemplated.

References

1. C. Basu, H. Hirsh, W. Cohen, Recommendation as classification: Using social and content-based information in recommendation. In *Proceedings of the Fifteenth National Conference on Artificial Intelligence* (1998), 714–720
2. D. Beckett, RDF/XML Syntax Specification (Revised). W3C Recommendation. (10 February 2004). Available at *http://www.w3.org/TR/rdf-syntax-grammar/*. Accessed 02/19/2005
3. T. Berners-Lee, Semantic Web Road map (1998). Available at *http://www.w3.org/DesignIssues/Semantic.html*. Accessed 02/15/2005
4. T. Berners-Lee, Semantic Web - XML2000 (2000). Available at *http://www.w3.org/2000/Talks/1206-xml2k-tbl/*. Accessed 02/13/2005
5. T. Berners-Lee, J. Hendler, O. Lassila, The Semantic Web: A new form of Web content that is meaningful to computers will unleash a revolution of new possibilities, *Scientific American* May (2001)
6. D. Bolchini, E.S.K. Yu, Modeling user requirements for Web application design, *Studies in Communication Sciences* 4 (2) (2004), 173–196
7. W. Brenner, R. Zarnekow, H. Witting, Intelligent Software Agent, Foundations and Applications, (Springer-Verlag. Berlin Heidelberg, 1998)
8. Y.A. Bishr, H. Pundt, W. Kuhn, M. Radwan, Probing the concept of information communities: a first step toward semantic interoperability. In M. Goodchild, M. Egen-Hofer, R. Fegeas, C. Kottman (eds), *Interoperating Geographic Information Systems*, (Kluwer Academic, 1999), 55–69
9. M. Claypool, A. Gokhale, T. Miranda, Combining content-based and collaborative filters in an online newspaper. In *Proceedings of the ACM SIGIR Workshop on Recommender Systems-Implementation and Evaluation* (1999)
10. M. Delgado, F. Herrera, E. Herrera-Viedma, M.J. Martín-Bautista, M.A Vila, Combining linguistic information in a distributed intelligent agent model for information gathering on the Internet. In P.P. Wang (ed), *Computing with Word*, (John Wiley & Son, 2001), 251–276
11. M. Delgado, F. Herrera, E. Herrera-Viedma, M.J. Martín-Bautista, L. Martínez, M.A. Vila, A communication model based on the 2-tuple fuzzy linguistic representation for a distributed intelligent agent system on Internet, *Soft Computing* 6 (2002), 320–328

12. Esperonto Project. Available at *http://esperonto.semanticweb.org*. Accessed 02/21/2005
13. B. Fazlollahi, R.M. Vahidov, R.A. Aliev, Multi-agent distributed intelligent system based on fuzzy decision making, *International Journal of Intelligent Systems* 15 (2000), 849–858
14. J. Ferber, Multi-Agent Systems: An Introduction to Distributed Artificial Intelligence (Addison-Wesley Longman, New York, 1999)
15. N. Good, J.B. Shafer, J.A. Konstan, A. Borchers, B.M. Sarwar, J.L. Herlocker, J. Riedl, Combining collaborative filtering with personal agents for better recommendations. In *Proceedings of the Sixteenth National Conference on Artificial Intelligence* (1999), 439–446.
16. T.R. Gruber, Toward principles for the design of ontologies used for knowledge sharing. In *International Journal of Human-Computer Studies* 43 (5-6) (1995), 907–928
17. N. Guarino, Formal ontology and information systems. In N. Guarino (ed) *Formal Ontology in Information Systems, Proceedings of FOIS'98.* Trento (Italy), (IOS Press, Amsterdam, 1998), 3–17.
18. R. Guha, R. McCool, E. Miller, Semantic search. 12th International World Wide Web Conference 2003 (WWW2003), Budapest (Hungary), (2003) 700–709
19. J. Hendler, Is there an intelligent agent in your future? (1999). Available at http://www.nature.com/nature/webmatters/agents/agents.html. Accessed 02/20/2005
20. J. Hendler, Agents and the Semantic Web, *IEEE Intelligent Systems*, March, April (2001), 30–37
21. F. Herrera, E. Herrera-Viedma, J.L. Verdegay, A model of Consensus in Group Decision Problems under Linguistic Assessments, *Fuzzy Sets and Systems* 78 (1996), 73–87
22. F. Herrera, E. Herrera-Viedma, J.L. Verdegay, Direct Approach Processes in Group Decision Making using Linguistic OWA operators, *Fuzzy Sets and Systems* 79 (1996), 175–190.
23. E. Herrera-Viedma, Modelling the Retrieval Process of an Information Retrieval System Using an Ordinal Fuzzy Linguistic Approach, *Journal of the American Society for Information Science and Technology (JASIST)* 52 (6) (2001), 460–475
24. E. Herrera-Viedma, E. Peis, Evaluating the informative quality of documents in SGML format using fuzzy linguistic techniques based on computing with words, *Information Processing and Management* 39 (2) (2003), 195–213
25. E. Herrera-Viedma, F. Herrera, L. Martínez, J.C. Herrera, A.G. López, Incorporating Filtering Techniques in a Fuzzy Multi-Agent Model for Gathering of Information on the Web, *Fuzzy Sets and Systems* 148 (1) (2004), 61–83
26. N. Jennings, K. Sycara, M. Wooldridge, A roadmap of agent research and development, *Autonomous Agents and Multi-Agents Systems* 1 (1998), 7–38
27. P. Maes, Agents that reduce the work and information overload, *Communications of the ACM*, 37 (7) (1994), 30–40
28. D.L. McGuinness, F. van Harmelen, OWL Web Ontology Language Overview. W3C Recommendation. 10 February 2004, (2004). Available at http://www.w3.org/TR/2004/REC-owl-features-20040210/. Accessed 02/16/2005
29. A. Moukas, G. Zacharia, P. Maes, Amalthaea and Histos: Multiagent systems for WWW sites and representation recommendations. In M. Klusch (ed), *Intelligent Information Agents* (Springer-Verlag, 1999), 293–322

30. Ontoknowledge Project. Available at http://www.ontoknowledge.org/. Accessed 02/18/2005
31. A. Popescul, L.H. Ungar, D.M. Pennock, S. Lawrence, Probabilistic models for unified-collaborative and content-based recommendation in sparse-data environments, *Proceedings of the Seventeenth Conference on Uncertainty in Artificial Intelligence (UAI)*, San Francisco (2001), 437–444
32. P. Reisnick, H.R. Varian, Recommender Systems. *Special issue of Comm. of the ACM* 40 (3) (1997)
33. J.B. Schafer, J.A. Konstan, J. Riedl, E-Commerce recommendation applications. *DataMining and Knowledge Discovery* 5 (1/2) (2001), 115–153
34. Semantic Web Advanced Development for Europe (SWAD-Europe). Available at http://www.w3.org/2001/sw/Europe/. Accessed 02/17/2005
35. SEWASIE Home. Available at http://www.sewasie.org. Accessed 02/17/2005
36. U. Shah, T. Finin, Y. Peng, J. Mayfield, Information Retrieval on the Semantic Web, *Proceedings of the 10*th *International Conference on Information and Knowledge Management* (2002), 461–468
37. B. Shapira, U. Hanani, A. Raveh, P. Shoval, Information filtering: A new two-phase model using stereotypic user profiling, *Journal of Intelligent Information Systems* 8 (1997), 155–165
38. Wonder Web. Available at http://wonderweb.semanticweb.org. Accessed 02/22/2005
39. R.R. Yager, Intelligent agents for World Wide Web advertising decisions, *International Journal of Intelligent Systems* 12 (1997), 379–390
40. L.A. Zadeh, The concept of a linguistic variable and its applications to approximate reasoning. Part I, In *Information Sciences* 8 (1975), 199–249. Part II, In *Information Sciences* 8 (1975), 301–357. Part III, In *Information Sciences* 9 (1975), 43–80
41. Y. Zou, T. Finin, L. Ding, H. Chen, R. Pan, Using semantic web technology in multi-agent systems: a case study in the TAGA trading agent environment. In *ACM International Conference Proceeding Series. Proceedings of the 5th international conference on Electronic commerce* (2003), 95–101

Fuzzy Concept-Based Models in Information Browsers

Elena García Barriocanal and Miguel-Ángel Sicilia

Computer Science Department, University of Alcalá, Ctra. Barcelona km. 33.600 – 28871, Alcalá de Henares, Madrid (Spain)
{elena.garciab, msicilia}@uah.es

Summary. The growing interest and expanding efforts oriented to create metadata associated to Web resources represents an opportunity for new models of information seeking that use highly–structured metadata records as a navigation *affordance*. In this context, Semantic Web technology provides formal languages for the description of metadata and for the linking of such metadata elements to shared conceptualizations called ontologies. Ontologies can be used to implement user seeking moves that support different search strategies, according to Bate's concepts. Nonetheless, many user moves require the consideration of some degree of imprecision or uncertainty that is either attributable to the ontology structure in itself or to the interpretation of the selections associated to the move by the system as a query criteria. This paper describes a collection of such fuzzy moves as a generic meta–model for the design of fuzzy concept browsers, extending previous research on ontology–based information seeking. In addition, a concrete design is provided as an illustration of the range of possible interaction designs.

1 Introduction

The recent evolution and growth of the *World-Wide-Web* has resulted in an increasing need for advanced search and filtering tools. Traditional text retrieval, document clustering and natural language processing were among the techniques that were first successfully applied [9]. These content–based methods were later complemented by the consideration of the subjective interest that is implicit to Web links. The *PageRank* algorithm [22] used in the *Google* search engine is the most popular of these methods. But nowadays, the vision of a Semantic Web [3] provides a new and complementary perspective on the problem of information filtering. The main idea behind the current approaches to the Semantic Web is that of encoding meta-information associated to digital resources in the form of ontology–based annotations. Formal ontologies are engineered artifacts aimed at representing a shared, consensual conceptualization of a given domain [15]. Description logics [1] and their extensions

E.G. Barriocanal and M.-Á. Sicilia: *Fuzzy Concept-Based Models in Information Browsers*, StudFuzz **197**, 121–134 (2006)
www.springerlink.com

are in turn the underlying representation framework for these ontologies, and specific Web-oriented markup languages (e.g. OWL) have been specified for the purpose of interoperability [18]. In consequence, ontology-based annotations are connected to existing conceptualizations that in some cases include large commonsense knowledge bases [20]. This is actually a very different substrate for information retrieval than the natural language texts used in previous approaches, and it requires novel access methods that exploit its capabilities.

A consequence of the eventual availability of large volumes of ontology-based metadata would be the requirement on new methods of information filtering that use ontologies as a mean for information browsing [11]. Research on *browsing* interfaces to filter information combines human-computer interaction (HCI) issues with system-oriented views of the process of information retrieval. Such trade-off was discussed by Bates [2] and it is also the focus of the discipline of Interactive Information Retrieval (IIR) [24]. Research on fuzziness in the content-oriented aspect of information seeking has produced a number of extended techniques and models, e.g. [4, 5]. Nonetheless, the use of fuzzy models in information browsing interfaces requires a different approach in which user *moves* [2] are interpreted as guiding criteria for an adaptive interface. Nonetheless, the interpretation of moves in the user interface is dependant on the conceptual relationships of the elements in the browsing context, and these relationships between concepts should account for the cognitive dimension of the process of search. An example of such cognitive aspects is described in [27], in which the "distance" between a term and its directly subsumed entities is used as a metric for deciding where to navigate. Recently, the work of Naeve [21] has resulted in a set of design principles for concept browsers, drawn from the experience in the development of the *Conzilla* browser [8]. These principles can be applied to the design on information-seeking interfaces, and the navigation of the contexts can be supported by fuzzy information in the decision of the conceptual neighborhood to be exposed.

This chapter attempts to provide a foundation for the principles and design criteria for concept browsers that account for the principal fuzzy modeling elements that have a direct influence in ontology-browsing behavior. The resulting meta-model can be used as a reference framework for the design of concrete browsing schemes that integrate domain-specific aspects or that are realizations of concrete sets or assumptions or hypotheses about the behavior of information seekers.

The rest of this chapter is structured as follows. A general model for fuzzy concept browsers is provided in Section 2. Then, the mapping of a concrete model of such fuzzy conceptual browsing to Semantic Web technology is described in Section 3. Finally, conclusions and future research directions are provided in Section 4.

2 Fuzziness in Concept Browsers

As pointed out by Bates [2], sometimes the user prefers to control the search process to some extent, according to certain search strategies and tactics that can be effectively supported in user interfaces. Tactics are defined by Bates as "one or a handful of moves made to further a search", and include, for example, changing a term by a more specific one or adding similar terms. Ontologies can be used to give direct support to tactics, as described by García and Sicilia [11]. Here we focus on the fuzzy aspects of moves and their interpretation. To do so, a basic model for resources annotated through straightforward fuzzy ontologies is provided in the first sub-section, and the specifics of a range of fuzzy tactics are described in the second sub-section.

2.1 A Basic Model for Resources and Their Annotations

Many existing approaches to annotating Web resources are based on extending markup elements or embedding metadata fragments in the hypermedia nodes (as are HTML elements), e.g. [16]. Other approaches use separate metadata files that use URIs to reference the elements described. Nonetheless, when using ontological knowledge bases to represent metadata, both of these approaches have the drawback of separating the representations of the resources from the element itself. This entails that hypermedia elements and their relationships are not represented as concepts and properties in the ontology, missing an opportunity for a richer level of representation. The approach described by Sicilia and Garcia [28] entails a representation of hypermedia wholes and their parts in the same ontological language as annotations, thus providing a coherent integration of both. When using Web-based ontology description languages [10], ontology instances of hypermedia elements become identified by URIs, thus coming to a flexibility in producing distributed and loosely-connected annotations for public or organizational hypermedia descriptions.

For the sake of simplicity, we will assume here a `Resource`[1] (crisp) class representing any kind of Web or hypermedia resource that is of interest in the resource seeking process. These resources are assimilable to `ComputerFileCopy` individuals according to the *OpenCyc* commonsense ontology[2]: "an information bearing thing that is identified as a unit by a unique name [...] every file is such that if some information were removed from it – if it were truncated in a certain respect – it would no longer be (the same) file", which in turn are InformationBearingThings (`IBTs`): "item(s) that contain information (for an agent who knows how to interpret it)".

Then, *annotations* are everything that predicates something about `Resources`. For example, if we say that a Web page `providesAnExample` of the *quicksort* algorithm, then the ontological representation of the algorithm

[1] Ontology classes and properties are in Courier font from here on

[2] The open source version of the *Cyc* knowledge base [20]

(let's say, as an instance of the `SortingAlgorithm` class) becomes a term directly reachable from the resource and vice-versa. In other words, semantic information in its most simple form is concerned with anything *reachable* from resources or ontology terms or instances. Expression (1) provides a notation to refer to everything that can be transitively reached through the traversal of any property (assimilable to the concept of contextual neighborhoods [21] if we consider "is-a" relationship as a *prop*). It should be noted that this includes all the properties applicable to classes subsuming the element x.

$$\forall x \in \top \;\; \mathcal{R}(x) = \{y | prop^{+}(x,y)\} - \{x\} \tag{1}$$

Reachability somewhat determines the scope of seeking in ontology-based browsers, except in the case that axioms or logical rules were used to broaden the notion of what is reachable to what is "inferable". We will restrict for simplicity to the scope of reachability as expressed in (1).

The problem of browser design from this representation can be stated in terms of two design elements:

i Which are the user interface moves and tactics allowed.
ii How the actual moves of the users are to be interpreted as query-formulating criteria.

A catalog of tactics for [i] is provided in the following section, and some HCI principles could be applied for this. Further, HCI techniques could be used for the evaluation as described in [12]. Nonetheless, the range of possibilities for element [ii] are still largely unexplored. In what follows, the scheme implemented in OntoIR [13] is provided as an illustration.

Expression 2 describes a simple interpretation that maps the set of selected terms $\mathcal{C}$ to a concept expression Ψ denoting a collection of Web resources.

$$r_{OntoIR} : \mathcal{C} \longmapsto \overbrace{\mathcal{R}_{\mathcal{D}} \sqcap (T_1 \sqcup T_2 \cdots \sqcup T_l)}^{\Psi} \tag{2}$$

Expression (2) describes matching items as a concept expression $\mathcal{R}_{\mathcal{D}}$ restricted to the selected types of documents $T_1, \ldots, T_n$ (e.g. articles, book, etc). $\mathcal{R}_{\mathcal{D}}$ is a combination of three sets of Web resources related with the sets of instances C, E and S. Expression (3) defines S in terms of the concepts $D_1, D_2, \ldots, D_h$ as the concept covering all the instances of the domain concepts in $\mathcal{C}$, and C is defined as a subset of S in which an arbitrary domain relationship R exists to other element of S. In the same expression, E is defined as a set of concepts external to S that are related to one or more concepts in S.

$$C \doteq S \sqcap \exists\, R.S \;\; and\; E \doteq \neg S \sqcap \exists\, R.S \;\; given\; that \;\; S \doteq D_1 \sqcup D_2 \cdots \sqcup D_h \tag{3}$$

$\mathcal{R}_{\mathcal{D}}$ can be determined by expression (4).

$$\mathcal{R}_{\mathcal{D}} \doteq Resource \;\sqcap (\exists\; about.C \;\sqcup \exists\; about.E \;\sqcup \exists\; about.S) \tag{4}$$

Where *Resource* is the top class of all kind of Web resources in the ontology. From expression (4), relevance criteria for resources can be implemented regarding to each of the three sets C, E and S. The OntoIR design follows a simple heuristic ($\exists\ about.C \succ S \sqcap \neg\exists\ about.C \succ \exists\ about.E$) favoring the presence of (intra-)relations for the relevance sorting of the results. This concrete design provides an approach to reachability spanning only two levels of traversal, that is, only two traversals at most are required to determine the sets that are used for retrieving the instances. Even though the formulation just presented operates at the level of instances (A-box in description logics jargon), relations between concepts (T-box) could be exploited similarly through *reification* of some concepts as instances, as described technically in [26].

Crisp interpretation schemes like that of OntoIR could be extended to the case of dealing with imprecision in two orthogonal directions:

- The moves in themselves may be considered as imprecise, and even degrees of imprecision may be derived from sequences of moves, e.g. when a selection is canceled or the user goes back in the navigation.
- The concepts, relations and instances that are dealt with in the interpretation of the moves may be considered to have imprecise relationships.

We have restricted ourselves to the second of the two directions, since the first one requires extensive research on user behavior that is not currently available. With regards to fuzziness inside the ontology, fuzzy description logics [17, 29, 30] can be used as a model. For example, the semantics of $\mathcal{ALC}_{FH}$ are based on the following interpretations, where η_M is a membership modifier (this category is intended to model fuzzy hedges like $VERY$ or $FAIRLY$):

$$
\begin{aligned}
\top^{\mathcal{I}}(d) &= 1 \quad \forall d \in \Delta^{\mathcal{I}} \\
\bot^{\mathcal{I}}(d) &= 0 \quad \forall d \in \Delta^{\mathcal{I}} \\
A^{\mathcal{I}} &: \Delta^{\mathcal{I}} \rightarrow [0,1] \\
R^{\mathcal{I}} &: \Delta^{\mathcal{I}} \times \Delta^{\mathcal{I}} \rightarrow [0,1] \\
(C \sqcap D)^{\mathcal{I}}(d) &= min\{C^{\mathcal{I}}(d), D^{\mathcal{I}}(d)\} \\
(C \sqcup D)^{\mathcal{I}}(d) &= max\{C^{\mathcal{I}}(d), D^{\mathcal{I}}(d)\} \\
(\neg C)^{\mathcal{I}}(d) &= 1 - C^{\mathcal{I}}(d) \\
(MA)^{\mathcal{I}}(d) &= \eta_M(A^{\mathcal{I}}(d)) \\
(\forall R.C)^{\mathcal{I}}(d) &= inf_{d' \in \Delta^{\mathcal{I}}}\{max\{1 - R^{\mathcal{I}}(d, d'), C^{\mathcal{I}}(d')\}\} \\
(\exists R.C)^{\mathcal{I}}(d) &= sup_{d' \in \Delta^{\mathcal{I}}}\{min\{R^{\mathcal{I}}(d, d'), C^{\mathcal{I}}(d')\}\}
\end{aligned}
$$

A (fuzzy) knowledge base Σ is the union of a finite set of fuzzy assertions Σ_A (often called "A-box") and a finite set of fuzzy terminological axioms Σ_T (often called "T-box"). Terminological axioms are either fuzzy specializations $A \tilde{\leq} B$ or fuzzy concept definitions $A \tilde{=} C$. Fuzzy assertions take the forms summarized in expression (5), where $n \in [0, 1)$. For example, $\langle u : VIP \geq 0.8\rangle$ may be interpreted as "user u belongs to the VIP user group to an extent greater or equal than 80%", and $\langle (u, p) : interestedIn \leq 0.3\rangle$ may be interpreted as "the interest of user u in product p is lower or equal than 30%".

$$\left\langle \left\{ \begin{array}{c} a : C \\ (a,b) : R \end{array} \right\} \left\{ \begin{array}{c} > \\ < \\ \leq \\ \geq \\ = \end{array} \right\} n \right\rangle \tag{5}$$

Given the above definitions, the interpretation of OntoIR can be extended in an straightforward way by considering $\mathcal{R}_\mathcal{D}$ a fuzzy concept characterized from the fuzzy assertions that connect *Resources* to instances of the set of concept selected $\mathcal{C}$. The computation process would proceed in two steps:

- First, the *about* relations with sets C, E and S is used to compute the membership in $\mathcal{R}_\mathcal{D}$. The difference in addition to the fuzzy semantics of the $\exists$ constraints lies in the addition of a new concept K that will provide an account for the strength and number of relationships inside the core term C, as detailed below (8). This is a numerical replacement for the original OntoIR preference for elements belonging to the set C.
- Then, the resource types selected would be considered. Here the categories T_i are crisp, but some degree of similarity between document forms may be considered. For example, when searching for `ScientificJournalArticles`, it can be considered that `ProceedingsArticles` could be an option. This would entail a replacement of the conjunctions and disjunctions in expression (2) to avoid the maximum-minimum semantics that would effectively result in crisp document types.

The final fuzzy interpretation is expressed in (6) and (7), where there are m document typed defined in the ontology, and the similarity class $\sigma(T_i)$ is used to represent the membership of each `Resource` to document type T_i defined logically by the combination of the similarities with all the document types to which the resource belongs (in a crisp sense). Those σ-terms are used for those document types not selected by the user, and they allow a resource to be selected even in the case that its type has not been explicitly selected.

$$r_{OntoIR'} : \mathcal{C} \longmapsto \overbrace{\mathcal{R}_\mathcal{D} \sqcap (T_1 \sqcup T_2 \cdots \sqcup \sigma(T_{m-1}) \sqcup \sigma(T_m))}^{\Psi} \tag{6}$$

$$\mathcal{R}_\mathcal{D} \doteq Resource \sqcap ((\exists\, about.C \sqcup K) \sqcup \exists\, about.E \sqcup \exists\, about.S) \tag{7}$$

In expression (7), the degree of membership of an instance ins belonging to C is computed by the expression (8), in which $PROP$ represents the set of all the properties in the considered fuzzy ontology, and $\mu(p, i_1, i_2)$ is used to denote the value associated to a property assertion as described above.

$$\mu(ind, K) = \prod_{prop_i \in PROP} \mu(prop_i, ind, ind') \;\; ind' \in S \tag{8}$$

Symbol $\prod$ in expression (8) stands for an strictly increasing aggregation operator that favors to some extent the amount of properties with non-zero grades in their assertion. A simple choice for this operator may be a simple arithmetic sum with a bound in full membership. Nonetheless, other choices that give less importance to the properties may be considered.

The interpretation scheme described is only one of the many design possibilities. Alternatives may also allow for the selection of instances or relations, and others may take into account the ordering of the set of elements that are part of the ongoing query. In any case, any of them would be subject to fuzzy interpretation, which provides a built-in relevance ranking as part of the mechanism that resolves the query.

2.2 Fuzzy Elements in the Implementation of Ontology-Guided Tactics

The description of the interpretation above deals with the retrieval of resources *after* the query has been formulated, in that case as a simple set of selected concepts. In ontology-browsing interfaces, these formulations come from interactive navigation of the ontology elements, called moves, which in turn are used in search tactics. Table 1 provides an overview of some well-known tactics and their interpretation in terms of the fuzziness represented inside the ontology.

The fuzzy measures for distance described in Table 1 are attached to classes (i.e. to elements in the *T-box*), so that they are not supported in fuzzy description logics as the one described above. Distance between terms can be represented explicitly [27] or be computed from the structure of the ontology whenever significant heuristics can be found.

3 A Semantic Web Implementation of a Fuzzy Generic Concept Browser

In this section, a concrete implementation of an ontology-based browser for a specific domain is described as an illustration of some of the fuzzy techniques sketched so far. The browser is intended to give access to a collection of Spanish graphical humor pieces, according to an ontology that provides two dimensions for the description of humor pieces: (a) frames in the sense used by Rojo-López [25] and (b) humor technique applications, based on the classification of Buijzen and Valkenburg [6].

Based upon the work of Fillmore [7], "frame semantics" allows the description of the mental structures that provide the common ground required to understand a concrete piece of humor. Frames are thus considered as any concept instance in the ontology and their reachable concepts, and the more specific "frame connecting element" is considered to be the concrete element in the humor piece that somewhat "triggers" the connection with the semantic frame(s). For example, as described in [14], a humor piece celebrating

Table 1. Fuzzy ontology support for search tactics

Tactic	**Bates' description** [2]	**Ontology-based implementation**	**Example**	**Fuzzy aspects**
SUPER	To move upward hierarchically to a broader (superordinate) term.	Change B by A, where B $\sqsubseteq$ A.	Move from Siamese_Cat to Cat	Use a fuzzy distance measure to select the term to change for
SUB	To move downward hierarchically to a more specific (subordinate) term.	Change A by B, where B $\sqsubseteq$ A	Move from Cat to Siamese_Cat	Use a fuzzy distance measure associated to the "is-a" relation to select the term to change for.
RELATE	To move sideways hierarchically to a coordinate term.	Change B by C, where there exist a concept A such that B $\sqsubseteq$ A, C $\sqsubseteq$ A ("directly")	Move from Siamese_ Cat to Angora_ Cat (through Cat)	Here two approaches can be used: (a) Use the fuzzy distance of the "is-a" relation to select the term that is "closer" by adding the distances up to A and down to C. (b) use instead a distance measure associated to the applicable properties
CONTRARY	To search for the term logically opposite from that describing the desired information	Change B by C, where B $\sqcap$ C = $\varnothing$ (at least)	Move from LateHarvest to EarlyHarvest (as wine categories)	The level of oppositeness of fuzzy terms can be measured by the overlapping of a term with the (fuzzy) opposite of the other. Even though antonyms are lexical relations, using such measures constitutes an alternative when such information is not available
PARALLEL	To make the search formulation broad (or broader) by including synonyms or otherwise conceptually parallel terms	Add term C where C is similar to any of the terms that form part of the current navigation	When searching for information on softwareInspection include also softwareReview	Similarity between concepts can be measured in fuzzy terms by induction from the actual instances or by explicitly incorporating similarity measures. Existing studies on the derivation of fuzzy lexical categories [23] can be used instead
PINPOINT	To make the search formulation precise by minimizing (or reducing) the number of parallel terms, retaining the more perfectly descriptive terms	The inverse of PARALLEL		Measures of similarity can be used for that purpose, retaining the fuzzy terms that has less internal entropy [19] as more specific

the Spanish Constitution Anniversary could have as frame "spanish democracy" as subsumed by the wider notion of "political system", and the frame connecting element may be for example, the depiction of the facade of the "Congress of the Deputies" that represents democracy in Spain. Since the declaration of the frame of a piece is subject to some degree of subjectiveness, the property `isFrameOf` (and its inverse `framedIn`) were used with the fuzzy description logics semantics described above. The same was applied to the `frameConnectingElement` declaration.

Fig. 1. Screenshot of the concept browser for humor pieces.

Figure 1 shows an screen capture for the semantic navigation browser, in which initially only instances referred as frames or frame connecting elements are showed (filtering those that have higher degrees in fuzzy assertions), and the terms to which they belong (again to a given degree) are also showed, and it is provided a "add class" link to move them to the list of selectable elements. In the interface, a number of tactics have been applied, all of them having a consideration of fuzziness:

- SUPER and SUB are explicitly provided for classes (using the arrows that are showed following term names). Here the choosing of super or sub-concepts is mediated by fuzzy term distance as described in [27], since the "closer" terms are suggested first to users.
- When clicking on an instance, other instances related to them are showed in the lower part of the interface. In this case, "closer" instances (as represented in fuzzy assertions) are showed first, and instances not reaching a threshold are hidden unless the user asks for them with the "show more" link. This may be considered a form of PARALLEL or RELATE.
- Instances with a similar "structure" are showed and allowed to be added to the query. Such similarity is computed by comparing property assertions of the two instances. If a degree of similarity in their fuzzy grade is found, the instance is considered to be similar enough. This again can be considered as a form of PARALLEL.

As an example of the navigation mediated by the tactics described, let's suppose that the user starts with the collection regarding the "25th

Fig. 2. Example humor piece.

Anniversary of the Spanish Constitution"[3]. Then, starting from a concept as `InstitutionalBuilding`, the traversal to an more specific category as `GovernmentBuilding` will be showed as a navigational option (of type SUB) due to a considerable degree of explicit association (let's say 0.8), while the navigation (of type SUPER) to the more general concept `Building` could instead be avoided by the navigation system due to a significant "distance", let's say, a proximity of 0.2. Those numbers are relative to the domain as described in [27], and they are only tentative estimations, but serve as a better estimation than simply considering every move to be at the same "distance". Following the example, let's consider the humor piece as the one depicted in Figure 2. This piece of humor uses the technique `Repartee` (i.e. verbal banter) as the essential technique, in this case playing with the senses of the Spanish word "labor" which means both "task" and also "the result of knitting clothes or textiles". Verbal banter is complemented with the `Absurdity` of a woman knitting a pullover inside the parliament, and with a reference to "Marcelino Camacho" (instance mc), a syndicalist and former communist party leader (very well-known in Spain). Such piece p may be described as

[3] http://www.constitucion.es/humor/

$\langle (p, mc) : about \geq 0.7 \rangle$. Then, we can consider the following assertions (a fragment of the entire knowledge base):

$$LeftWingOrganization \sqsubseteq DemocraticOrganization$$
$$DemocraticOrganization \sqsubseteq \exists operatesIn.DemocraticSystem$$
$$\langle ccoo : LeftWingOrganization \geq 0.9 \rangle$$
$$\langle (ccoo, mc) : drivenBy \geq 0.8 \rangle$$

where *ccoo* is the representation of a Spanish union of which Camacho was a leader. If the navigation through *DemocraticSystem* leads to *LeftWingOrganization* (LWO), the user may select it as the query formulating criterion. In such case, the instance *ccoo* will be included in the core set C, and mc will be included in set E by virtue of the property *drivenBy*. In consequence, p will become a query result with the degree 0.7. According to the formulation described above, the fuzzy degree that connect mc to an element inside S is not considered, but a simple modification in the matching scheme would do so. This provides a flexibility in the representation of semantic relations, similar to that of element K.

Humor technique applications are the representation of a resource used by the humor author or present in a humor piece that contributes to producing the humorous interpretation, e.g. absurdity, exaggeration, verbal banter and the like. Another page in the interface allows for a selection of the techniques that are of interest in the current search.

Once both classes or instances and humor technique applications are selected in a series of user moves, the search will proceed to locate relevant instances of `HumorPiece` by the following algorithm:

1 First, the specified humor techniques (if any) are used to collection H_0 of humor pieces ordered by the aggregation of the degrees associated to the `appliedIn` assertions relating humor pieces with the techniques selected.
2 Then, a procedure of recursive search of humor pieces included in H_0 traversing the relationships from the instances selected is carried out, limiting a number of levels of recursion. The complete path of assertions is recorded. Then, the final paths are sorted by an aggregation of the partial degrees of each property assertion, penalizing longer ones. This results in the final H_1 list of results. An additional filtering step by collection can be carried out, but this is no relevant to our present discussion.

In the case that the user selects a class, a subset of the instances with higher degrees of membership are selected, and step [2] is applied to each of them.

The use of fuzzy techniques has been added as a way to reduce the number of elements in the interface according to the numeric representations of fuzziness integrated in the ontology. This eases interaction by reducing cognitive overload with a sound justification on the contents of the knowledge base.

4 Conclusions and Future Work

Fuzziness in ontology-based information seeking provides both a degree of flexibility in query interpretation and also a built-in criteria for the ranking of query results. Approaches that take subsets of the reachable resources can make use of existing fuzzy description logics and other imprecision-representation elements both in the generation of the moves available to the user at a given instant, and also for the generation of results. A generic model has been described, and tentative interpretations for well-known moves have been provided. In addition, a concrete prototype demonstrating some of the points of this kind of retrieval tools has been described.

The design of ontology-based browsers is still an unexplored area, and in consequence, fuzzy extensions have still not been studied yet. Further research in the area would require a balanced approach between the amount and richness of the fuzzy extensions used and their impact in the usability of the final user interfaces.

References

1. Baader F, Calvanese D, McGuinness D, Nardi D, Patel-Schneider P (eds.) (2003) The Description Logic Handbook: Theory, Implementation and Applications. Cambridge Press.
2. Bates M J (1990) Where Should the Person Stop and the Information Search Interface Start?. Information Processing and Management 26: 575–591.
3. Berners-Lee T, Hendler J, Lassila O (2001). The Semantic Web. Scientific American 284(5): 34–43.
4. Bookstein A (1981) Fuzzy requests: an approach to weighted Boolean searches. Journal of the American Society for Information Science 31.
5. Bordogna G, Pasi G (1993) A Fuzzy Linguistic Approach Generalizing Boolean Information Retrieval: A Model and Its Evaluation. Journal of American Society for Infomation Science 44(2): 70–82.
6. Buijzen M, Valkenburg P M (2004) Developing a typology of humor in audiovisual media. Media Psychology 6: 147–167.
7. Fillmore C J (1976) Frame semantics and the nature of language. In: Annals of the New York Academy of Sciences: Conference on the Origin and Development of Language and Speech 280: 20–32.
8. Nilsson M, Palmér M (1999) Conzilla – Towards a concept browser. CID-53, TRITA-NA-D9911, Department of Numerical Analysis and Computer Science, KTH, Stockholm.
9. Faloutsos C, Oard D W (1995) A survey of Information Retrieval and Filtering Methods. Technical Report CS-TR3514, Dept. of Computer Science, Univ. of Maryland.
10. Fensel D (2002) Language Standardization for the Semantic Web: The Long Way from OIL to OWL. In: Proc. of the 4th International Workshop on Distributed Communities on the Web, DCW 2002: 215–227.
11. García-Barriocanal E, Sicilia M A (2003) User Interface Tactics in Ontology-Based Information Seeking. Psychology e-journal 1(3): 243–256.

12. García-Barriocanal E, Sicilia, M A, Diaz P, Aedo I (2003) An interactive ontology-based query formulation approach for exploratory styles of interaction. In: Proceedings of the 10th International Conference on Human – Computer Interaction (HCI 2003). Lawrence Erlbaum Associates.
13. García-Barriocanal E, Sicilia M A (2003) Designing ontology-based interactive information retrieval interfaces. In: R. Meersman and Z. Tari, (eds) Proceedings of OTM Workshops LNCS 2889: 152–165, Springer.
14. García-Barriocanal E, Sicilia M A, Palomar D (2005) A graphical humor ontology for contemporary cultural heritage access. In: Proceedings of the European Conference on Information Systems (to appear).
15. Gruber T. (1995) Towards principles for the design of ontologies used for knowledge sharing. International Journal of Human-Computer studies 43(5/6): 907–928.
16. Heflin J D, Hendler J A (2001) A Portrait of the Semantic Web in Action. IEEE Intelligent Systems and their applications 16(2): 54–59.
17. Hölldobler S, Dinh Khang T, Störr H P (2002) A fuzzy description logic with hedges as concept modifiers. In: Hoang Phuong, N., Nguyen, H.T., Cat Ho, N., Santiprabhob, P. (eds.): Proceedings of InTech/VJFuzzy'2002: 25–34.
18. Horrocks I, Patel-Schneider P F, van Harmelen, F (2003) From SHIQ and RDF to OWL: The making of a web ontology language. Journal of Web Semantics 1(1): 7–26.
19. Klir G, Folger T (1988) Fuzzy Sets, Uncertainty and Information. Prentice-Hall, Englewood Cliffs.
20. Lenat D D (1995) Cyc: A Large-Scale Investment in Knowledge Infrastructure. Communications of the ACM 38(11): 33–38.
21. Naeve A (2001) The Concept Browser – a new form of Knowledge Management Tool. In: Proceedings of the 2nd European Web-based Learning Environments Conference (WBLE 2001): 24–26.
22. Page L, Brin S, Motwani R, Winograd T (1998) The PageRank citation ranking: Bringing order to the web. Technical report, Stanford Digital Library Technologies Project, 1998. Paper SIDL-WP-1999-0120.
23. Rieger B B (1994) Fuzzy Computational Semantics. In: Japanese-German-Center Berlin (Eds.) Joint Japanese-European Symposium on Fuzzy Systems 1992 [Publications of the JGCB: Series 3 Vol. 8]: 197–217. Berlin (JDZB)
24. Robins D (2000) Interactive Information Retrieval: Context and Basic Notions. Informing Science Journal 3(2): 57–62.
25. Rojo-Lopez A M (2002) Frame semantics and the translation of humour. Babel 48(1): 34–77.
26. Sicilia M A, García-Barriocanal E, Díaz P, Aedo I (2003) A Literature-Based Approach to the Annotation and Browsing of Domain-Specific Web Resources. Information Research 8(2).
27. Sicilia M A, García-Barriocanal E, Díaz P, Aedo I (2003) Representation of Concept Specialization Distance through Resemplance Relations. In: Advances in Soft Computing - Engineering, Design and Manufacturing: 173–182. Springer.
28. Sicilia M A, García-Barriocanal E (2005) On the Convergence of Formal Ontologies and Standardized e-Learning. Journal of Distance Education Technologies 3(2): 13–29.

29. Straccia U (2001) Reasoning within fuzzy description logics. Journal of Artificial Intelligence Research 14: 137–166.
30. Tresp C B, Molitor R (1998) A Description Logic for Vague Knowledge. In: Proceedings of the 13th biennial European Conference on Artificial Intelligence (ECAI'98): 361–365. J. Wiley and Sons, Brighton, UK.

Evaluation of Term-based Queries using Possibilistic Ontologies

Yannick Loiseau, Mohand Boughanem, and Henri Prade

Institut de Recherche en Informatique de Toulouse, 118 route de Narbonne, 31062 Toulouse cedex 4
{loiseau,bougha,prade}@irit.fr

Summary. In multiple-source information systems, attribute values are often assessed in linguistic terms belonging to different vocabularies. The request itself, which may include preferences, may be expressed using terms of another vocabulary, raising the problem of matching the query and the information in a semantic manner. The fuzzy pattern matching framework allows us to compute matching degrees between queries and data represented by fuzzy sets, even if they do not perfectly match. The qualitative pattern matching no longer requires a fuzzy set representation thanks to the use of ontologies for computing similarity degrees between terms. This allows us to deal with information querying in face of heterogeneous sources of information. This chapter presents this tool and its application to database and textual information retrieval on two examples.

1 Introduction

In information retrieval (IR) as well as in databases systems, the vocabularies used to state the linguistic information are often heterogeneous when documents come from different sources. The terms used to express categories associated to a concept by one of the sources do not always correspond to the terms used by the other sources. For instance, documents written by different authors, from different journals, even if they deal with the same field, may use slightly different vocabularies. Moreover, even in the case of a unique source, or if the different sources use the same vocabulary, queries specifying user needs may be stated in another vocabulary, since the user does not know a priori which terms are used in the documents collection. This issue is still more obvious when using documents written in different languages, as it is the case in multilingual information retrieval [1]. This raises the problem of putting into correspondence the terms used in the different vocabularies. Moreover, the user may want to express preferences, under the form of flexible queries (e.g. [2–4]), about her/his requirements that have to be taken into account in the query evaluation process.

Y. Loiseau et al.: *Evaluation of Term-based Queries using Possibilistic Ontologies*, StudFuzz **197**, 135–160 (2006)
www.springerlink.com

The fuzzy pattern matching framework [5] provides a tool for evaluating flexible queries in face of possibly imprecise data, where each linguistic label, both in the data or in the query, is represented by a fuzzy set. When these fuzzy sets are defined on the same domain, they can be compared using two indices called possibility and necessity measures. However, fuzzy pattern matching requires fuzzy set-based representations, usually associated with terms on numerical domains. In order to deal with more general linguistic terms, i.e. evaluating queries on documents where both of them are described by linguistic keywords that cannot be easily represented by fuzzy sets on a domain, the fuzzy pattern matching idea has been adapted to symbolic labels [6, 7] recently. Then, the matching between query terms and data no longer require the identity of terms, but rather a *semantic similarity* is computed between these terms, in a qualitative matching process. The relations between terms are now provided by means of an ontology estimating approximate synonymy and hypernymy relations between terms by means of possibility and necessity degrees. This *possibilistic ontology* is then used to evaluate relations that are not directly specified between terms of the ontology. This leads to an implicit query expansion. Thus it is possible to match a query and a piece of information taking into account preferences, even though terms do not match perfectly. This can partially solve the heterogeneous vocabularies issue.

The approach may appear to be reminiscent of thesaurus or terms association-based IR (e.g. [8]), where co-occurrence degrees are attached to pairs of terms. However, fuzzy pattern matching uses two degrees that respectively assess to what degree terms are possibly synonymous, or to what degree one is a specialization of the other. These two estimates differ from thesauri degrees. Indeed, terms such as *database* and *query* may be found co-occurrent, while *query* and *request* are not, even though the two latter terms are synonymous, while the two former are not.

The key notions of the fuzzy pattern matching used as a conceptual basis for the approach are first recalled. Its qualitative counterpart is then presented, where the semantic proximity of terms is assessed by means of possibility and necessity relations and propagated within possibilistic ontologies. Related works on weighted ontologies are also briefly surveyed. This chapter illustrates the use of this method on practical information sources, such as a tourism database and a collection of document titles. Finally, an extension of this model to full-text IR is outlined.

2 From Fuzzy to Qualitative Pattern Matching

2.1 Fuzzy Pattern Matching

The fuzzy pattern matching method has been developed in the fuzzy sets framework [5, 9, 10]. It is used to formulate flexible queries using fuzzy sets, evaluated on imprecise or fuzzy data also represented by fuzzy sets. This technique estimates to what extent it is possible and to what extent it is certain that data represented by imprecise attributes fulfill a flexible query representing the user needs and preferences.

A pattern is a set of elementary requirements represented by labels encoding properties defined on the attributes domains. As an example, the pattern "cheap and large" associated with a database of houses, states that the "price" and the "size" attribute values of the searched house should be respectively compatible with *cheap* and *large*. Each label such as *cheap* and *large* is associated to a fuzzy set membership function which restricts values that are more or less compatible with the label meaning. These values belong to the domains of the corresponding attributes. Here, *cheap* and *large* are associated with membership functions defined on the price and size domains, which can be numerical ranges or discrete sets of values. Moreover, data are also described by sets of labels associated to fuzzy sets. These sets are possibility distributions representing imprecise or fuzzy data.

Let R and D be a pattern label and a piece of data respectively, that must be compared. They belong to the same domain U. Let μ_R be the membership function associated to the label R and π_D be the possibility distribution corresponding to D, which are functions from U to $[0, 1]$. Let u be an element of U. $\mu_R(u)$ is the compatibility degree between the value u and the meaning of R. $\mu_R(u) = 1$ means a complete compatibility with R and $\mu_R(u) = 0$ means a complete incompatibility with R. In the same way, $\pi_D(u)$ estimates the possibility that u is the value of the considered attribute describing the data. D is a fuzzy set of *possible* values, among which only one is the true value of the uncertain piece of data, whereas R is a fuzzy set of *more or less* compatible values. More precisely, $\pi_D(u) = 1$ means that u is totally possible, whereas $\pi_D(u) = 0$ means that u is not at all a possible value for the element. However, distinct values u and u' such that $\pi_D(u) = \pi_D(u') = 1$ may exist. In the following, μ_R and π_D are normalized, that is $\max_U \mu_R(u) = 1$ and $\max_U \pi_D(u) = 1$.

Two measures are defined to estimate the compatibility between a request element R and its counterpart D in the data: a possibility degree $\Pi(R; D)$ and a necessity degree $N(R; D)$ defined as [9]:

$$\Pi(R; D) = \sup_{u \in U} \min(\mu_R(u), \pi_D(u)) \ , \tag{1}$$

$$\begin{aligned} N(R; D) &= 1 - \Pi(\overline{R}; D) \\ &= \inf_{u \in U} \max(\mu_R(u), 1 - \pi_D(u)) \ . \end{aligned} \tag{2}$$

The possibility measure $\Pi(R; D)$ estimates to what extent it is possible that R and D refer to the same value u. It represents the *intersection* of the fuzzy set of values compatible with R with the fuzzy set of possible values for D. The necessity $N(R; D)$ measures to what extent it is certain that the value corresponding to D is compatible with R. It is an *inclusion* degree of the possible values for D into the set R of values compatible with the query label.

This measure $N(R;D)$ reflects the asymmetry between the user needs and the information itself.

It can be shown that $\Pi(R;D) \geq N(R;D)$. Note that if D is precise, that is: $\exists t', \pi_D(t') = 1$ and $\forall u \neq t', \pi_D(u) = 0$, i.e. $D = \{t'\}$, then $\Pi(R;\{t'\}) = N(R;\{t'\}) = \mu_R(t')$. Note also that if $\mu_R = \pi_D$, $\Pi(R;D) = 1$, and if R is a fuzzy set, $1 \geq N(R;D) \geq \frac{1}{2}$. In the case of continuous membership functions on real domains, (2) gives $N(R;D) = \frac{1}{2}$ if $\mu_R = \pi_D$. This reflects that there are values that are possible at degree 0.5 that are not in the 0.5 level-cut of R, i.e. in $\{u, \mu_R(u) \geq \frac{1}{2}\}$.

Classical fuzzy pattern matching recalled here is used in the following as a basis for defining *qualitative pattern matching*. The idea of qualitative pattern matching is to state fuzzy semantic relations between linguistic terms, using possibility and necessity degrees, now obtained from *possibilistic ontologies*.

2.2 Possibilistic Ontology

Fuzzy sets can interface numerical values with linguistic terms, using membership functions, and the comparison of terms can be evaluated by fuzzy pattern matching. However, the terms used for describing information need to be represented by a fuzzy set on a clearly identified domain, which is a severe limitation for information retrieval purposes. Similarity measures between words have been extensively studied in information retrieval literature, using for instance distance between nodes in a taxonomy, or based on a common information probability (e.g. [11]). Besides, a usual method when a query fails is to replace it by similar queries generated using ontologies or thesauri (e.g. [12]). Qualitative pattern matching remedies limitations of fuzzy pattern matching by enlarging the meaning of a term through the fuzzy set of its similar terms.

In *qualitative pattern matching*, a set of terms $\mathcal{T}_i$ is assumed to be associated to each information domain i. $\mathcal{T}_i$ is the set of labels that can be used to describe information in domain i. More precisely, $\mathcal{T}_i = \{t_i^j, j = 1, n(i)\}$, where t_i^j is a label (e.g. *hotel*), that can be used in order to describe a piece of information (here, an accommodation place). Contrary to standard fuzzy pattern matching, terms, or more generally symbolic expressions, are not associated with fuzzy sets, but their meanings are related through *possibilistic ontologies* O_i for each domain i, where relations in O_i are modeled by possibility and necessity degrees. For two labels t_i^j and t_i^k:

- $\Pi(t_i^j, t_i^k) = \Pi(t_i^k, t_i^j)$ represents to what extent t_i^j and t_i^k can describe the same thing. A zero possibility means that the two labels never represent the same thing. A positive possibility lesser than 1 expresses that the two terms may mean the same thing, but it is not always the case.
- $N(t_i^j, t_i^k)$ estimates to what extent it is certain that t_i^k is a specialization of t_i^j. Moreover, $N(t_i^j, t_i^k) = N(t_i^k, t_i^j) = 1$ represents genuine synonymy. If t_i^k is a perfect specialization of t_i^j, then $N(t_i^j, t_i^k) = 1$. However, a zero value

only means a total lack of certainty for the specialization relation between these terms. This relation is asymmetric.

These measures must satisfy the following properties:

- Reflexivity: $\Pi(t_i^j, t_i^j) = 1$.
- Symmetry: $\Pi(t_i^j, t_i^k) = \Pi(t_i^k, t_i^j)$.
- $\Pi(t_i^j, t_i^k) \geq N(t_i^j, t_i^k)$, since specialization entails that the meanings overlap.
- $N(t_i^j, t_i^k) > 0 \Rightarrow \Pi(t_i^j, t_i^k) = 1$. If it is somewhat certain that t_i^k specialize t_i^j, then it must be fully possible that they are used for referring to the same thing.
- If labels are precise, we have $N(t_i^j, t_i^j) = 1$. If they are vague, we will suppose $N(t_i^j, t_i^j) \geq \frac{1}{2}$, according to fuzzy pattern matching. This expresses the uncertainty that the query and the data represent "really" the same thing. As an example, two people do not have necessarily the same idea of the price of something found *expensive* by both of them.

The degrees specified in the ontology are actually only defined on a subset of the Cartesian product of the vocabulary $\mathcal{T}_i \times \mathcal{T}_i$. They can be completed using previous properties and the two following forms of transitivity:

$$N(t_i^j, t_i^h) \geq \min\left(N(t_i^j, t_i^k), N(t_i^k, t_i^h)\right) , \tag{3}$$

$$\Pi(t_i^j, t_i^h) \geq N(t_i^j, t_i^k) * \Pi(t_i^k, t_i^h) . \tag{4}$$

where $*$ is defined as:

$$a * b = \begin{cases} b & \text{if } b > 1 - a , \\ 0 & \text{otherwise.} \end{cases}$$

Equation (3) represents the specialization transitivity [13]. The "hybrid transitivity" (4) states that if t_i^k specializes t_i^j and if t_i^k and t_i^h may refer to the same thing, then the meanings of t_i^j and t_i^h should overlap as well; see [14] for a proof of (3-4).

The certainty degree of the synonymy of t_i^j and t_i^h $Syn(t_i^j, t_i^h)$ can be computed as $\min(N(t_i^j, t_i^h), N(t_i^h, t_i^j))$. Using (3), it can be shown that this degree is max-min transitive. Let $Syn(t, t') = \min(N(t, t'), N(t', t))$. We have $\forall t'', Syn(t, t') \geq \min(Syn(t, t''), Syn(t'', t'))$. Indeed:

$$\begin{aligned} &\min(Syn(t, t''), Syn(t'', t')) \\ &\quad = \min(\min(N(t, t''), N(t'', t)), \min(N(t'', t'), N(t', t''))) \\ &\quad = \min(\min(N(t, t''), N(t'', t')), \min(N(t', t''), N(t'', t))) \\ &\quad \leq \min(N(t, t'), N(t', t)) \\ &\quad \leq Syn(t, t') \end{aligned}$$

Therefore, values that are not specified can be deduced from existing ones using the previous properties and relations. Values that cannot be inferred are supposed to be zero. The fact that default possibility is zero corresponds to a closed world hypothesis, since it is supposed that two terms cannot overlap if it is not specified. From a practical point of view and to simplify the use of the degrees, evaluations will be estimated "at worst", and the $\geq$ will be generally taken as an equality. It is therefore possible to estimate the relevance degrees between the data and the query even if the searched terms are not directly present in the information representation without using any explicit query expansion stage, as usually proposed in IR.

The extreme binary cases illustrate four situations:

1. the terms are genuine synonyms: $\Pi(t_i^j, t_i^k) = N(t_i^j, t_i^k) = N(t_i^k, t_i^j) = 1$;
2. one of the terms specializes the other term: $\Pi(t_i^j, t_i^k) = N(t_i^j, t_i^k) = 1$ or $\Pi(t_i^j, t_i^k) = N(t_i^k, t_i^j) = 1$;
3. the two meanings overlap, but are not true synonyms nor specializations: $\Pi(t_i^j, t_i^k) = 1$ and $N(t_i^j, t_i^k) = N(t_i^k, t_i^j) = 0$;
4. The meanings are clearly distinct: $\Pi(t_i^j, t_i^k) = N(t_i^j, t_i^k) = N(t_i^k, t_i^j) = 0$.

Intermediary values refine these distinctions. However, only few values in $[0, 1]$ will be generally used, to distinguish the case where a term always certainly specializes another, i.e. $N(t_i^j, t_i^k) = 1$, from the case where it is only *generally* true, which is expressed by $1 > N(t_i^j, t_i^k) > 0$.

As a matter of illustration, Fig. 1 presents a fragment of a simple ontology for accommodation places. This graph is a simplified representation of how an agent may perceive similarity relations between terms. Since it is possible to deduce implicit values for relations between terms, using properties and constraints of the used degrees, only direct links need to be given and are represented here.

In Fig. 1, note that words like *lodge* and *inn* are only considered as *possible* synonyms, or as entity that can provide the same services. Nothing can be inferred for the necessity from the possibility degree, and it can exist *lodges*

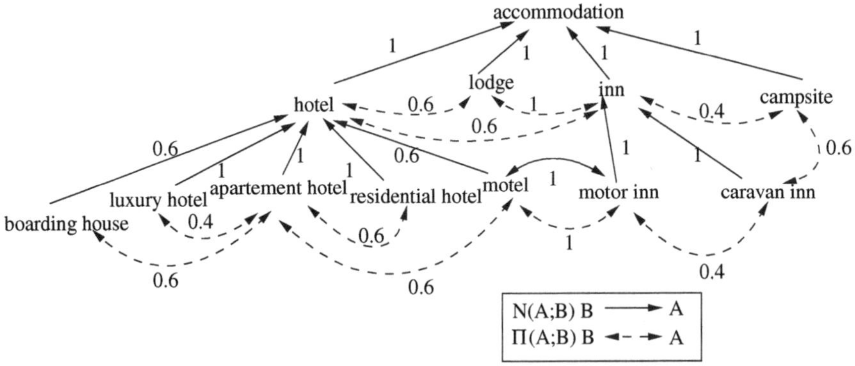

Fig. 1. Accommodation ontology.

that are not *inns*. On the other hand, both necessity degrees between *motel* and *motor inn* being 1, these terms are considered as genuine synonyms.

The values of degrees present in such ontologies are qualitative in nature, and estimates semantic relations between terms. As an example, $N(hotel, motel) = 0.6$ means that it may exist *motels* that are not considered as *hotel*, but that *generally*, *motels* are a kind of *hotels*. Thus, despite the use of numerical values, only the relative ordering between these values is significant, as the purpose is to rank-order query results and not to assess an absolute similarity degree. Practically, only few levels should be used, e.g. $\{0, 0.4, 0.6, 1\}$. These values may be associated for convenience with linguistic labels, such as "very similar", "rather similar", etc. to specify the strength of the relations.

Building such ontologies is a complex task, specially if done manually even on a task domain. Ontologies such as WordNet [15] can be used as a starting point. For instance, typed relations used in these ontologies, such as hypernymy can be matched with necessity degrees. Other relations such as "being a part of", e.g. a *room* and a *hotel*, can be interpreted in terms of possibility degrees. Besides, statistical ontologies can also be built from corpora analysis, extracting relations from terms co-occurrences (e.g. [16]). This may provide a basis for assessing values in possibilistic ontologies using both crisp semantic ontologies and possibilistic rescaling of probabilities (as used in Sect. 5). Subparts of general ontologies can be identified in order to obtain domain specific ontologies.

2.3 Qualitative Pattern Matching

We now consider the case of composed queries. Let Ω be the set of ontologies used to describe the different data domains:

$$\Omega = \{O_i | i = 1, \ldots, n\} ,$$
$$O_i = \{t_i^j \in \mathcal{T}_i\}, \forall i \in [\![1, n]\!] ,$$

with $\mathcal{T}_i$ the vocabulary associated with the ith domain. Formally, a piece of information (i.e. a document) is modeled as a set of terms (i.e. keywords), each term in the set belonging to a different ontology:

$$D = \bigwedge_i D_i \text{ and } \exists t_i' \in \mathcal{T}_i, D_i = \{t_i'\} .$$

Queries are conjunctions of disjunctions of (possibly weighted) terms. Thus a query R may be viewed as a conjunction of fuzzy sets R_i representing a conjunction of flexible user needs.

$$R = \bigwedge_i R_i, \text{ where } R_i = \bigvee_j (\lambda_i^j, t_i^j), t_i^j \in \mathcal{T}_i .$$

Note that weighted disjunctions allows us to define new concepts. As an example, a user can specify its own definition of a *cosy lodging* as:

$$(0.5, lodge) \vee (0.7, motel) \vee (0.8, apartment\ hotel) \vee (1, luxury\ hotel) .$$

The weights $\lambda_i^j \in [0,1]$ reflects how satisfactory this term is for the user (i.e. how well it corresponds to his/her request). It is assumed that $\max_j \lambda_i^j = 1$, i.e. at least one query term reflects the exact user requirement. Moreover, importance levels could be introduced between query elements, as described in [7].

The possibilistic query evaluation consists in retrieving all documents D such that the possibility of relevance $\Pi(R, D)$ or the necessity of relevance $N(R, D)$ are non zero. These two relevance degrees are computed as:

$$\Pi(R, D) = \min_i \max_j \min(\lambda_i^j, \Pi(t_i^j, t_i')) , \tag{5}$$

$$N(R, D) = \min_i \max_j \min(\lambda_i^j, N(t_i^j, t_i')) . \tag{6}$$

The max parts are weighted disjunctions corresponding to those in the query (where a fuzzy set of more or less satisfactory labels expresses a disjunctive requirement inside the same domain).

In the same way, as the query is a conjunction of elementary requirements pertaining to different domains, the min operator is used in the final aggregation. Note that if R contains a disjunction of redundant terms, that is $R = t \vee t'$ and $N(t, t') = 1$ in the ontology, it can be checked that evaluating t and $t \vee t'$ leads to the same result.

$\Pi(R, D)$ and $N(R, D)$ values estimate to what extent the document D corresponds possibly and certainly to the query R. Results are sorted first using decreasing values of $N(R, D)$, then decreasing values of $\Pi(R, D)$ for pieces of information having the same necessity value.

This matching process can be applied to databases, classical or fuzzy ones, or adapted to information retrieval for collections of sentences or keywords (see Sect. 4), or more generally to documents, as discussed in Sect. 5.

2.4 Other Approaches Using Ontologies

Statistical analysis of texts are used to estimate similarities between terms to define thesauri, maybe interpreted in a fuzzy way, as presented in [17], where fuzzy sets of terms similar to a given term are viewed as representing concepts. These thesauri are used to reformulate queries in order to retrieve more relevant documents. For instance, the *Ontoseek* information retrieval system [18] uses WordNet to expand queries. Ontologies can also be used to index collections [19], as presented in Sect. 5.

In databases, ontologies are used to extend queries to linguistically valued attributes, or to compute similarities between query terms and attribute values. For instance, [20] uses a thesaurus to match queries with a fuzzy database. Imprecise terms are defined as a fuzzy set of terms, and the fuzzy pattern matching is used in the matching process. However, even though links between terms are weighted, this ontology is not a possibilistic one as defined here, and links are sorted as in traditional ontologies.

A recent approach considers relevance rather than similarity between terms [21], where degrees representing terms specialization and generalization are introduced. These degrees are asymmetric, generalization being less favored (from a relevance point of view) than specialization. As an example, *poodle* specializes *dog* at 0.9 whereas *dog* generalizes *poodle* at 0.4. In the approach presented here, two kinds of degrees are used as well, but with a different meaning. The possibility degree is symmetrical, and a positive necessity for $N(t_i^j, t_i^k)$ implies nothing for $N(t_i^k, t_i^j)$, contrary to specialization and generalization degrees of two reversed pairs which are simultaneously strictly positive as in the above example. Moreover, the product used as transitivity operator in [21] leads to a weakening of association weights between terms with the distance in the ontology. Here, the *min* operator implies that the matching is independent of the ontology granularity (inserting a new term between two terms in the ontology cannot change their similarity).

The approach presented in [22] uses ontologies to represent the documents contents, and queries are stated as weighted sets of ontology nodes. Conjunctive queries evaluation is done by comparing minimum subgraphs containing query and document nodes. This comparison is based on a multi-valued degree of inclusion of the document graph in the query graph. Moreover, the documents description takes into account semantic equivalence between expressions, assuming that if a document strongly includes terms, it deals with more general concepts as well, which is equivalent to expand the query with more general terms.

Another approach [23] uses a weighted multilingual ontology to exploit multilingual documents in a translation and search process. In this system, the multilingual ontology is used to translate and expand the query. The query is stated as a subgraph of the ontology, by selecting concepts judged to be relevant. The relevance of each concept can be weighted by the user. The matching is done by computing the inclusion of the query representation in the document representation. A similar approach is presented in [24], where authors use an automatically built fuzzy ontology to expand the user query. The ontology is built using WordNet to extract keywords from a documents collection, the fuzzy relations being computed as in [17]. The ontology is then pruned to eliminate redundant relations.

In [25], a fuzzy ontology is used to summarize news. This ontology is built by fuzzifying an existing ontology using a fuzzy inference mechanism, based on several similarity measures between terms. These measures are computed by textual analysis of corpora. The fuzzy inference inputs are a *part-of-speech distance*, a *term word similarity* that counts the number of common Chinese ideograms in expressions or phrases, and a *semantic distance similarity* based on the distances in the crisp ontology.

Ontologies are also used in [26] to improve clustering of users profiles. These ontologies represent knowledge on the users' domains of interest. The users profiles are then linked through the ontology, which is used to compute a similarity measure between them, allowing a more accurate clustering system.

3 Using Qualitative Pattern Matching on a Database

The qualitative pattern matching framework as previously presented can be applied to classical databases. Possibilistic ontologies, defined on the domain of each linguistic attribute of the database, are thus used to exploit the ontological knowledge about the vocabulary and to make queries on these attributes more flexible. Moreover, as the approach is compatible with the fuzzy pattern matching framework, it can be combined with the evaluation of fuzzy criteria on numerically valued attributes. In this section, some experiment results are presented for illustration purposes. They are carried out on a small but realistic database implemented in the PRETI[1] platform.

Database attribute values are considered as fuzzy ones, and can therefore be represented by linguistic terms associated to fuzzy sets when such a representation exists (e.g. on numerical ranges), or by natural language terms from a known vocabulary stated in a fuzzy ontology as presented in Sect. 2.2. Attribute values represented by terms can be fuzzily matched, exploiting the knowledge from the ontology. Queries will be conjunctions attribute requirements, the conditions for each attribute being stated as a weighted disjunction of acceptable elements of the attribute domain, as presented in Sect. 2.

Description of the PRETI Platform

PRETI[1] is an experimental system used in IRIT laboratory. It contains about 600 records about houses to let in the Aude French department. In this example, only a small subset of the available attributes is used for the sake of simplicity: an identifier, the house location described by one linguistic term, the comfort level encoded by an integer in $[\![0,4]\!]$, and the price being a real interval giving the minimum and maximum prices.

Used Ontologies

To illustrate the previous framework, fuzzy data and ontologies are needed. Different geographical partitions are used to build the ontology used to represent the location attribute. Districts (French *communes*) are the leafs of the area hierarchies, and are the values actually specified in the database. The sub-ontologies that are used are the following:

cantons: There are 35 *cantons* represented as sets of *communes* (city districts). Their labels start with *c_*. As this classification is crisp, relations in

[1] `http://www.irit.fr/PRETI/accueil.en.php`

the ontology are pure inclusions ($N = 1$). However, a *canton* and its main *commune* have a reverse necessity of 0.6, since a user may mean a *canton* using the *commune* label. For instance, we have $N(c\_limoux, limoux) = 1$ and $N(limoux, c\_limoux) = 0.6$, even if the *canton* strictly contains the city.

arrondissement: the department is split in three administrative districts (French *arrondissement*), described as sets of *cantons*: Carcassonne ($a\_carcassonne$), Narbonne ($a\_narbonne$) and Limoux (a_limoux). These relations are also crisp inclusions ($N = 1$). As for *canton* and *commune*, a reverse necessity is defined between an *arrondissement* and its main *canton*. However, since a user is less likely to use a *canton* as an *arrondissement*, the necessity degree is 0.5 (e.g $N(a\_limoux, c\_limoux) = 1$ and $N(c\_limoux, a\_limoux) = 0.5$).

micro-regions: They can be more or less associated to historical or cultural areas, and are also stated as sets of *communes*. Since some *communes* are classified in several micro-regions, the intersection of some micro-regions is not empty. For instance, *narbonne* commune pertains to micro-regions: *cabardes*, *corbieres*, *lauragais*, *narbonnais* and *razes-limouxin*.

Altogether these different administrative districts belong to the same ontology. It is completed by terms referring to physical geography and induced by the fuzzy distinction between mountain, seaside and other areas. The membership of *communes* to these fuzzy terms are given respectively by the average distance to the coast and the average altitude. Moreover, the *pyrenees* micro-region is stated as being included with $N = 0.8$ in *mountain*.

Term mountain

The membership of communes to the *mountain* concept is computed according to their average altitude (between 0 and 1200 meters). The membership function is thus like the one presented in Fig. 2(a), where the "mountainness" of an area is estimated as a percentage of the maximal altitude. It induces the spacial repartition presented in Fig. 2(b). The altitude of communes in black in not known. This will be interpreted as $N = 0, \Pi = 1$, which corresponds to a total lack of information.

Term seaside

The *seaside* term is defined using the maximum and minimum distance to the coast. Here, if no distance information is given, the degrees are $N = 0$ and $\Pi = 1$ also (the proximity to other communes whose distance to the sea is known is not used). Otherwise, degrees are computed using the membership function described in Fig. 3(a). In the evaluation, a threshold of 50 km has been used; see in Fig. 3(b).

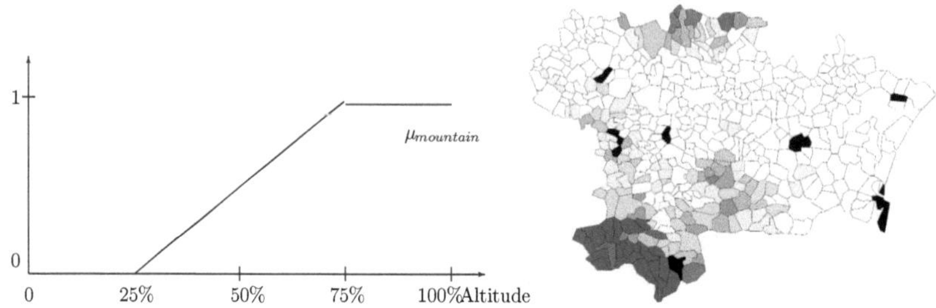

(a) Distribution (b) Repartition

Fig. 2. Term *mountain*.

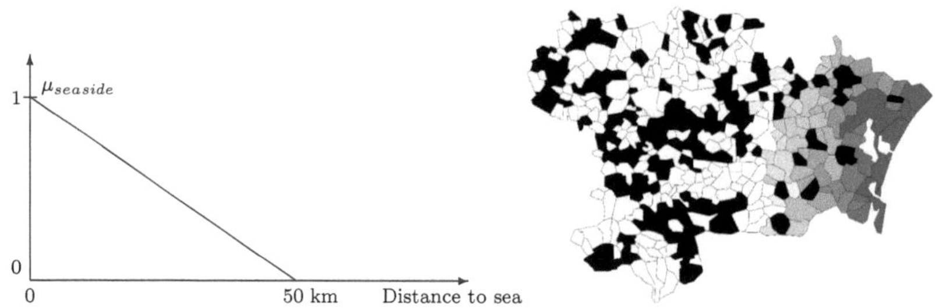

(a) Distribution (b) Repartition (50 km threshold)

Fig. 3. Term *seaside*.

Examples of Queries

Let us consider an elementary query: $R_1 = corbieres$. Evaluating this query on the database without the ontology returns an empty result. Actually, *corbieres* is a micro-region, and the attribute *location* only refers to communes. Using the ontology including micro-regions gives the results reported in Table 1.

Results with $N = 1$ are due to houses located in *communes* that are included is the searched micro-region. The results at $N = 0.6$ are obtained with a transitivity through *cantons*. For instance, some houses are in the *albieres* commune, located in the *cabardes* micro-region. According to the ontology, these terms are not directly related. This degree is obtained in several steps:

Table 1. Results for the query on "*corbieres*"

Number of results	Π	N
309	1	1
72	1	0.6
16	1	0

1. The micro-regions sub-ontology gives: $N(corbieres, mouthoumet) = 1$, since *mouthoumet* is in *corbieres*.
2. The canton sub-ontology gives: $N(mouthoumet, c\_mouthoumet) = 0.6$ and $N(c\_mouthoumet, albieres) = 1$, since this commune pertains to this canton.
3. Using transitivity, we thus have $N(corbieres, albieres) = 0.6$.

Results having $N = 0$ and $\Pi = 1$ pertain to communes in a micro-region that intersect the *corbieres* micro-region. The global results make sense since the system returns first houses in the requested area, then houses in *cantons* related with the requested micro-region, and lastly houses in a micro-region connected with the requested one (since they intersect it). The size of the area that is put into relation with *corbieres* increases when the relevance decreases, and so does the chance that the retrieved house is in an interesting area for the user. Thus, ontologies allow us therefore to extend the searchable domain of textual attributes without a direct expansion of the query by restating it.

Let us now consider a query that involves preferences, namely $R_2 = pyrenees \wedge comfort\{(0.7, 2) \vee 3\}$, expressing that the user is looking for a house to let in Pyrénées, a micro-region, with a comfort level of 3 or possibly 2. The evaluation of the comfort requirement is made in a standard way, namely: $N((0.7, 2) \vee (1, 3), 2) = \Pi((0.7, 2) \vee (1, 3), 2) = 0.7$; $N((0.7, 2) \vee (1, 3), 3) = \Pi((0.7, 2) \vee (1, 3), 3) = 1$. If the ontology is not used, no results are retrieved, whereas using the ontology gives 73 houses (Table 2). The first 13 results pertain to the *pyrenees* micro-region and have a comfort

Table 2. Results for R_2

Number of results	Π	N
13	1	1
47	0.7	0.7
2	1	0.6
7	0.7	0.6
3	1	0
1	0.7	0

of 3. The 47 following houses are also in *pyrenees*, but have a comfort of 2 only. Results with $N = 0.6$ are obtained though their *canton*, as in R_1, and have a comfort of 3 and 2 respectively, depending on the possibility degree. The following 3 results have a comfort of 3, but are in communes having only a non zero possibility degree of matching with *pyrenees*, since they pertain to cantons having a non empty intersection with *pyrenees*. The last results are in the same situation, but with a comfort level of 2.

Table 3. Results for R_3

Number of results	Π	N
7	1	1
1	1	0.9
7	1	0.8
55	0.7	0.7
2	1	0.6
2	0.7	0.6
2	1	0.5
1	0.7	0.5
130	1	0.4
151	0.7	0.4
2	0.7	0.3
8	1	0.2
2	0.7	0.2
3	1	0.1
9	0.7	0.1
5	1	0
8	0.7	0

Let us now consider the query $R_3 = mountain \wedge comfort\{(0.7, 2) \vee 3\}$. The criterion is the same for the comfort attribute, but now uses a fuzzy term for specifying the location. More results are therefore retrieved (see Table 3).

In the ontology, $N(mountain, pyrenees) = 0.8$. The first group of 13 houses found with R_2 are now discriminated into the three first groups of results of Table 3. The two first groups pertains to *mountain* with a necessity $N \geq 0.9$ and with a comfort of 3. The 7 houses with $N = 0.8$ either pertains to *pyrenees* or have a necessity with *mountain* at least equal to 0.8 given by the altitude. The same analysis can be done for other results. The possibility value 0.7 is induced by a comfort of 2, as previously. For necessity degrees with a value less than 0.7, the final degree value is determined by the inclusion degree of the commune in *mountain*, since it is the minimum of this value and the one implied by the comfort level, which is at least 0.7. The system returns here 395 houses.

In this case, the use of a fuzzy term, with a broader sense than the area name (*pyrenees*) leads to more results, and provides a better granularity in the result ordering.

Let us consider another query involving disjunction between symbolic labels, namely $R_4 = (seaside \vee c\_carcassonne) \wedge [0, 1500]$ (a house near the coast or in the carcassonne *canton* and with a price lower than 1500).

The query price requirement is evaluated in the following way: $\Pi = N = 1$ if the attribute value interval is included in the query one; $\Pi = 1, N = 0$ if the two intervals overlap; $\Pi = 0, N = 0$ if the two intervals are disjoint.

Table 4. Results for R_4

Number of results	Π	N
1	1	1
2	1	0.9
3	1	0.8
1	1	0.7
7	1	0.6
36	1	0.5
1	1	0.4
8	1	0.3
237	1	0

R_4 leads to the results presented in Table 4. Most of the non-integer values for the necessity degrees are induced by the evaluation of the term *seaside*, since the requirement $c\_carcassonne \wedge [0, 1500]$ leads only to results with $\Pi = 1, N = 0$ and $\Pi = 1, N = 0.5$ due to a transitivity through *arrondissement*. Again, the presence of a fuzzy term in the ontology leads to a more refined ranking. This is even more effective when using preference weights in the disjunction of symbolic terms as in the following query.

By weighting *carcassonne*, $R_5 = (seaside \vee (0.7, c\_carcassonne)) \wedge [0, 1500]$ states that the user prefers an house closed to the coast, even if one in the historical city of Carcassonne would be still acceptable (Carcassonne is far from the coast, $N(seaside, c\_carcassonne) = 0$). The difference between R_4 and R_5 is introduced by the weighting of *carcassonne*. In Table 5, the 36 houses that had $\Pi = 1, N = 0.5$ in Table 4 are discriminated in 17 results with $\Pi = 1, N = 0.5$, which correspond to houses with a 0.5 degree with *littoral*, and 19 with $\Pi = 0.7, N = 0.5$, corresponding to houses in the Carcassonne *canton*. In the same way, the 237 last houses, of which 53 are obtained through

Table 5. Results for R_5

Number of results	Π	N
1	1	1
2	1	0.9
3	1	0.8
1	1	0.7
7	1	0.6
17	1	0.5
19	0.7	0.5
1	1	0.4
8	1	0.3
184	1	0
53	0.7	0

a relation with *c_carcassonne*, have now a possibility reduced to 0.7 due to the weighting in the query.

Qualitative pattern matching with databases containing linguistic labels, allows the semantic evaluation of flexible queries that also use linguistic terms. The evaluation of the semantic similarity of terms is done by means of possibilistic ontologies, but may also use fuzzy set based representations, especially for terms referring to numerical scales, since the approach is fully compatible with standard fuzzy pattern matching. As shown by the above illustration, the evaluation process does not look for a strict matching between identical terms, which avoids the reformulation of the query.

These ideas apply to information retrieval as well, since the data are then textual. This is the topic of the next sections.

4 Retrieving Titles Using Qualitative Pattern Matching

In this illustration, a collection of titles of articles is considered. Titles are viewed as set of keywords, obtained by lemmatizing their significant terms and forgetting the stop-words. Therefore, the information does no longer refer to distinct attributes (with their own domain) as in the database example of the previous section. Keywords correspond to a unique multiple-valued attribute, in which terms pertain to the same global domain $\mathcal{T}$.

In the following, it is assumed that all the terms used in the query and in the titles are in the ontology. This can be practically achieved by enforcing the user to choose query terms in the ontology, and by making sure that terms appearing in representative titles are indeed in the ontology.

Queries are still conjunctions of disjunctions of possibly weighted terms, but all terms are now in the same domain (i.e. vocabulary). Moreover, weights are also introduced at the conjunctions level in order to express the relative importance of the elementary requirements of the query. Queries are thus weighted Boolean expressions on keywords. Namely,

$$D = \{t_i', t_i' \in \mathcal{T}\}\ , R = \bigwedge_k \left(\omega_k, \bigvee_j (\lambda_k^j, t_k^j) \right)\ , \text{ with } t_k^j \in \mathcal{T}\ .$$

This *importance* weighting obey the same constraint as the weights λ_k^j. In practice, disjunctions are between terms which are more or less interchangeable for the user (the weight λ_k^j expressing his/her preference between them). The weight ω_k expresses how compulsory is each elementary requirement in the conjunction.

In this fuzzy context, conjunctions are still evaluated by the *min* operator, and disjunctions by the *max* operator to compute the possibility and necessity degrees. The evaluation equations (5)–(6) are therefore rewritten as:

$$\Pi(R, D) = \min_k \max_{i,j} \min(\lambda_k^j, \Pi(t_k^j, t_i')) \ , \tag{7}$$

$$N(R, D) = \min_k \max_{i,j} \min(\lambda_k^j, N(t_k^j, t_i')) \ , \tag{8}$$

to acknowledge the multiple-valued aspect of the keywords attribute. Since the ω_k's are importance weights, this leads to the more general weighted min formula (the above formulas (7–8) are retrieved when all ω_k are equal to 1):

$$\Pi(R, D) = \min_k \max \left(1 - \omega_k, \max_{i,j} \min(\lambda_k^j, \Pi(t_k^j, t_i'))\right) \ ,$$

$$N(R, D) = \min_k \max \left(1 - \omega_k, \max_{i,j} \min(\lambda_k^j, N(t_k^j, t_i'))\right) \ .$$

Therefore, having an importance weight ω_k less than 1 leads to retrieve results violating the corresponding elementary requirement with a degree at most equal to $1 - \omega_k$.

Experimentation Protocol

In this experimentation, the TREC protocol is followed, defining queries and their corresponding relevant documents to compute precision values for the retrieval system. However, the experiments reported below are not a real evaluation, as in TREC campaigns, but rather an illustration of the potentials of the approach and its application to textual information retrieval.

4.1 Data Description

The collection contains about 200 titles of computer science articles, mainly in English but some in French, from artificial intelligence and information retrieval fields. In order to index these "documents" and to evaluate queries using qualitative pattern matching, a simple ad hoc ontology corresponding to titles terms is used. The ontology has been built a posteriori to fulfill the assumption that all documents and queries terms must be in the ontology. First, terms are generalized by their stem, given by the Porter algorithm ($N(stem, term) = 1$). To represent the few cases where different terms lead to the same stem, this relation is not considered as genuine synonymy, and the reverse necessity is set to $N(term, stem) = 0.9$. Other relations are introduced by translating French terms in English and considering a term and its translation as synonyms. Moreover, some compound expressions, such as *fuzzy set*, as well as the weights associated to the links between the terms, are added manually. A fragment of this ontology is shown in Fig. 4.

4.2 Examples of Queries

Queries used in the illustration are:

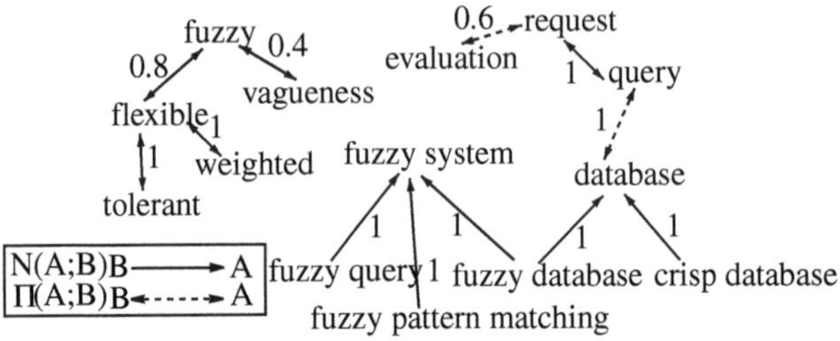

Fig. 4. Ontology fragment for the titles collection.

1. *nutrition*∨(*repas*∧*équilibré*), reformulated as (*nutrition*∨*repas*)∧(*nutrition*∨*équilibré*) to fit the query format.
2. (*nutrition* ∨ *meal*) ∧ (*nutrition* ∨ *balanced*), which is a translation of the previous one.
3. $fuzzy \wedge information$
4. *model* ∧ (*reasoning* ∨ *decision*)

To help interpreting the precision values presented in the following, the number of relevant documents to each query is given in Table 6.

Table 6. Number of relevant documents for each query

Query	Relevant doc.
1	6
2	6
3	15
4	41

4.3 Evaluation and Results

In this experiment, results obtained by just looking if each query term is present or not in the titles, are compared with those obtained by using the ontology. Tables 7(a) and 7(b) show the precision values of the two evaluations respectively.

For the two first queries, one of the relevant documents is in French. In the first case, the system cannot translate the query terms to match the index and does not retrieve the French document that does not contains *nutrition* (title 218, see Table 8). Only 5 among the 6 relevant documents are retrieved. that is 5 over 6, thus the P5 value of 1 and P10 of 0.5. As *nutrition* has the same writing in English as in French, the 5 other documents are retrieved. On the other hand, the translation is possible using the ontology, and the

Table 7. Precision values for queries the evaluations

(a) without ontology

Query	P5	P10	P15	AvgPr
1	1.00	0.50	0.33	0.83
2	1.00	0.50	0.33	0.83
3	0.20	0.10	0.07	0.07
4	1.00	0.70	0.47	0.17

(b) using ontology

Query	P5	P10	P15	AvgPr
1	1.00	0.60	0.40	1.00
2	1.00	0.60	0.40	1.00
3	0.40	0.60	0.53	0.34
4	1.00	0.90	0.87	0.39

Table 8. Relevant titles for queries 1 and 2

135	Nutri-Expert, an Educational Software in Nutrition
218	Nutri-Expert et Nutri-Advice, deux logiciels d'aide à la construction de repas équilibrés pour l'éducation nutritionnelle
234	Balancing Meals Using Fuzzy Arithmetics and Heuristic Search Algorithms
237	Multicenter randomized evaluation of a nutritional education software in obese patients
238	Expert system DIABETO and nutrition in diabetes
239	Evaluation of microcomputer nutritional teaching games in 1876 children at school

system retrieves all relevant documents, thus both average and P5 precisions are equal to 1. Relevant titles for the two first queries are given in Table 8.

Without ontology, query 3 selects only one title: "Fuzzy sets and fuzzy information granulation theory", whereas with the ontology this query retrieves 9 more titles presented in Table 9. In the ontology, *possibilistic logic* IS A kind of *fuzzy logic*, which IS *fuzzy*, and therefore, $N(fuzzy, possibilistic\ logic) = 1$ (title 264). Moreover, the ontology considers *flexible* and *fuzzy* as 0.8 synonyms, as well as *data* and *information*; *database* being a specialization of *data*, giving degrees for titles form 287 to 114 in Table 9. The 0.5 degree is given by a 0.5 necessity between terms *information* and *knowledge* in the ontology. Note that titles that are not closely related to the query, but still weakly relevant, such as titles 195 and 128 are also retrieved at the end of the list, but with a positive possibility degree only (the possibility weights come from the application of (4). However, since the qualitative pattern matching does not require a perfect match between the query and the data, more relevant titles are retrieved and the precision for this query is thus improved.

Weighted Queries

Consider now weighted versions of the previous queries 3 and 4:

3. $(0.3, fuzzy) \wedge (1, information)$
4. $(0.7, model) \wedge (1, ((0.8, reasoning) \vee (1, decision)))$

Table 9. Detailed results for query 3 *with* ontology

Doc.	Π	N	Title
223	1	1	Fuzzy sets and fuzzy information granulation theory
259	1	1	Quasi-possibilistic logic and its measures of information and conflict
264	1	1	Practical Handling of Exception-tainted rules and independence information in possibilistic logic
287	1	0.8	Fuzzy logic techniques in multimedia database querying
156	1	0.8	Fuzzy logic techniques in Multimedia database querying: a preliminary investigation of the potentials
129	1	0.8	Flexible queries in relational databases - The example of the division operator
114	1	0.8	Semantics of quotient operators in fuzzy relational databases
236	1	0.5	Fuzzy scheduling: Modelling flexible constraints vs. coping with incomplete knowledge
216	1	0.5	Uncertainty and Vagueness in Knowledge-Based Systems
137	1	0.5	Checking the coherence and redundancy of fuzzy knowledge bases
195	0.5	0	Handling locally stratified inconsistent knowledge bases
128	0.5	0	Some syntactic approaches to the handling of inconsistent knowledge bases

For query 3, greater importance is given to the term *information* w.r.t. *fuzzy*, while in query 4, weight 0.7 expresses that retrieving *model* or an equivalent term in the title is less important than satisfying the disjunctive part of the query. Weight 0.8 reflects lower preference for *reasoning* compared to *decision*. Results of these queries are presented in Tables 10 and 12 respectively.

Table 10. Results of weighted queries *without* ontology

Query	P5	P10	P15	AvgPr
3	0.20	0.30	0.20	0.11
4	1.00	1.00	0.87	0.73

Without using the ontology, results of the weighted version of query 3 are improved as the average precision raises from 0.07 to 0.11. The non-weighted version retrieved only one document. Lowering the importance of *fuzzy* leads to retrieve more titles as detailed in Table 11

Even if quite a lot of documents have a debatable relevance (average precision 0.11), a few more relevant ones are retrieved. This improvement is due to the collection itself that contains mainly documents on *fuzzy* topics. Therefore, this term is not as discriminant as *information* in this particular collection, and its importance in the query can be lowered.

Table 11. Results of weighted query 3 *without* ontology

223	1	1	Fuzzy sets and fuzzy information granulation theory
292	0.7	0.7	Internet-based information discovery: Application to monitoring science and technology
291	0.7	0.7	TétraFusion: Information Discovery on the Internet
288	0.7	0.7	Information discovery from semi-structured sources Application to astronomical literature
284	0.7	0.7	On using genetic algorithms for multimodal relevance optimisation in information retrieval
264	0.7	0.7	Practical Handling of Exception-tainted rules and independence information in possibilistic logic
260	0.7	0.7	On the use of aggregation operations in information fusion processes
259	0.7	0.7	Quasi-possibilistic logic and its measures of information and conflict
251	0.7	0.7	Logical representation and fusion of prioritized information based on guaranteed possibility measures: Application to the distance-based merging of classical bases
193	0.7	0.7	Possibilistic merging and distance-based fusion of propositional information

Table 12. Results of fuzzy queries using ontology

Query	P5	P10	P15	AvgPr
3	0.40	0.30	0.27	0.34
4	1.00	1.00	1.00	0.76

In the same way, for R_4, more relevant documents are retrieved in the 10 first results due to the lower importance of *model*, thus the increasing of P10.

The weighting of query terms allows therefore to exploit knowledge about the collection, lowering the importance of terms known not to be discriminant.

The impact of the combined use of the weights and the ontology is difficult to analyze on this small experiment. Query 4 performance is improved by both the weighting and the ontology. Nevertheless, weighting *fuzzy* in query 3 with the ontology decreases the precision. Indeed, a 0.3 weight leads to retrieve titles that are not linked with the concept of fuzziness, but having possibility and necessity degrees of 0.7 (which correspond to the fact that it is not very important to have *fuzzy* in the title). Therefore, they obtain a better rank than titles linked with this concept with a 0.5 necessity degree, which would actually be more relevant. This raises the issue of the commensurateness of the scales used for assessing the weights in the query and in the ontology.

Generally, the implicit query expansion by means of the ontology improves the system performances. This simple illustration shows that the ontology is an important aspect of the system efficiency, and therefore that this approach depends on the quality of the ontology used. By introducing weights in the query, the user can represent its preferences and priorities, as well as take into account some knowledge about the collection. However, the impact of the weighting is a difficult aspect to evaluate, since some values can improve results and some can worsen them, specially when using the ontology which gives a additional fuzzification of the final relevance degree. This experiment cannot be considered as an evaluation of a real system, due to the limited number of titles in the collection and the few queries used. However, it illustrates the approach, showing its possibilities and limitations.

5 Toward an Extension of the Approach to Full-text IR

In the previous section, even if no database attribute is considered, the illustration cannot be seen as genuine full-text IR. Indeed, no statistical analysis is done to compute terms importance in the document. In this section, possibilities of extension of the model to full-text IR, by using statistical analysis to estimate possibility and necessity degrees between the ontology terms and the documents are explored.

5.1 Possibilistic Indexing

To be homogeneous with the ontology model, the association between documents and the ontology nodes must be stated using the same possibility and necessity degrees, taking into account the statistical weights of the terms in the documents. Classically, the significance weight ρ_i^j for a given term t_i w.r.t. a document D_j is computed by combining its frequency tf_{ij} in D_j and its inverse frequency $idf_i = \log(d/df_i)$, where df_i is the number of documents containing t_i and d is the number of documents in the collection. The weights ρ_i^j are assumed to be rescaled between 0 and 1. The document D_j is therefore represented by the fuzzy set of its significant terms [4,27]: $D_j = \{(\rho_i^j, t_i), i = 1, n\}$, where n is the number of terms in the ontology.

Assuming that the ρ_i^j is an intermediary degree between the possibility and the necessity that the term describes the document, the possibility and necessity degrees can be computed as follows [28]:

$$\begin{cases} \Pi(t_i, D_j) = 2\rho_i^j \ ; \ N(t_i, D_j) = 0 & \text{if } \rho_i^j < \frac{1}{2} \,, \\ \Pi(t_i, D_j) = 1 \quad ; \ N(t_i, D_j) = 2\rho_i^j - 1 & \text{otherwise} \,. \end{cases} \tag{9}$$

The intuition underlying (9) is that a sufficiently frequent term in the document is necessarily somewhat relevant, while a less frequent term is only possibly relevant.

The ontology model agrees with the *synset* concept in WordNet. A synset is a set of synonymous terms such as: $S = \{t_i \in \mathcal{T}\}$ such that $\forall(i,j), t_i, t_j \in S \iff t_i \neq t_j$, $\Pi(t_i, t_j) = 1$ and $N(t_i, t_j) = N(t_i, t_j) = 1$. This allows us to take synsets as ontology nodes, thanks to the transitivities properties (3)–(4). Indeed, a term (used in a given sense) belongs to only one synset and

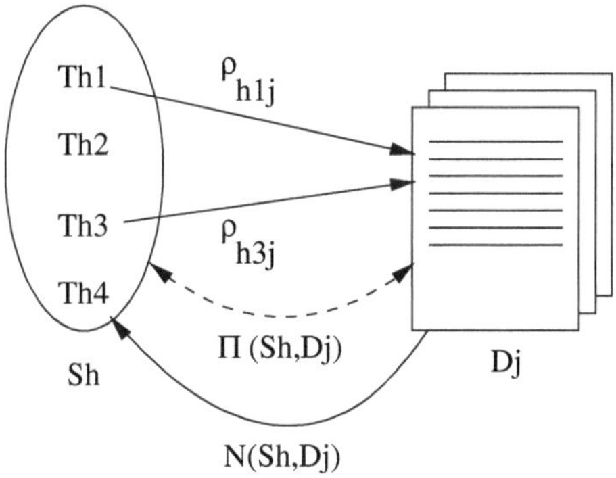

Fig. 5. Links between document and synset.

is a synonym of all other synset terms. We have to estimate to what extent a synset $S = \{t_i, i = 1, p\}$ describes a document D_j, that is to compute $\Pi(S, D_j)$ and $N(S, D_j)$. Since all synset terms are supposed to describe the document equally, we have $\Pi(S, D_j) = \max_{i,t_i \in S}(\Pi(t_i, D_j))$ and $N(S, D_j) = \max_{i,t_i \in S}(N(t_i, D_j))$ (see Fig. 5).

Indexing Example

As an example, let us consider the document D represented by the index given in Table 13.

Table 13. Index example

Term	ρ	Π	N
Database	0.6	1	0.2
Artificial Intelligence	0.2	0.4	0
AI	0.7	1	0.4
Machine learning	0.8	1	0.6

This suggests that this document deals with artificial intelligence, more specially with machine learning, applied to databases. Notice that despite *artificial intelligence* and *AI* have exactly the same meaning, their weights are different, since from a statistical point of view, the term *AI* is more frequent in the document than *artificial intelligence*. Thus, the (Π, N) degrees between the synset $\{ArtificialIntelligence, AI\}$ and D is $(\Pi, N) = (\max(0.4, 1), \max(0, 0.4)) = (1, 0.4)$.

5.2 Query Evaluation

Given a collection of documents indexed using an ontology, the query evaluation can be done similarly as described in Sect. 5. However the significance degrees between query terms and documents are no longer supposed to be 1 or 0. Taking into account the possibility and the certainty of significance as given by (9), leads for a query R and a document D, to the following relevance status value (rsv):

$$rsv(R, D) = (\Pi(R, D), N(R, D)) \ ,$$

where degrees are given by:

$$\Pi(R, D) = \min_k \max_{i,j} \min(\lambda_k^j, \Pi(t_k^j, t_i), \Pi(t_i, D)) \ ,$$
$$N(R, D) = \min_k \max_{i,j} \min(\lambda_k^j, N(t_k^j, t_i), N(t_i, D)) \ ,$$

The importance weights of the elementary requirement in a query (ω_k) can be added, which leads to:

$$\Pi(R, D) = \min_k \max\left(1 - \omega_k, \max_{i,j} \min(\lambda_k^j, \Pi(t_k^j, t_i), \Pi(t_i, D))\right) \ ,$$
$$N(R, D) = \min_k \max\left(1 - \omega_k, \max_{i,j} \min(\lambda_k^j, N(t_k^j, t_i), N(t_i, D))\right) \ .$$

The above expressions provide bases for the extension of the approach to general documents information retrieval.

Besides, in the above formulas, the aggregation of the evaluations associated with each elementary requirement is performed by means of the conjunction *min*. It is well known in information retrieval that the minimum operation is often too restrictive in practice, and is usually outperformed by other operations such as the sum. However, it has been shown in a recent work [29], that it is possible to refine the minimum operation (using a leximin ordering on ordered sets of values to be compared), and to obtain results as good or even better than with the sum. Such a refinement could be applied also in the above approach.

6 Conclusion

The approach described in this chapter is an adaptation of fuzzy pattern matching to purely linguistic terms. The main idea is to retrieve information containing terms that may not match exactly those of the query. To cope with this point, a *possibilistic ontology* is used, where the relations between therms are stated by the possibility and the certainty that their meanings refer to the same thing. This allows us to specify semantic relations, such as synonymy or specialization and generalization of meanings. Thanks to the transitivity properties of possibilistic ontologies, relations that are not explicitly stated can be deduced. A property of this model is the independence of the similarity with respect to the hierarchical distance of terms in the ontology, and therefore to the granularity of the vocabulary.

The application of qualitative pattern matching to databases allows the evaluation of flexible queries on linguistic terms, in agreement with the more standard handling of queries and data represented by fuzzy sets. Its use in information retrieval systems, avoiding query reformulation owing to the a priori vocabulary knowledge contained in the ontology. Since the matching is qualitative in nature, results can be rank-ordered even though no document matches exactly the query. The experiments undertaken separately in a database and in a textual data collection show that the approach is viable for both fields. Indeed, results are improved by the use of the possibilistic ontology and prioritized queries.

References

1. G. Grefenstette, editor. *Cross-Language Information Retrieval.* Kluwer Academic, Boston, 1998.
2. G. Bordogna and G. Pasi. A fuzzy linguistic approach generalizing boolean information retrieval: a model and its evaluation. *Journal of the American Society for Information Science*, 44(2):70–82, 1993.
3. T. Andreasen, H. Christiansen, and H. L. Larsen, editors. *Flexible Query Answering Systems.* Kluwer, 1997.
4. D. Kraft, G. Bordogna, and G. Pasi. Fuzzy set techniques in information retrieval. In *Fuzzy Sets in Approximate Reasoning and Information Systems*, chapter 8, pages 469–510. Kluwer Academic Publishers, 1999.
5. D. Dubois, H. Prade, and C. Testemale. Weighted fuzzy pattern matching. *Fuzzy Sets and Systems*, 28:313–331, 1988.
6. M. Boughanem, Y. Loiseau, and H. Prade. Graded pattern matching in a multilingual context. In *Proc. 7th Meeting Euro Working Group on Fuzzy Sets*, pages 121–126. Eurofuse, Varena, 2002.
7. Y. Loiseau, H. Prade, and M. Boughanem. Qualitative pattern matching with linguistic terms. *Ai Communications, The European Journal on Artificial Intelligence (AiCom)*, 17(1):25–34, 2004.
8. G. Salton. Experiments in automatic thesaurus construction for information retrieval. In *IFIP Congress*, pages 115–123, 1971.
9. M. Cayrol, H. Farreny, and H. Prade. Fuzzy pattern matching. *Kybern.*, 11:103–16, 1982.
10. D. Dubois and H. Prade. Tolerant fuzzy pattern matching: an introduction. In P. Bosc and J. Kacprzyk, editors, *Fuzziness in Database Management Systems*, pages 42–58. Physica-Verlag, 1995.
11. P. Resnik. Semantic similarity in a taxonomy: an information-based measure and its application to problem of ambiguity in natural language. *J. Artif. Intellig. Res.*, 11:95–130, 1999.
12. A. Bidault, C. Froidevaux, and B. Safar. Similarity between queries in a mediator. In *Proc. 15th European Conference on Artificial Intelligence*, pages 235–239. ECAI'02, Lyon, July 2002.

13. J.P. Rossazza, D. Dubois, and H. Prade. A hierarchical model of fuzzy classes. In R. De Caluwe, editor, *Fuzzy and Uncertain Object-Oriented Databases*, pages 21–62. World Pub. Co., 1997.
14. D. Dubois and H. Prade. Resolution principles in possibilistic logic. *Int. Jour. of Approximate Reasoning*, 4(1):1–21, 1990.
15. G.A. Miller, R. Beckwith, C.Fellbaum, D. Gross, and K.J. Miller. Introduction to wordnet: An on-line lexical database. *Journal of Lexicography*, 3(4):235–244, 1990.
16. C.J. Crouch. An approach to the automatic construction of global thesauri. *Information Processing and Management*, 26(5):629–640, 1990.
17. S. Miyamoto. *Fuzzy sets in Information Retrieval and Cluster Analysis.* Kluwer Academic Publisher, 1990.
18. N. Guarino, C. Masolo, and G. Vetere. Ontoseek : content-based access to the web. *IEEE Intelligent Systems*, 14(3):70–80, 1999.
19. Mustapha Baziz, Mohand Boughanem, Nathalie Aussenac-Gilles, and Claude Chrisment. Semantic cores for representing documents in ir. In *SAC'2005- 20th ACM Symposium on Applied Computing.* Santa Fe, New Mexico, USA., 13–17 mars 2005.
20. N. Mouaddib and P. Subtil. Management of uncertainty and vagueness in databases: the FIRMS point of view. *Int. Journal of Uncertainty, Fuzziness and Knowledge-Based Systems*, 5(4):437–457, 1997.
21. H. Bulskov, R. Knappe, and T. Andreasen. On measuring similarity for conceptual querying. In *Flexible Query Answering Systems, LNAI 2522*, pages 100–111. Springer, 2002.
22. M. Boughanem, G. Pasi, and H. Prade. Fuzzy set approach to concept-based information retrieval. In *10th International Conference IPMU*, pages 1775–1782. IPMU'04, Perugia (Italy), July 2004.
23. V. Cross and C.R. Voss. Fuzzy ontologies for multilingual document exploitation. In *Proc. of the 18th Conference of NAFIPS*, pages 392–397. New York City, IEEE Computer Society Press, June 1999.
24. D.H. Widyantoro and J. Yen. A fuzzy ontology-based abstract search engine and its user studies. In *FUZZ-IEEE*, pages 1291–1294, 2001.
25. C-S Lee, Z-W Jian, and L-K Huang. A fuzzy ontology and its application to news summarization. *IEEE Transactions on Systems, Man and Cybernetics*, 35(5):859–880, October 2005.
26. A. Smirnov, M. Pashkin, N. Chilov, T. Levashova, A. Krizhanovsky, and A. Kashevnik. Ontology-based user and requests clustering in customer service management system. In V. Gorodetsky, J. Liu, and V. Skormin, editors, *Autonomous Intelligent Systems: Agent and Data Mining*, pages 231–246. Int. Workshop , AIS-ADM 2005, Springer-Verlag, 2005.
27. D.A. Buell. An analysis of some fuzzy subset applications to information retrieval systems. *Fuzzy Sets and Systems*, 7(1):35–42, 1982.
28. H. Prade and C. Testemale. Application of possibility and necessity measures to documentary information retrieval. *LNCS*, 286:265–275, 1987.
29. M. Boughanem, Y. Loiseau, and H. Prade. Rank-ordering documents according to their relevance in information retrieval using refinements of ordered-weighted aggregations. In *3rd International Workshop on Adaptive Multimedia Retrieval.* AMR'05, Glasgow (UK), July 2005.

Part III

Web Information Retrieval

Formal Theory of Connectionist Web Retrieval

Sándor Dominich[1], Adrienn Skrop[1], and Zsolt Tuza[1,2]

[1] University of Veszprém, Department of Computer Science, Egyetem u. 10, 8200 Veszprém, Hungary
[2] Hungarian Academy of Sciences, Computer and Automation Research Institute, Budapest, Hungary
{dominich,skrop,tuza}@dcs.vein.hu

Summary. The term soft computing refers to a family of techniques consisting of methods and procedures based on fuzzy logic, evolutionary computing, artificial neural networks, probabilistic reasoning, rough sets, chaotic computing. With the discovery that the Web is structured according to social networks exhibiting the small world property, the idea of using taxonomy principles has appeared as a complementary alternative to traditional keyword searching. One technique which has emerged from this principle was the "web-as-brain" metaphor. It is yielding new, associative, artificial neural networks- (ANN-) based retrieval techniques. The present paper proposes a unified formal framework for three major methods used for Web retrieval tasks: PageRank, HITS, I^{2}R. The paper shows that these three techniques, albeit they stem originally from different paradigms, can be integrated into one unified formal view. The conceptual and notational framework used is given by ANNs and the generic network equation. It is shown that the PageRank, HITS and I^{2}R methods can be formally obtained from the generic equation as different particular cases by making certain assumptions reflecting the corresponding underlying paradigm. The unified formal view sheds a new light upon the understanding of these methods: it may be said that they are only seemingly different from each other, they are particular ANNs stemming from the same equation and differing from one another in whether they are dynamic (a page's importance varies in time) or static (a page's importance is constant in time), and in the way they connect the pages to each other. The paper also gives a detailed mathematical analysis of the computational complexity of WTA-based IR techniques using the I^{2}R method for illustration. The importance of this analysis consists in that it shows that (i) intuition may be misleading (contrary to intuition, a WTA-based algorithm yielding circles is not always "hard"), and (ii) this analysis can serve as a model that may be followed in the analysis of other methods.

1 Introduction

The term soft computing (SC) refers to a family of techniques consisting of methods and procedures based on fuzzy logic, evolutionary computing, artificial neural networks, probabilistic reasoning, rough sets, chaotic computing.

S. Dominich et al.: *Formal Theory of Connectionist Web Retrieval*, StudFuzz **197**, 163–194 (2006)
www.springerlink.com

In many applications, uncertainty (the lack of precise and exact rules connecting causes and goals) and imprecision (partial or noisy knowledge about the real world) are important features. Also, in many applications the computational effort required by exact conventional methods (also referred to as hard computing) can make the problem at hand intractable or is unnecessary (when the precision of the output is overweighed by other properties, e.g., more economical or feasible [13].

The application of soft computing techniques to Information Retrieval (IR) aims at capturing aspects that could hardly be satisfactorily modeled by other means. One example for such a technique is based on fuzzy sets theory, which allows for expressing the inherent vagueness encountered in the relation between terms and documents, and fuzzy logic makes it possible to express retrieval conditions by means of formulas in the weighted Boolean model; an overview can be found in [35]. Connectionism represents another approach. Basic IR entities (documents, terms) are represented as an interconnected network of nodes. Artificial neural networks (ANNs) and semantic networks (SN) are two techniques used for this; they are modeled mathematically using graphs. ANN learning allows to model relations between documents as well as documents and terms. It was used with the principal aim to increase the accuracy of document-term weights [3, 4, 5, 15, 23, 37, 39, 55]. ANNs were also applied to query modification aiming at enhancing retrieval performance [14], to retrieval from legal texts [44,45], and to clustering for IR [12,21,30,31,44, 53].

Three major methods used for Web retrieval tasks are: PageRank, Hubs and Authorities (HITS), Interaction Information Retrieval (I^2R). They are based on different paradigms originally: PageRank: normalized importance; HITS: mutual re-enforcement; I^2R: winner-take-all. But are really different and independent of one another? The paper shows that, albeit they stem originally from different paradigms, they can be integrated into one unified formal framework.

Computational complexity (CC) is an under-treated topic in IR. Most IR researchers and developers understand CC as meaning physical running time. This interpretation of the term CC is justified by and important in practice. However, a correct mathematical justification for the complexity of IR algorithms can be obtained only by properly analyzing complexity. In an ANN, the winner-take-all (WTA) strategy can yield circles, and so one may be tempted to assume – based on the result that finding circles in a graph is "hard" – that an ANN-based IR algorithm may not always tractable (i.e., its CC is not polynomial). But is it really and necessarily the case? The paper shows that intuition may be misleading, and at the same time the CC analysis of the WTA-based retrieval method can serve as a model that may be followed in the analysis of other methods.

2 Connectionist Web Information Retrieval

The present section aims at giving a short but comprehensive "state-of-the-art" of connectionist Web IR, namely of applying ANNs to retrieval tasks on the Web.

Part 2.1 contains a compact description of the basic ANN concepts and results used. In part 2.2, typical applications of ANNs to IR (text retrieval, categorization) are briefly presented. This is followed, in part 2.3, by a concise literature overview of Web retrieval techniques and applications using ANNs.

2.1 Artificial Neural Network

The underlying idea of ANNs goes back to [29], where it is stated, as a fundamental principle, that the amount of activity of any artificial neuron depends on its weighted input, on the activity levels of artificial neurons connecting to it, and on inhibitory mechanisms. This idea gave birth to a huge literature and many applications, especially due to the results obtained in, e.g., [22, 25, 27]. An *artificial neuron* is a formal processing unit abstracted from real, biological neurons (Fig. 1a). An artificial neuron ν has inputs $I_1, \ldots, I_n$, these can be weighted by the weights $w_1, \ldots, w_n$. The total input I is a function of the inputs and their weights, usually a linear combination of them, i.e., $I = \sum_i I_i w_i$. The total input I stimulates the neuron which can thus be activated and can take on a *state* z. The state z, also referred to as an *activation level*, is a function g of I, $z = g(I)$. For example, the function g can be a

(a) threshold function: $z = 1$ if $I > k$, and $z = 0$ if $I \leqslant k$, where k is a threshold value;
(b) identity function: $z = I$.

The artificial neuron produces an output O via its *transfer function* f depending on its state z, i.e., $O = f(z)$. The transfer function f can take on several forms, e.g.,

(c) identity function: $O = f(z) = z$,
(d) sigmoid function: $O = f(z) = \frac{1}{1+e^{-z}}$.

Artificial neurons can be connected to each other thus forming an *Artificial Neural Network* (*ANN*; Fig. 1b). Given two interconnected neurons ν_i and ν_j in an *ANN*, the *output* $f_j(z_j)$ of ν_j can be transferred to ν_i via the connection between them which can alter $f_j(z_j)$ by a weight w_{ij}. The quantity $w_{ij} \cdot f_j(z_j)$ reaches the artificial neuron ν_i for which it is an input.

The following general differential equation can be derived for the state z_i of neuron ν_i [16]:

$$\frac{dz_i(t)}{dt} = -z_i(t) + \sum_{j=1}^{n} f_j(w_{ij}, z_j(t), z_i(t)) + I_i(t) \tag{1}$$

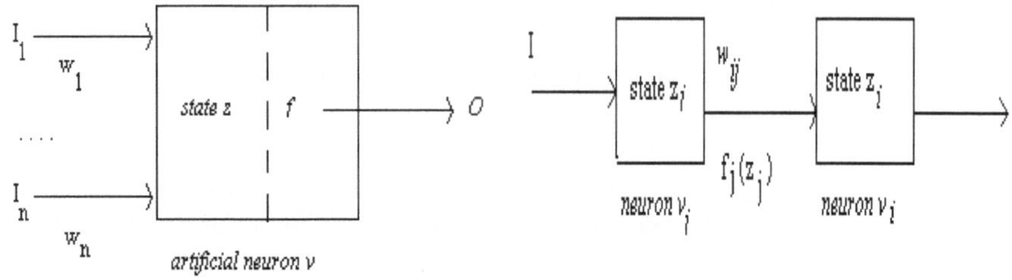

a) artificial neuron b) artificial neural network

Fig. 1. (**a**) An artificial neuron. A linear combination of the weighted (w_i) inputs (I_i) activates the neuron ν which takes on a state z and produces an output O via its transfer function f. (**b**) ANN. Interconnected artificial neurons (ν_j, ν_i). I is an input to neuron ν_j, and $f_j(z_j)$ is its output, which is an input to neuron ν_i weighted by the quantity w_{ij}, i.e., it is $w_{ij} \cdot f_j(z_j)$.

t denotes time, $z_i(t)$ denotes the activity level of the ith artificial neuron, w_{ij} denotes the weight of a link from the jth to the ith artificial neuron, $I_i(t)$ denotes external input to the ith artificial neuron, $f_j(z_j(t), w_{ij}, z_i(t))$ denotes the influence of jth artificial neuron upon the ith artificial neuron. Equation (1) is a generic equation, and can have different forms depending on the choice of I_i, f_j, w_{ij} corresponding to the particular case or application where the ANN is being used. For example, when applied to neurons, z_i denotes membrane voltage, I_i means an external input, w_{ij} is interpreted as a weight associated to the synapse, whereas f_j takes the from of a product between the weight and z_j. For analogue electric circuits, z_i denotes the potential of a capacitor, the left hand side of the equation is interpreted as a current charging a capacitor to potential z_i, whereas the summed terms mean potentials weighted by conductance. Because eq. (1) can be written for every $i = 1, 2, \ldots, n$, we have a *system of differential equations.* The study of an *ANN* is done assuming that *initial states* z_0 are known at some initial point t_0. It can be shown that in a small enough vicinity $|z - z_0|$ of z_0 and $|t - t_0|$ of t_0, the system (1) has a unique solution. From a practical point of view, the existence of solutions of (1) can be answered positively due to the Cauchy-Lipschitz theorem [41], and is stated here without proof:

Theorem 1. *Let*

$$F(t, z) = \frac{1}{\mu_i}(I_i(t) - z_i(t) + \sum_j f_j(z_j(t), w_{ij}, z_i(t)))$$

where μ_i *is a coefficient. Consider the initial condition* $z(t_0) = t_0$. *If the function* $F(t, z)$ *is continuous in a region* $\Omega \subset \mathbb{R}^2$ ($\mathbb{R}^2$ *denotes the real plane*), *and the function* $F(t, z)$ *is a local Lipschitz contraction, i.e.,* $\forall P \in \Omega \exists K \subset \Omega$ *and* $\exists L_K > 0$ *constant such that* $|F(t, z_1) - F(t, z_2)| \leqslant L_K |z_1 - z_2|, \forall (t, z_1), (t, z_2) \in$

K, then there exists a vicinity $V_0 \subset \Omega$ of the point (t_0, z_0) in which the equation has a unique solution satisfying the initial condition $z\ (t_0) = t_0$, which can be obtained by successive numeric approximations. □

Equation (1) gives the state of every neuron at time t. By letting time t to evolve, a sequence $z_i(t), i = 1, \ldots, n$, of states results. This is referred to as the *operation* of ANN. Normally, an ANN evolves in time towards a state that does not change anymore. This is called an *equilibrium* and is defined as $dz_i/dt = 0$, $i = 1, 2, \ldots, n$. One important mode of operation of an ANN is referred to as the *winner-take-all* (*WTA*) strategy which reads as follows: only the neuron with the highest state will have output above zero, all the others are "suppressed". In other words, WTA means to select the neuron that has maximum state and deactivate all the others; formally the WTA can be expressed as follows:

$$(z_i = 1 \text{ if } z_i = \max_j\ z_j) \wedge (z_k = 0 \text{ if } z_k \neq \max_j\ z_j).$$

Artificial neurons can be grouped together to form structures called *layers*. A collection of layers forms a *processing structure* to *solve a problem*. Usually, there are the following types of layers (Fig. 2):

(a) *input layer*: it accepts the problem input data which is called an *input pattern*; there is usually one input layer,
(b) *output layer*: it delivers a solution (which is called an *output pattern*) to the problem to be solved, there is usually one output layer,
(c) *hidden layer*(s): there may be several such layers; they act as intermediary processing ensembles.

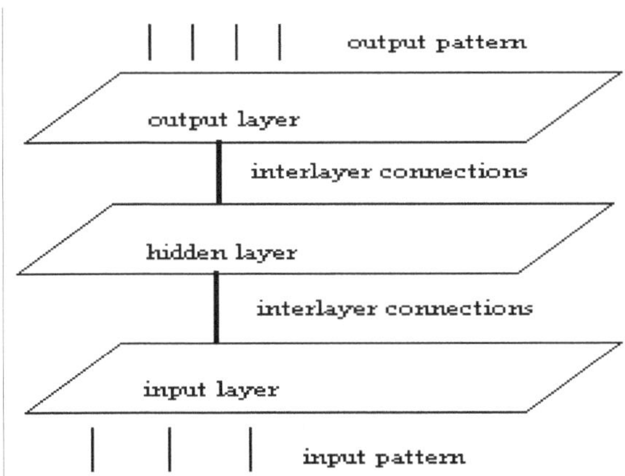

Fig. 2. Typical ANN layered architecture: input, hidden and output layers. The input of the architecture accepts the input pattern, whereas ANN's output is the output pattern. Hidden layers perform intermediary processing.

Artificial neurons within the same layer usually have the same transfer function, and obey the same *learning rule* (an algorithm according to which the weights change). Typically, learning rules are derived from the *Hebb's Rule* which reads as follows: if two neurons are simultaneously active, the weight between them is increased by a quantity $k \cdot f_i \cdot f_j$. Examples of learning rules derived from Hebb's Rule are the *Delta Rule*, *Kohonen Rule*, etc.. Learning rules are based on *learning methods* such as:

(i) *supervised learning*: both the input pattern and the solution are known; the actual output pattern, produced during the operation of the network, is compared to the solution, and the weights are changed accordingly;
(ii) *unsupervised learning*: the network develops its own classification rules;
(iii) *reinforcement learning*: the net is rewarded if the actual output pattern is accepted.

Globally, in a layered ANN architecture, there is an activation flow between layers:

(a) *feed-forward propagation*: the activation flows from the input layer towards the output layer;
(b) *feed-backward propagation*: the activation flows from the output layer towards the input layer;
(c) *interactive propagation*: the activation flows both for- and backward;
(d) *equilibrium*: the network relaxes.

2.2 Information Retrieval Using Multi-Layered Artificial Neural Networks

In [36], a typical application of ANNs to IR is described: ANNs are used to model a probabilistic IR model using single terms as document components. A 3-layer network (Q, T, D) is considered, Q and D denote the layer of queries Q_k and documents D_i, respectively. T denotes the layer of index terms t_j which are connected bi-directionally with documents and queries (Fig. 3).

Intra-layer connections are disallowed. The net is considered to be initially blank. As each document neuron is created and fed into the network, associated index term neurons are also created. From the document neuron D_i to the index term neuron t_j a new connection is created and its strengths w_{ij} is set to $w_{ij} = f_{ij}/L_i$, where f_{ij} denotes the number of occurrences of term t_j in document D_i, and L_i denotes the length (e.g. total number of index terms) of that document. From a term neuron t_j to a document neuron D_i a connection is created if that term is new, and its strengths w_{ji} is set to a special form of inverse document frequency as follows:

$$w_{ji} = \log \frac{p}{1-p} + \log \left(\frac{1 - s_{ji}}{s_{ji}} \right) \qquad (2)$$

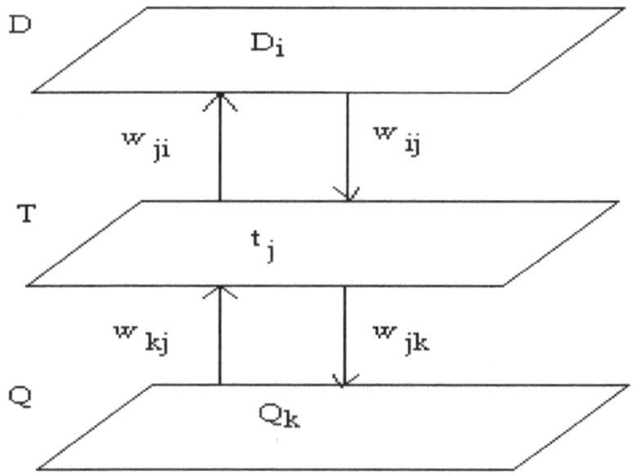

Fig. 3. A 3-layer ANN architecture for IR. There is a query, a document and a terms layer (Q, D, T). There are weighted connections between terms on the one hand, and documents and queries on the other hand.

where p is a small positive constant (the same for all terms), and s_{ji} is the proportion F_j/N, where F_j represents the document collection number of occurrences of the jth term, and N is the total number of terms in the document collection.

Queries Q_k are treated as if they were documents. Let us consider a query neuron Q_k, and clamp its activity to 1. Any term neuron t_j affected by this query will receive as input $1 \cdot w_{kj}$, and outputs the same value. From each of the terms affected by the query some documents D_i (those having terms in common with the query) are affected receiving input $I_i = \sum_j w_{ji} w_{kj}$, and output the same value I_i. The documents so affected turn back the activation to term neurons which receive input $I_j = \sum_i I_i w_{ij}$, and which output the same value I_j to query neurons. These output values can then be used for ranking purposes.

Example. Let us consider the document $(i = 1, p = 0.5)$: D_1 = Bayes' Principle: The principle that, in estimating a parameter, one should initially assume that each possible value has equal probability (a uniform prior distribution). Consider the following terms $(j = 2)$: t_1 = Bayes' Principle, t_2 = probability. The connection strengths are as follows: w_{ij} for $i = 1, j = 1, 2$: $w_{11} = 0.5$, $w_{12} = 0.5$; w_{ji} for j = 1,2, $i = 1$: $w_{11} = 0$, $w_{21} = 0$. Let the query be Q_k = Bayes' Principle $(k = 1)$. The corresponding weights are: w_{kj}, for $k = 1$, $j = 1$, 2: $w_{11} = 1$, $w_{12} = 0$; w_{jk}, for $j = 1$, 2, $k = 1$: $w_{11} = -\infty$, $w_{21} = 0$. Activation spreading: from Q_k an activation with value $1 \cdot w_{kj} = 1 \cdot w_{11} = 1$ is started, it is spread over to D_1 having value $w_{kj} w_{ji} = w_{11} w_{11} = 1$. From D_1, the activation is turned back to the term it has started from, having values 0.5.

The above model only operates with inter-layer interactions, and does not take into account intra-layer interactions such as, for example, possible relationships between documents or terms. The model was further developed in [38]. The methods for computing term associations can be divided into two categories. One can estimate term relationships directly from term co-occurrence frequencies. On the other hand, one can infer term associations from relevance information through feedback. In the first approach, semantic relationships are derived from the characteristics of term distribution in a document collection. These methods are based on the hypothesis that term co-occurrence statistics provide useful information about the relationships between terms. That is, if two or more terms co-occur in many documents, these terms would be more likely semantically related. In [54], a method for computing term associations by using a three-layer feed-forward neural network is presented. Term associations are modeled by weighted links connecting different neurons. Each document is represented by a node. Likewise, each query is also represented by a node. Document vectors are input to the network. The nodes in the input layer represent the document terms. These term nodes are connected to the document nodes with individual weights. The nodes in the hidden layer represent query terms. These nodes are connected to the document term nodes. a_{ij} is a weight between the ith document term node and the jth query term node. A value q_j, which represents the importance of term s_j in query q, is used as a scaling factor. The output layer consists of just one node, which pools the input from all the query terms. In this neural network configuration, the weights of the input and output layers are fixed. This makes the task of training the network much easier, because only the weights of the hidden layer are adjustable. The training method was tested using the ADINUL collection, the maximum document length is 95, and the maximum query length is 27, where length is the number of terms in a vector, This means that there are many terms with zero weights in the document and query vectors. It is therefore possible to reduce the dimension of the term-association matrix by eliminating the zero weighted terms. Those terms which are absent in all the query vectors were eliminated. The remaining terms were used to describe the document vectors. In this way, the dimension of the term-association matrix was reduced to 200×200 from an initial dimension of 1217×1217.

Text categorization, also known as automatic indexing, is the process of algorithmically analyzing an electronic document to assign a set of categories (or index terms) that succinctly describe the content of the document. This assignment can be used for classification, filtering, or retrieval purposes. Text categorization can be characterized as a supervised learning problem. We have a set of example documents that have been correctly categorized (usually by human indexers). This set is then used to train a classifier based on a learning algorithm. The trained classifier is then used to categorize the target set. Ruiz and Srinivasan [47] present the design and evaluation of a text categorization method based on a multi-layer ANN. This model uses a "divide and conquer" principle to define smaller categorization problems based on a predefined hierarchical structure. The final classifier is a hierarchical array of neural networks. The method was evaluated using the UMLS Metathesaurus as the underlying hierarchical structure, and the OHSUMED test set of MEDLINE records. The method is scalable to large test collections and vocabularies because it divides the problem into smaller tasks that can be solved in shorter time.

2.3 Artificial Neural Network-based Web Retrieval – A Literature Overview

Web retrieval owes a considerable part of its success to classical IR and AI. IR has gone through major periods of development from the Boolean Model though the Vector Space Model and Probabilistic Model to present day Language Model. Also, IR has gained a formal theoretical foundation and has thus become an applied formal (mainly mathematical) discipline too [19, 51]. In the 1980s, knowledge-based expert systems were developed to generate user models in order to assist searching. In the 1990s, ANN-based techniques appeared in IR. Since the Internet and the World Wide Web have become robust and reliable depositing a huge amount of valuable information, Web retrieval has become an important part of everyday life. Popular search engines, like Google, Altavista, Yahoo!, are spidering the Web indexing and retrieving Web pages. However, these search engines suffer from information overload (too many pages to crawl and to show to the user) and low precision. Also, they deviate more and more from their original searching technology roots, and are becoming "media companies" stressing advertising [8, 9].

There are major differences between classical IR and Web retrieval, they are summarized below:

(a) Most Web documents are in HTML (Hypertext Mark Up Language) format, containing many tags. Tags can provide important information about the page; for example, a bold typeface markup ($\langle b \rangle$) usually increases the importance of the term it refers to. Tags are usually removed in the indexing phase.
(b) In traditional IR, documents are typically well-structured (e.g., research papers, abstracts, newspaper articles) and carefully written (grammar, style). Web pages can be less structured and are more diverse: they can differ in language, grammar, style and length; they can contain a variety of data types including text, image, sound, video, executable. Many different formats are used, such as HTML, XML, PDF, MSWord, mp3, avi, mpeg, etc..
(c) While most documents in classical IR tend to remain static, Web pages are dynamic: they can be updated frequently, they can be deleted or added, they can be dynamically generated.
(d) Web pages can be hyperlinked, this generates a linked network of Web pages (referred to as Web graph). An URL (Universal Resource Locator) from a Web page to another page, anchor text, the underlined, clickable text can provide additional information about the importance of the target page.

(e) The size of the Web, i.e., the number of Web pages and links between them, is orders of magnitudes larger than the size of corpuses and databases used in classical IR.
(f) The number of users in the Web is much larger, and they are more diverse in terms of interest, search experience, language, and so on.

These characteristics represent challenges to Web searching. Web retrieval should be able to address these characteristics, to cope with the dynamic nature of the Web, and to scale up with size. Thus far, there have been proposed just a few Web search tools using ANNs (they are not commercial search engines). The typical ones are briefly described below.

Sir-Web is a connectionist Web search system [42]. It is based on a client (Java)-server (Horb) architecture, indexes 32,760 files with 463,616 keywords, and operates with AND-ed words queries. It adjusts weights during operation based on a user feedback by clicking on words offered on a list. Unfortunately, there is not much information available as regards exact operation of Sir-Web, further the Sir-Web server broke the connection many times with time out error. Also, the files Sir-Web indexes seem to come from mainly Japanese sites (based on the hit lists returned).

In [14], a prototype search system, called WebCSA (Web Search by Constraint Spreading Activation), using Constrained Spreading Activation (CSA) is proposed for retrieving information from the Web. The activation is spread in an ANN subject to constraints such as

(i) distance constraint which places a constraint on how far from a source neuron the activation should spread;
(ii) fan-out constraint which means that the spreading of activation should stop when it reaches a neuron with a large fan-out (outdegree);
(iii) path constraint: activation should spread along preferred paths (e.g., along paths having some predefined semantics); activation constraint: a threshold value is defined for the activation level of a neuron (i.e., activation spreads over from a neuron if its activation level exceeds the threshold, value).

WebCSA uses an ostensive retrieval approach, and is intended to complement and enhance search engines' results. A user performs a search using a commercial search engine first, and marks the Web pages that are thought to be relevant. WebCSA then downloads these pages, indexes them, and creates a list of weighted terms from their content (search profile). A navigation, controlled by constraints, in the Web follows using the URLs found. Each selected page is downloaded and indexed. A cosine similarity value is computed between them and the search profile, and an activation level is assigned to the pages. The navigation process is continued from active pages, and stops when the user decides so, or when some pre-set termination condition is met. Experiments were carried out to evaluate WebCSA's relevance effectiveness involving graduate students. A 30% improvement was obtained over search engines.

In [46], an automatic clustering method is proposed and applied to Web retrieval using SOM (Self Organising Maps; [34]). SOM is an unsupervised ANN resistant to noisy input. SOM's biological motivation was the human sensory and motor maps in the brain cortex. SOM consists of a one-layer ANN with artificial neurons arranged in a 2-D grid (alternatively 3-D or spherical structures can also be used), Each neuron i is assigned an n-dimensional vector m_i, e.g., representing objects' features. In each training iteration t a vector $x(t)$ is selected at random, and presented to the neurons. The activation function can be the Euclidean distance (other measures also exist, e.g., scalar product, Manhattan distance) between the vectors $x(t)$ and m_i, $Euc(x(t),m_i)$, $i = 1, \ldots, n$. The neuron for which this distance is lowest will be the winner. The vector m_i of the winner as well as those of its "neighbors" are adapted (there are several definitions of who are the "neighbors") according to the learning rule $m_i(t+1) = m_i(t) \cdot \alpha \cdot h \cdot (x(t) - m_i(t))$, where α is the learning rate and h is the neighborhood parameter. Thus, the objects are mapped on a 2-dimensional map (grid of ANs) in such a way that similar objects are placed close to each other. SOM is fully connected, i.e., every map-node is connected to every input node with some connection weight. The input pattern is presented several times, the network operates and develops its own classification automatically. The paper proposes a prototype as layer between the user and a commercial Web search engine (AltaVista). The user submits a query to the prototype. The top 200 pages returned by the commercial search engine are fetched, and then presented to SOM which clusters them and assigns a label to each cluster. These are then shown to the user.

In [10], a meta-search engine called MetaSpider is proposed. It performs real-time post-retrieval document clustering. MetaSpider sends out the user query to multiple search engines and collects the results. The corresponding Web pages are downloaded and indexed using the Arizona Noun Phraser based on key phrases that occur in pages. Phrases are presented to the user in descending order of frequency. Clicking on a phrase, the user can view a list of pages containing that phrase. All the phrases are sent to a SOM for automatic categorization. Thus, the user can have an overview of the set of pages collected.

i2rMeta is a Web meta-search engine using the Interaction Information Retrieval (I^2R) paradigm [20], which is implemented using a WTA-based ANN. The query is entered as a set of terms (separated by commas), they are Porter-stemmed, and then currently sent to six commercial spider-based Web search engines as HTTP requests. The hit lists returned by the Web search engines are considered, the corresponding Web pages are downloaded in parallel. Each Web page undergoes the following processing: tags are removed, words are identified, stop-listed and Porter-stemmed. The thus resulted Web pages and the query are processed according to I^2R: they are viewed as artificial neurons, weighted connections are created between them. An activation is started at the query-neuron, it is spread according to WTA until a circle develops; the pages forming the circle will be retrieved.

3 Formal Theory of Connectionist Web Retrieval

Firstly, three major techniques – PageRank, HITS, I^2R – used in Web retrieval tasks are briefly recalled as they were originally proposed (parts 3.1, 3.2, 3.3). They are based on different paradigms:

(a) PageRank: normalized importance (the importance of a Web page depends on the importance of Web pages linked to it);
(b) HITS: mutual re-enforcement (the importance of hub and authority pages mutually depend on each other);
(c) I^2R: winner-take-all (the importance of a Web page is given by its activity level within a reverberatice circle).

The aim of the present section is to show that the above three techniques, albeit they stem originally from different paradigms, can be integrated into one unified formal view. The conceptual and notational framework used in our approach is given by ANNs and the generic network equation. It will be shown (in parts 3.4, 3.5, 3.6) that the PageRank, HITS and I^2R methods can be formally obtained from the generic equation as different particular cases by making certain assumptions reflecting the corresponding underlying paradigm.

3.1 PageRank

In the PageRank method, a Web page's importance is determined by the importance of Web pages linking to it. Brin and Page [6], extending the Garfield static citation principle [24], define the PageRank value R_i of a Web page W_i using the following equation:

$$R_i = \sum_{W_j \in \mathrm{B}_i} \frac{R_j}{L_j} \tag{3}$$

where L_j denotes the number of outgoing links from the page W_j, B_i denotes the set of pages W_j pointing to W_i. Equation (3) is a homogenous and simultaneous system of linear equations, which always has trivial solution (the null vector), but which has nontrivial solutions too if and only if its determinant is equal to zero. Let $G = (V, A)$ denote the directed graph of the Web, where the set $V = \{W_1, W_i, \ldots, W_N\}$ of vertices denotes the set of Web pages. The set A of arcs consists of the directed links (given by URLs) between pages. Let $M = (m_{ij})_{N \times N}$ denote a square matrix attached to the graph G such that $m_{ij} = {}^1/_{L_j}$ if there is a link from W_j to W_i, and 0 otherwise (Fig. 4).

Because the elements of the matrix M are the coefficients of the right hand side of eq. (3), this can be re-written in a matrix form as $M \times R = R$, where R denotes the vector of PageRank values, i.e., $R = [R_1, \ldots, R_i, \ldots, R_N]^T$. If the graph G is strongly connected, the column sums of the matrix M are equal to 1. Because the matrix M has only zeroes in the main diagonal, the

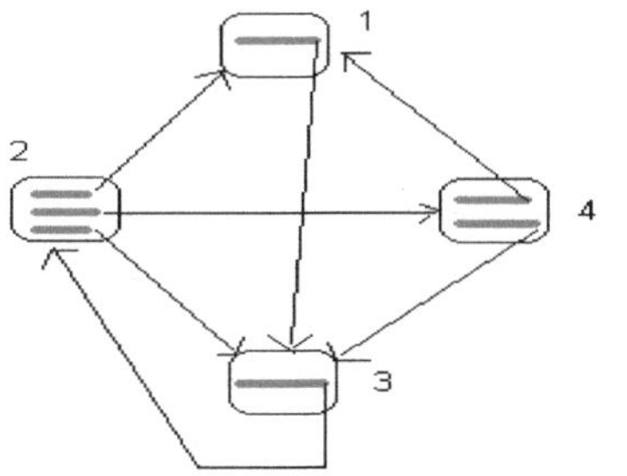

Matrix M					PageRank
0	1/3	0	1/2		0.325
0	0	1	0		0.651
1	1/3	0	1/2		0.651
0	1/3	0	0		0.217

Fig. 4. A small Web graph G with four pages: 1, 2, 3 and 4. The *horizontal* bars within each page symbolise URLs to other pages as shown by the *arrows*. The elements of the matrix M are also shown, they were computed as follows: $m_{ij} = 1/L_j$ (see text). The PageRank values, i.e., the eigenvector corresponding to the eigenvalue 1, were computed using the Mathcad command "eigenvec(M, 1)".

matrix $M - I$ has zero column sums (I denotes the unity matrix). Let D denote its determinant, i.e., $D = |M - I|$. If every element of, say, the first line of D is doubled we get a new determinant D', and we have $D' = 2D$. We add now, in D', every other line to the first line. Because the column sums in D are null, it follows that $D' = 2D = D$, from which we have $D = 0$. The matrix $M - I$ is exactly the matrix of eq. (3), which thus has nontrivial solutions too. The determinant $|M - I|$ being equal to 0 is equivalent to saying that the number 1 is an eigenvalue of the matrix M. The PageRank values are computed in practice using some numeric approximation procedure to calculate the eigenvector R corresponding to the eigenvalue 1.

Another equation to compute PageRank values in practice is the following:

$$R_i = (1 - d) + d \cdot \sum_{W_j \in \mathrm{B}_i} \frac{R_j}{L_j} \tag{4}$$

where $0 < d < 1$ is a damping factor (e.g., set to 0.85 in practice).

3.2 Authorities and Hubs

A method for computing hubs and authorities is suggested in [33]. Two types of Web pages are defined first: hubs and authorities. They obey a mutually reinforcing relationship, i.e., a Web page is referred to as an authority if it is pointed to by many hub pages, and a hub if it points to many authoritative pages (Fig. 5.a)). Given a page p, an authority weight $x^{\langle p \rangle}$ and a hub weight $y^{\langle p \rangle}$ is assigned to it. If p points to pages with large x-values, then it receives large y-values, and if p is pointed to by pages with large y-values, then it should receive a large x-value.

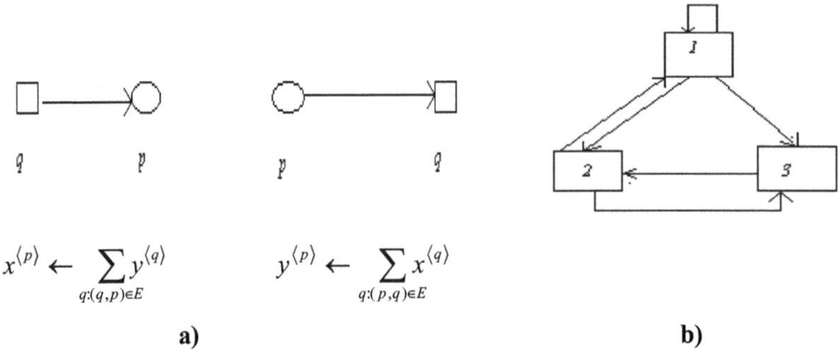

Fig. 5. (**a**) Illustration of operations for computing hubs and authorities. (**b**) Mini-Web (example).

The following iterative operations can be defined:

$$\begin{aligned} x^{\langle p\rangle} &\leftarrow \sum_{q:(q,p)\in E} y^{\langle q\rangle} \\ y^{\langle p\rangle} &\leftarrow \sum_{q:(p,q)\in E} x^{\langle q\rangle} \end{aligned} \tag{5}$$

where E denotes the set of arcs of the Web graph. To find "equilibrium" values for the weights of n linked Web pages, a sufficient number of iterations are repeated starting from the initial values $x_0 = (1, \ldots, 1)$ and $y_0 = (1, \ldots, 1)$ for both x and y. It can be shown that x is the principal eigenvector of $M^T M$, and y is the principal eigenvector of MM^T.

Example. Let $M = \begin{pmatrix} 1 & 1 & 1 \\ 0 & 0 & 1 \\ 1 & 1 & 0 \end{pmatrix}$ denote the adjacency matrix of the mini-Web graph of Fig. 5.b). Perform the operations $x_{i+1} = M^T y_i$ and $y_{i+1} = M x_i$ (after every such iteration length normalize both vectors) until the vectors x and y do not change significantly (convergence). In this example, after four steps the following values are obtained: $x = (0.628; 0.628; 0.46)$ and $y = (0.789; 0.211; 0.577)$.

3.3 Interaction Information Retrieval

In the Interaction Information Retrieval (I^2R) method [18] each Web page o_i is viewed as an artificial neuron, and is associated an n_i-tuple of weights corresponding to its identifiers (e.g., obtained after stemming and stoplisting) $t_{ik}, k = 1, \ldots, n_i$. Given now another page o_j. If identifier $t_{jp}, p = 1, \ldots, n_j$, occurs f_{ijp} times in o_i then there is a link from o_i to o_j, and this has the following weight:

$$w_{ijp} = \frac{f_{ijp}}{n_i} \tag{6}$$

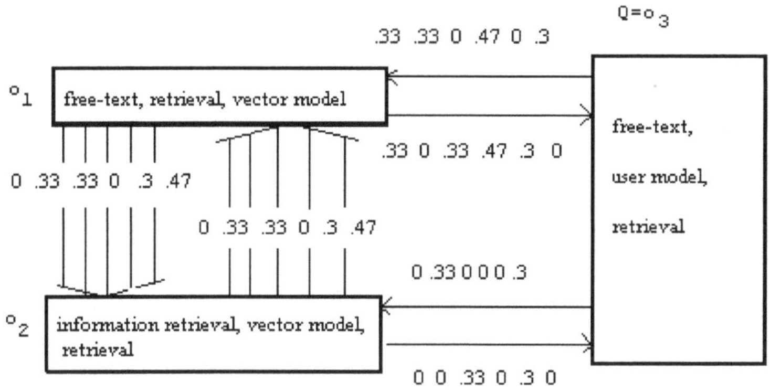

Fig. 6. Retrieval in Interaction Information Retrieval (I^2R). All links having the same direction between Q and o_1, and Q and o_3 are shown as one single arrow to simplify the drawing. The activation starts at Q, and spreads over to o_1 (total weight $= 0.33 + 0.33 + 0.47 + 0.3 = 1.43$) from which to o_2, and then back to o_1. o_1 and o_2 form a reverberative circle, and hence o_1 and o_2 will be retrieved in response to Q.

If identifier t_{ik} occurs f_{ikj} times in o_j, and df_{ik} denotes the number of pages in which t_{ik} occurs, then there is a link from o_i to o_j, and this has the following weight:

$$w_{ikj} = f_{ikj} \cdot \log \frac{2N}{df_{ik}} \tag{7}$$

The total input to o_j is as follows:

$$\sum_{k=1}^{n_i} w_{ikj} + \sum_{p=1}^{n_j} w_{ijp} \tag{8}$$

The other two connections – in the opposite direction – have the same meaning as above: w_{jik} corresponds to w_{ijp}, while w_{jpi} corresponds to w_{ikj}. The query Q is considered to be a page, too. The process of retrieval is as follows (Fig. 6 shows and example). A spreading of activation takes place according to a WTA strategy. The activation is initiated at the query $Q = o_j$, and spreads over along the strongest total connection thus passing on to another page, and so on. After a finite number of steps the spreading of activation reaches a page which has already been a winner earlier thus giving rise to a loop (referred to as a reverberative circle). This is analogous to a local memory recalled by the query. Those objects are said to be retrieved which belong to the same reverberative circle.

3.4 Interaction Information Retrieval: Particular Case of the Generic Equation

Let (Fig. 7):

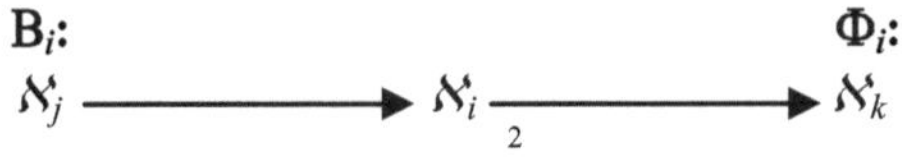

Fig. 7. Objects and links as viewed in I^{2}R. $\aleph_1, \aleph_2, \ldots, \aleph_i, \ldots, \aleph_N$ form an artificial neural network, $\Phi_i = \{\aleph_k | k = 1, \ldots, n_i\}$ denotes the set of artificial neurons that are being influenced by $\aleph_i$, $\mathrm{B}_i = \{\aleph_j | j = 1, \ldots, m_i\}$ denote the set of artificial neurons that influence $\aleph_i$.

(i) $\Delta = \{O_1, O_2, \ldots, O_i, \ldots, O_N\}$ denote a set of Web pages under focus, each page O_i is assigned an artificial neuron $\aleph_i, i = 1, \ldots, N$; thus we may consider $\Delta = \{\aleph_1, \aleph_2, \ldots, \aleph_i, \ldots, \aleph_N\}$,
(ii) $\Phi_i = \{\aleph_k | k = 1, \ldots, n_i\}$ denote the set of artificial neurons that are being influenced (i.e., synapsed, pointed to by) by $\aleph_i$, $\Phi_i \subseteq \Delta$,
(iii) $B_i = \{\aleph_j | j = 1, \ldots, m_i\}$ denote the set of artificial neurons that influence (i.e., synapse to, point to) $\aleph_i$, $B_i \subseteq \Delta$.

The I^{2}R technique can be formally derived from the generic eq. (1) as follows.

Because the objects to be searched are IR objects, e.g, pages, no external input (i.e., from outside the Web) can be assumed, so we take $I_i(t) = 0$. One way to define f_j is to conceive the influence of an object j upon another object i as being determined by the strengths of the connections which convey this influence, i.e., weights w_{ij} of the links between them. Equation (1) thus reduces to the following equation:

$$\frac{dz_i(t)}{dt} = -z_i(t) + \sum_{\aleph_j \in \mathrm{B}_i} w_{ij} \tag{9}$$

In order to simplify the writing, let us introduce the following notation: $\sum_{\aleph_j \in \mathrm{B}_i} w_{ij} = \Sigma^{(i)}$. It is known from the theory of differential equations [41] that the solution of eq. (9) has the following general form

$$z_i(t) = C \cdot e^{-t} + \Sigma^{(i)} \tag{10}$$

where C is a constant depending on the initial condition. When the I^{2}R network operates for retrieval activation spreading is taking place according to WTA strategy. At any time step t_u, $u = 0, 1, \ldots$, exactly one neuron $k \in \{1, \ldots, N\}$, i.e., the winner, is active, all the other neurons $s \in \{1, \ldots, k-1, k+1, \ldots, N\}, s \neq k$, are deactivated, i.e., $z_s(t_u) = 0$. Taking into account this initial condition the activity level of any non-winner neuron s is given by the following function:

$$z_s(t) = (1 - e^{t_u - t})\Sigma^{(s)} \tag{11}$$

If time t is let to increase the activity level $z_s(t)$ tends to stabilize on the total input value $\Sigma^{(s)}$ of that neuron s:

$$\lim_{t \to \propto} z_s(t) = \Sigma^{(s)} \tag{12}$$

At the next time step t_{u+1}, of these neurons s the winner will be that neuron p whose activity level z_p exceeds the activity level z_s of any other neuron s, i.e., $z_p \geqslant z_s$. This re-writes as follows:

$$(1 - e^{t_u - t})\Sigma^{(p)} \geqslant (1 - e^{t_u - t})\Sigma^{(s)} \tag{13}$$

Because $t > t_u$ we have $e^{t_u - t} < 1$, and so $(1 - e^{t_u - t})$ is strictly positive. Hence, the winner-condition $z_p \geqslant z_s$ becomes equivalent to $\Sigma^{(p)} \geqslant \Sigma^{(s)}$. In other words, the neuron with the highest total input (given by eq. (8)) will be the winner. Figure 8 shows example – and typical – plots of activity levels $z_s(t)$ for four neurons, it can be nicely seen how the activity levels reach asymptotically their limit which is equal to the corresponding total input values 1, 5, 3, 6.

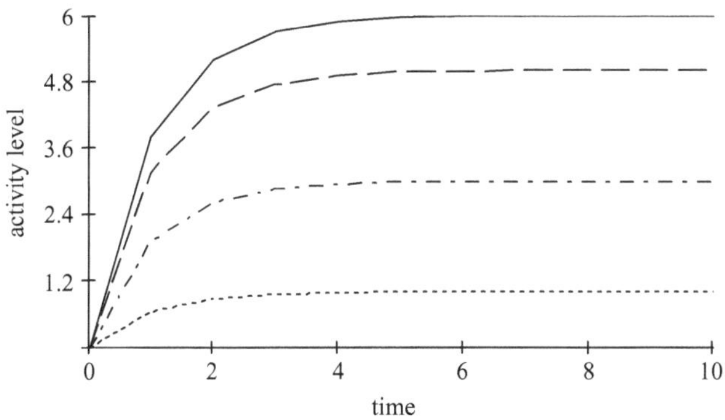

Fig. 8. Typical plots of activity levels for four neurons in I^{2}R. It can be nicely seen how the activity levels reach asymptotically their limit which is equal to the corresponding total input. The highest will be the winner.

3.5 PageRank: Particular Case of the Generic Equation

The importance levels of Web pages are viewed as the activity levels of associated artificial neurons. Once computed, they remain constant while being used in the retrieval process. Thus, the activity level $z(t)$ may be viewed as a particular case, namely constant in time. Then eq. (1) has a null in its left hand side (the derivative of a constant is zero), and is thus asking for finding the equilibrium as a solution; it becomes (see Fig. 7 for notations):

$$0 = I_i - z_i + \sum_{\aleph_j \in \mathrm{B}_i} f_j(z_j, w_{ij}, z_i) \tag{14}$$

No external (i.e., from outside the Web) inputs to Web pages are assumed in PageRank, hence we take $I_i(t) = 0$. Taking into account the principle of

PageRank, the function f_j does not depend on z_i, but it depends on z_j (the citation level of a Web page depends on the citation levels of the pages linking to it). The function f_j is taken as the dot product of the vector of activity levels and corresponding weights of the pages pointing to it, i.e., $f_j = z_j w_{ij}$. Thus, eq. (14) re-writes as follows:

$$0 = -z_i + \sum_{\aleph_j \in \mathrm{B}_i} z_j w_{ij} \tag{15}$$

Let w_{ij} be defined as a weight meaning the frequency of URL_i(i.e., page j contains the URL of page i) in page j, i.e., $w_{ij} = 1/n_j$, where n_j denotes the number of URLs occurring in page j, and multiple occurrences of an URL are considered as single occurrences, then eq. (15) becomes:

$$z_i = \sum_{\aleph_j \in \mathrm{B}_i} \frac{z_j}{n_j} \tag{16}$$

which is the same as PageRank's eq. (3). If an external input I_i is assumed in eq. (10) and defined as $I_i = 1 - d$, where $0 < d < 1$, further the function f_j is defined as $f_j = d \cdot z_j \cdot w_{ij}$, then eq. (4) of PageRank can be obtained:

$$z_i = 1 - d + d \cdot \sum_{\aleph_j \in \mathrm{B}_i} \frac{z_j}{n_j} \tag{17}$$

3.6 Hubs and Authorities: Particular Case of the Generic Equation

If the activity level $z(t)$ of eq. (1) is conceived as corresponding to an authority or hub weight x or y, respectively, of eq. (5), and is viewed as being constant in time (when being used; similar to PageRank), then eq. (1) has a null in its left hand side (the derivative of a constant is zero), and $z(t)$ does not depend on time, i.e., $z(t) = z$. Thus, eq. (1) becomes (see Fig. 7 notations):

$$0 = I_i - z_i + \sum_{\aleph_j \in \mathrm{B}_i} f_j(z_j, w_{ij}, z_i) \tag{18}$$

No external inputs to Web pages are assumed, hence we take $I_i = 0$. It can also be seen that the authority of a Web page is influenced only by the authorities of the pages pointing to it. Thus, the function f_j under the summation of eq. (18) becomes $f_j = z_j$. Writing now eq. (18) also for all neurons k pointed to by neurons i, the following two particular cases of the generic equation are obtained:

$$\begin{aligned} 0 &= -z_i + \sum_{\aleph_j \in \mathrm{B}_i} z_j \\ 0 &= -z_k + \sum_{\aleph_k \in \Phi_i} z_i \end{aligned} \tag{19}$$

which are identical with eq. (5).

4 Computational Complexity in Connectionist Web Retrieval

Computational complexity (CC) is an under-treated topic in IR. Due to a variety of reasons, most IR researchers and developers understand CC as meaning physical running time, and instead of the term CC that of scalability is being used meaning the variation of the physical running time and memory usage as a function of data volume. No doubt, this interpretation of the term CC is justified by and important in practice. However, a systematic mathematical treatment of CC of different IR algorithms would be welcome, because

(i) the CC of different IR algorithms used contribute to the physical running time of applications and experiments;
(ii) in this way, a correct mathematical justification for the complexity of IR algorithms can be obtained (beside the empirical one, as it is usually the case at present).

The aim of the present section is to give a detailed mathematical analysis of the CC of WTA-based IR techniques. The importance of this analysis is given by the following reasons:

(i) intuition may be misleading: because the WTA strategy can yield circles, one may be tempted to assume – based on the result that finding circles in a graph is "hard" – that a WTA-based algorithm is not always tractable (i.e., its CC is not polynomial); the mathematical analysis will reveal that intuition may be misleading;
(ii) at the same time, this analysis can serve as a model that may be followed in the analysis of other methods.

Firstly, in part 4.1, the basic CC concepts used are briefly presented (a good introduction can be found in [52]. This is followed by a literature overview of the CC of different soft computing-based IR tasks (part 4.2). Then, part 4.3 contains an extensive analysis of the CC of WTA techniques (using the I^2R method as illustration).

4.1 Basic Concepts

An *algorithm* can be defined as a finite set of well-defined instructions or steps to be performed in order to accomplish some task. It is commonly accepted that what is referred to as a "well-defined" procedure can be emulated on a special *abstract model of computation* called a *Turing machine* (this is referred to as the Church-Turing hypothesis). Algorithms play a central role in computer science because any computer program is essentially a representation (an encoding) of an algorithm.

When implementing algorithms in the form of computer programs it is important to have an estimate of how much physical time the execution of the program will take. A measure for this is referred to as *computational complexity* (CC), sometimes – and somewhat incorrectly – referred to as simply time complexity. CC aims at estimating numerically the "volume" or "quantity" of operations (calculations) in an algorithm. The practical importance of CC consists in that it is one of the factors influencing the physical running time of the algorithm under consideration for a given set of real data. The CCs of different algorithms can be compared to each other, so the "faster" (i.e., having smaller CC) can be chosen for a given problem. CC may be conceived as being an *upper bound* for the *number of steps an algorithm performs in order to obtain the desired result as a function of its input size.* A step usually means an important and relevant operation in the algorithm. In what follows, CC will be expressed by $O(S)$, pronounced as "big-Oh" of S, where S is an expression depending on the input size N. If S is a polynomial in N then CC is said to be of *polynomial* type. For example, we need to make at most N comparisons in order to find a given name in a list of N names, so the CC of this type of searching is $O(N)$, and is polynomial because N is a first-degree polynomial. The practical importance of an algorithm's polynomial CC is that the problem at hand will always be tractable with that algorithm for virtually any input size (as opposed to those class of algorithms whose CC is not a polynomial of input size but an exponential function, e.g., the Traveling Salesman Problem which is intractable even for relatively low values of the number of cities).

The following remarks are also in place here. The CC of ANN-based algorithm can be estimated by means of the O notation at its most detailed and basic level of each and every one computational step. However, because an ANN is a statistical construct implementing fault-tolerant soft computing, at the design and running level one speaks about layers, neurons, connections, weights, activation levels, transfer functions, learning rules. Thus, next to using solely the traditional O measure, a need for novel complexity measures is emerging based on computationally important aspects such as network topology (number of layers, connectivity, number of neurons), signal propagation (e.g., feed-forward, -backward, recurrent), transfer function (threshold, identity, etc.), learning rules (used to compute the weights). A survey of complexity in neural computation using Perceptron and Spiking Neuron ANNs can be found in [43, 50].

4.2 Computational Complexity in Soft Computing-based Information Retrieval – A Literature Overview

In [28], an activation spreading network based on a Hopfield Net is proposed for collaborative filtering (CF). CF is concerned with making a personalized recommendation by aggregating the experiences of similar users. A key aspect in CF is identifying users similar to the one who needs a recommendation. Users are represented by the items they purchase or rate. The value of 1 indicates a purchase, and 0 indicates that no purchase has occurred. Thus, a consumer-product matrix can be generated. This matrix is the adjacency matrix representation of a graph with consumer-nodes and item-nodes. If an item was purchased by a consumer there is link between the corresponding consumer-node and item-node. Let us assume that there three consumers, c_1, c_2 and c_3, and four items, p_1, p_2, p_3, p_4. We know that both consumers c_1 and c_2 have purchased the items p_2 and p_4, and that consumer c_2 has also purchased item p_3. Then, we can recommend consumer c_1 item p_2. The graph is a bipartite graph with two groups of nodes:

(i) consumer-nodes, there are no links between them,
(ii) item-nodes, there are no links between them.

There are links between corresponding consumer- and item-nodes. The association between a consumer- and an item-node is determined by the existence of a path between them and also by the length of this path. The higher the number of distinct paths the higher the association between them. The degree of association $a(c, p)$ between a consumer-node c and an item-node p is defined as the sum of the weights w of all distinct paths. The weight w is computed as $w = \alpha^x$, where $\alpha \in (0; 1)$ is a constant, and x is the length of the path. An activation is started at the node c, it is then spread to the nodes p linked to it. From these nodes, the activation is spread to their neighbors, and so on. A Hopfield Net is applied to support the spreading o activation until there is no significant change in the activation levels. Computational complexity was evaluated in terms of physical running time only using a test collection, and no formal analysis was carried out.

In [7], a query-concept learner (Maximizing Expected Generalization Algorithm, MEGA) for relevance feedback is proposed that learns query criteria through an intelligent sampling process. MEGA models query concepts in k-CNF, i.e., in conjunctive normal form with at most k literals over Boolean variables. The time complexity for learning a k-CNF is of the order $O(M^k)$, where M is the number of features to depict an object. To solve this curse of dimensionality problem so that a concept can be learned quickly the paper proposes a divide-and-conquer method to reduce learning time of MEGA by dividing the learning task into G subtasks to achieve a speedup of $O(G^{k-1})$ folds. Experiments using both synthesized data and real images showed that MEGA converges to a target concept much faster (in terms of both wall-clock time and number of user iterations) than traditional schemes.

In [49], a semantics-based clustering and indexing approach (*SemQuery*) is presented to support visual queries on heterogeneous features of images. To improve the efficiency in finding the closest images to the query image, spatial data indexing methods were used. The time complexity of an index tree is $O(\log N)$, where N is the number of images that are indexed. If images are retrieve based on the closeness of feature vectors l index trees are needed corresponding to the l feature classes in the set of feature classes $F = \{f_1, \ldots, f_l\}$. The time complexity of such retrieval will be $l \cdot O(\log N)$. In semantic clustering, for each semantic cluster an index tree is needed, where the number of images in that cluster is n and $n < N$. The computational complexity to choose the semantic cluster is $O(t)$, where t is the number of templates representing clusters. The number of templates is generally far less than the number of images, that is $t \ll N$, the overall time complexity of retrieval using semantic clustering will be $O(\log N)$. The paper also proposes a multi-layer perceptron neural network, NeuroMerge, to merge the results obtained from the heterogeneous features. Experimental analysis is conducted and presented to demonstrate the efficiency of the proposed approach. The required time to merge the similarity measurements after the network is trained was computed. On average, NeuroMerge took 0.01 milliseconds to merge the measurements and linear merging took 0.004 milliseconds. NeuroMerge's computation to merge the measurements does not depend on the number of images in the database.

In [48], a pattern matching query language for XML is presented where the interpretation of a query is founded on cost-based query transformations, and XML documents are modeled as labeled trees. Two polynomial-time algorithms are presented that efficiently find the best n answers to the query. *primary* finds all approximate results, sorts them by increasing cost, and prunes the result list after the nth entry. All functions used by *primary* have a polynomial time complexity with respect to the number of nodes in the data tree. The overall time complexity of the algorithm is $O(n^2 rsl)$, where n is the number of query selectors, r is the maximal number of renamings per selector, s is the maximal number of data nodes that have the same label, and l is the maximal number of repetitions of a label along a path in the data tree. *secondary* uses a structural summary – the schema – of the database to estimate the best k transformed (second–level) queries, which in turn are executed against the database. The maximal time needed to generate k second-level queries is $O(n^2 r s_s l k^2 \log k)$, where s_s is the selectivity in the schema. Let s_d denote the selectivity in the data tree, i.e., the maximal number of instances of a node class. The maximal time needed for a single iteration of the algorithm is $O(k(n^2 r s_s l k \cdot \log k + s_d m))$. All the parameters of the formula are typically small numbers. Experiments were carried out to evaluate the efficiency of the algorithms in terms of the evaluation time of a query pattern. Tests results show that the schema-driven query evaluation is faster than the direct evaluation if n, the number of results, is small. For some queries, the schema-based algorithm is faster even if all the results are requested.

In [11], a signature clustering algorithm (compact representations of video clips) is proposed to improve video retrieval performance. In the first phase of the algorithm an MST (minimum spanning tree) is constructed based on the Kruskal algorithm. The order of complexity is $O(e \cdot \log e)$ where $e = |E(P(V, \mu))|$, i.e. the number of edges of the threshold graph $P(V, \mu)$ with a vertex set V, and there is an edge between any two vertices whose corresponding signature distance is less than a fixed distance threshold μ. In the second phase, similar clusters are identified by repeatedly setting the thresholds to the length of the edges of the MST, and checking the pre-computed edge density for each newly-formed connected component. For each threshold tested, all the information required to identify similar clusters is pre-computed and the complexity is simply $O(1)$. Thus, the computational complexity of the second phase is $O(|V|)$, the same order as the maximum number of edges in the MST. The combined complexity of the two phases in the implementation is thus $O(e \cdot \log e) + O(|V|)$.

In [10,46], an automatic clustering method is proposed and applied to Web retrieval using SOM. Scaling and computational performance of the prototype based on this method were evaluated by measuring the physical time for users (the time that the user spends searching). Experiments are reported along with results. User time had an upper limit of 20 minutes. The users spent between 13 minutes up to 20 minutes searching.

In [1,26], efficient techniques are discussed for computing PageRank. It is showed that PageRank can be computed for very large sub-graphs of the web (up to hundreds of millions of nodes) on machines with limited main memory. An implementation is described using the principle underlying the block nested-loops-join strategy that efficiently controls the cost per iteration even in low memory environments. Running time measurements on various memory configurations are presented for PageRank computation. Several methods are discussed for analyzing the convergence of PageRank based on the induced ordering of the pages. In [32], it is noted that there is a drawback of the original PageRank algorithm and an improved PageRank algorithm is presented that computes the PageRank values of the Web pages correctly. Implementation issues of the improved algorithm are presented and it is showed that the implementation of the algorithm does not require a large amount of spatial and computational overhead.

In [2], an adaptive method is presented (Metric Similarity Modeling, MSM) to automatically improve the performance of ranked text retrieval systems. MSM represents documents as vectors in a multidimensional space. There are two costs to using the MSM algorithm. The firs cost is paid only once during the indexing, the second is paid for each retrieval using the new representations. The indexing cost is the cost of calculating $W^{(k)}$. The matrix $W^{(k)}$ is a matrix having k rows and t columns, t is the number of terms and represents objects in the term vector space as objects in the k dimensional semantic space. This cost is determined by the cost of computing the Singular Value Decomposition which is typically of order $O(n^3)$ where n is the number of rows or columns of the matrix. In the case of MSM, n is the number of documents d. Certain steps in the indexing algorithm also require the matrix multiplication of document matrices, these operations are of order $O(d^2t)$ and $O(t^2d)$. The cost at retrieval time to compute the semantic representation for the query takes $O(kt_q)$, where t_q is the number of terms in the query. Finding the documents which are most similar to the query vector requires $O(kd)$ time.

4.3 Computational Complexity of Winner-Take-All-based Retrieval

As it could be seen, in IR models using ANNs a large number of weights should be computed, even repeatedly. Thus, the question of tractability and hence complexity of such a computation arises. In what follows, the computational complexity of the I^2R method is treated in detail, which may serve as a model for analyzing other connectionist IR methods.

The complexity of weights calculation in I^2R is given by the following:

Theorem 2. *The complexity of weights computation is polynomial.*

Proof. As it could be seen (eq. (6) and (7)), there are $2 \times (n_i + n_j)$ number of weights between every pair (o_i, o_j), $i \neq j$, of which there are C_M^2, hence $2 \times (n_i + n_j) \times C_M^2$ has complexity $O(M^2N)$, and $N = \max_{i,j}(n_i, n_j)$, i.e., the largest of object lengths. The computation of the sums of weights (eq. (8)) between a given object o_i and all the other objects o_j, of which there are $M - 1$, takes time $(n_i + n_j) \times (M - 1)$, and thus an upper bound for the computation of all sums in the network is $(n_i + n_j) \times (M - 1)^2 = O(NM^2)$ because i can vary, too, at most $M-1$ times. Hence an overall upper bound for weights computation is $O(NM^2) + O(NM^2) = \mathrm{O}(NM^2) = O(K^3)$, where $K = \max(N, M)$. □

In other words the computation of weights is tractable. The complexity of the retrieval process is given by the following results.

Theorem 3. *The links of maximum weights from all o_q can be found in $O(M^2)$ time.*

Proof. For each o_q we open a stack $S(q)$ and a real variable $m(q)$. Initially $S(q)$ is empty and $m(q) = 0$. Scanning the weights w_{iq} for all the $M-1$ values of i, we do nothing for $w_{iq} < m(q)$, but put o_i into $S(q)$ if $w_{iq} = m(q)$. Finally, if $w_{iq} > m(q)$, we first empty $S(q)$, then put o_i into $S(q)$ and redefine $m(q)$: $= w_{iq}$. At the end of this procedure, the contents of $S(q)$ tells precisely the maximum-weight links at o_q. □

Theorem 4. *One cycle can be retrieved in $O(M)$ steps.*

Proof. Open a block $\underline{a} = a_1a_2 \ldots a_M$, of M zeroes. Starting at o_q (that represents the query), rewrite $a_q = 1$. When at o_i, choose one $o_j \in S(i)$ (the top element). If $a_j = 1$, a cycle has been found. Otherwise we rewrite $a_j := 1$ and continue the search there. It is clear that a nonzero entry in $\underline{a}$ (and hence also a cycle) will be reached in at most M steps. □

We define the graph $G_{\max}$ with vertex set $\{\nu_1, \nu_2, \ldots, \nu_M\}$ and arcs $\nu_i\nu_j$ where $\nu_j \in S(i)$. By definition, a cycle is feasible with respect to the retrieval process if and only if it corresponds to a cycle in $G_{\max}$. We denote by $G'_{\max}$ the sub-graph of $G_{\max}$ containing those arcs which lie in at least one cycle.

Theorem 5. *The sub-graph $G'_{\max}$, and also a family of cycles covering all edges of $G'_{\max}$, can be found in polynomial time.*

Proof. Selecting any one arc $\nu_i\nu_j$, it can be tested in polynomial (in fact, linear) time whether there exists a directed path from ν_j to ν_i in $G_{\max}$ (as a necessary and sufficient condition for $\nu_i\nu_j$ to lie in a cycle), and also such a path of minimum length can be found, applying Breadth-First Search. □

After weights summation there are $M-1$ weighted links $s_1, \ldots, s_{i-1}, s_{i+1}, \ldots, s_M$ from an object o_i to all the other objects. Because i varies from 1 to M there are at most $M \times (M-1) = O(M^2)$ links to be evaluated in all (in a search). Depending on the multiplicity (i.e., unique, double, triple maximum, or higher) of the maximum of the sequence $s_1, \ldots, s_{i-1}, s_{i+1}, \ldots, s_M$ the number of reverberative circles can increase. The number of retrieved objects depends on two factors:

(a) the number of reverberative circles,
(b) the number of objects a reverberative circle contains.

In order to render the influence of the first parameter simulations were carried out using MathCAD. M was taken 100,000, and different number of sequences of weights were generated at random (random number generator with uniform distribution, six digit accuracy). In each of these cases the maximum and its multiplicity was determined (Table 1).

The value of the empirical density function on every interval Δx = (0, 1), (1, 2), (2, 3), (3, 4) is calculated separately using the usual ratio:

$$\frac{NoOfVals}{Length \times TotalVals} \tag{20}$$

(where *NoOfVals* denotes the number of values in the interval, whereas *TotalVals* is the total number of values) for each of the three cases (Table 1), and then the corresponding values are averaged. The empirical density function can be well approximated by, e.g., the function:

Table 1. Simulation of the multiplicity of maxima. (In 985 sequences out of 1000 sequences there was a unique maximum, in 14 cases there were double maxima, in 3 cases there was one triple maximum, and there were no maxima with multiplicity 4)

	Multiplicity of maxima Number of sequences			
Number of sequences	1	2	3	4
1 000	985	14	1	0
2 500	2469	30	1	0
10 000	9532	439	26	2

$$f(x) = u^2 e^{-u^{0.7}x} \tag{21}$$

After curve fitting (calculations carried out using standard Mathcad curve fitting) this becomes:

$$f(x) = 3.864 \cdot e^{-1.605x} \tag{22}$$

which is an estimated density function, and thus the probability to have maximum with multiplicity a or b, Prob(a, b), in a random sequence $s_1, \ldots, s_{i-1}$, $s_{i+1}, \ldots, s_M$ of weights can be estimated using the usual definition from probability:

$$\int_a^b 3.864 e^{-1.605x} dx \tag{23}$$

Thus, we obtain the following probabilities: Prob(1,2) = 0.386; Prob(2,3) = 0.078; Prob(3,4) = 0.016. The simulation results show that there always are a few multiple maxima, their proportion is not high, but the probability to have maxima with higher multiplicity increases with the number of sequences. (*Note.* To the best of our knowledge the probability of the multiplicity of maximum in a random sequence is an open, interesting and difficult mathematical problem.)

We give now an asymptotic estimate of the probability that, assuming uniform weight distribution, the maximum value is unique (that is, the retrieval procedure is continued in a unique direction). Let $S = s_1 s_2 \ldots s_k$ be a randomly chosen sequence of length k where $s_i \in \{1, \ldots, n\}$ and $\text{Prob}(s_i = j) = 1/n$ for every i and j, independently for all i. (To simplify notation, we write k for $M - 1$; and n denotes the number of possible weight values w_{ij}.) Suppose that the maximum occurs at a unique element of S, say $s_i = m$ and $s_j < m$ for all $j \neq i (1 \leqslant j \leqslant k)$. Each of the k positions in S is equally likely to occur as this particular i, and for every j we have $\text{Prob}(s_j < m) = (m - 1)/n$. Thus,

$$\text{Prob}(S \text{ has unique maximum}) = P(n, k) = \frac{k}{n} \sum_{m=2}^{n} \left(\frac{m-1}{n} \right)^{k-1} \tag{24}$$

In order to obtain fairly tight asymptotic estimates, we write the right-hand side of (24) in the form

$$P(n, k) = \frac{k}{n} \sum_{q=1}^{n} \left(1 - \frac{q}{n} \right)^{k-1} \tag{25}$$

and apply the inequalities

$$e^{-\frac{uv}{1-u}} < (1 - u)^v < e^{-uv} \tag{26}$$

As for an upper bound, we immediately obtain

$$P(n,k) < \frac{k}{n}\sum_{q\geqslant 1} e^{-\frac{q(k-1)}{n}} = \frac{k}{n}\frac{1}{e^{\frac{k-1}{n}}-1} \tag{27}$$

Introducing the notation $x = k/n$, a convenient estimate seems to be

$$f(x) = \frac{x}{e^x - 1} \tag{28}$$

Lower bounds are somewhat more complicated. Since the leftmost and rightmost sides of (26) are not far from each other only if u is near zero, for the purpose we can split the sum in (27) into two parts, for q small and q large, respectively. To do this, a convenient threshold $b = \lceil c(n/k) \cdot \log n \rceil$ is taken, where c is a suitably chosen constant. Then, for $q > b$, all of the summands are smaller than n^{-c} and hence become negligible, depending on c that can be fixed with respect to the accuracy required. Assuming $q \leqslant b$, and applying the lower bound in (28), we obtain

$$P(n,k) > \sum_{q=1}^{b} e^{-q(k-1)/n\left(1-\frac{i}{n}\right)} > \sum_{q\rangle 0}\left(e^{-k+1/n\left(1-\frac{1}{b}\right)}\right)^q - \sum_{q\rangle b}\left(e^{-k+1/n\left(1-\frac{1}{b}\right)}\right)^q \tag{29}$$

Introducing the notation

$$y = e^{-k+1/n\left(1-\frac{1}{b}\right)} \tag{30}$$

for the expression occurring in the parentheses in (29), we obtain

$$P(n,k) > \frac{y}{1-y} - \frac{y^b+1}{1-y} \tag{31}$$

We note that the negative term in eq. (31) is again $O(n^{-c})$, by the choice of b, i.e., it may become negligible, and then the formula simplifies to $y/(1-y)$. If both k and n/k are large, the formula (28) seems to be quite convenient for use. Having checked for the simulation values, i.e., for $k = 100000$ and $n = 1000000$ (that is, six-digit accuracy in weights), we obtain the that the probability to have multiple maxima is $\approx$0.05 that happens to compare acceptably well with the average value of $\approx$0.1 of the probabilities for multiple maxima obtained in the simulations.

5 Conclusions

The main contribution of this paper is the construction of a unified formal framework for three major methods used for Web retrieval tasks: PageRank, HITS, I^2R. They are based on different paradigms originally: PageRank (normalized importance), HITS (mutual re-enforcement), I^2R (winner-take-all). The paper shows that the above three techniques, albeit they stem originally from different paradigms, can be integrated into one unified formal view. The conceptual and notational framework used is given by ANNs and the generic network equation. It was shown that the PageRank, HITS and I^2R methods can be formally obtained from the generic equation as different particular cases by making certain assumptions reflecting the corresponding underlying paradigm.

This unified formal view sheds a new light upon the understanding of these methods: it may be said that they are only seemingly different from each other, they are particular ANNs stemming from the same equation and differing from one another in whether they are dynamic (a page's importance varies in time) or static (a page's importance is constant in time), and in the way they connect the pages to each other.

At the same time, the existence of solutions (i.e., the convergence of the algorithms) in all three methods is given by Theorem 1. This is a simple and unified justification of their convergence as opposed to those given thus far (based on quite sophisticated linear algebra considerations).

The creation of unified formal frameworks for different Web retrieval methods is not entirely new. Ding et al. [17] gave a unified formal framework for PageRank and HITS using linear algebra techniques. Lempel and Moran [40] proposed a method called SALSA to define hubs and authorities based on Markov chains. They also gave a meta-framework for both SALSA and HITS. However, to the best of our knowledge, the contribution of the present paper is entirely new in that it seems to be the first to give a unified formal framework for all three methods.

Another contribution of the present paper is a detailed mathematical analysis of the computational complexity of WTA-based IR techniques using the I^2R method for illustration. The importance of this analysis consists in that it shows that (i) intuition may be misleading (contrary to intuition, a WTA-based algorithm yielding circles is not always "hard"), and (ii) this analysis can serve as a model that may be followed in the analysis of other methods.

Acknowledgements

The authors would like to thank the anonymous reviewers for their helpful comments, and to acknowledge the support of grants OTKA T 037821 and OTKA T049613 of the National Foundation for Scientific Research, Hungary.

References

1. Arasu, A. (2002). PageRank Computation and the Structure of the Web: Experiments and Algorithms. *Proceedings of the World Wide Web 2002 Conference*, Honolulu, Hawaii, USA, 7–11 May, http://www2002.org/CDROM/poster (visited: 4 Nov 2002)

2. Bartell, B. T. (1994). Optimizing Ranking Functions: A Connectionist Approach to Adaptive Information Retrieval. *Ph.D. Thesis*, Department of Computer Science and Engineering, University of California, San Diego, 1994. http://www.cs.ucsd.edu/groups/guru/publications.html (visited: 10 May 2004)
3. Belew, R.K. (1987). A Connectionist Approach to Conceptual Information Retrieval. *Proceedings of the International Conference on Artificial Intelligence and Law* (pp. 116–126). Baltimore, ACM Press.
4. Belew, R.K. (1989). Adaptive information retrieval: Using a connectionist representation to retrieve and learn about documents. *Proceedings of the SIGIR 1989* (pp. 11–20). Cambridge, MA, ACM Press.
5. Bienner, F., Giuvarch, M. and Pinon, J.M. (1990). Browsing in hyperdocuments with the assistance of a neural network. *Proceedings of the European Conference on Hypertext* (pp. 288–297). Versailles, France.
6. Brin, S., and Page, L. (1998). The Anatomy of a Large-Scale Hypertextual Web Search Engine. *Proceedings of the 7th World Wide Web Conference*, Brisbane, Australia, 14-18 April, pp: 107–117
7. Chang, E. and Li, B. (2003). MEGA – The Maximizing Expected Generalization Algorithm for Learning Complex Query Concepts. *ACM Transactions on Information Systems*, 21(4), pp: 347–382.
8. Chen, H. (2003a). Introduction to the JASIST special topic section on Web retrieval and mining: a machine learning perspective. *Journal of the American Society for Information Science and Technology*, vol. 54, no. 7, pp: 621–624.
9. Chen, H. (2003b). Web retrieval and mining. *Decision Support Systems*, vol. 35, pp: 1–5.
10. Chen, H., Fan, H., Chau, M., Zeng, D. (2001). MetaSpider:Meta-Searching and Categorisation on the Web. *Journal of the American Society for Information Science and Technology*, vol. 52, no. 13, pp: 1134–1147.
11. Cheun, S. S. and Zakhor, A. (2001). Video Similarity Detection with Video Signature Clustering. *Proceedings of the 8*th *IEEE International Conference on Image Processing*, vol. 1. pp: 649–652.
12. Cohen, P., and Kjeldson, R. (1987). Information retrieval by constrained spreading activation in semantic networks. *Information Processing and Management*, 23, 255–268.
13. Cordon, O., Herrera-Viedma, E. (2003). Editorial: Special issue on soft computing applications to intelligent information retrieval. *International Journal of Approximate Reasoning*, vol. 34, pp: 89–95.
14. Crestani, F., Lee, P. L. (2000). Searching the web by constrained spreading activation. *Information Processing and Management*, vol. 36, pp: 585–605.
15. Cunningham S.J., Holmes G., Littin J., Beale R., and Witten I.H. (1997). Applying connectionist models to information retrieval. In Amari, S. and Kasobov, N. (Eds.) *Brain-Like Computing and Intelligent Information Systems* (pp 435–457). Springer-Verlag.
16. De Wilde, Ph. (1996). *Neural Network Models*. Springer Verlag.
17. Ding, C., He, X., Husbands, P., Zha, H., Simon, H.D. (2002). PageRank, HITS, and a unified framework for link analysis. *Proceedings of the ACM SIGIR 2002*, Tampere, Finland, pp: 353–354.

18. Dominich, S. (1994). Interaction Information Retrieval. *Journal of Documentation*, 50(3), 197–212.
19. Dominich, S. (2001). *Mathematical Foundations of Information Retrieval.* Kluwer Academic Publishers, Dordrecht, Boston, London.
20. Dominich, S. (2004). Connectionist Interaction Information retrieval. *Information Processing and Management*, vol 39, no.2, pp: 167–194
21. Doszkocs, T., Reggia, J., and Lin, X. (1990). Connectionist models and information retrieval. *Annual Review of Information Science & Technology*, 25, 209–260.
22. Feldman, J.A., and Ballard, D.H. (1982). Connectionist models and their properties. *Cognitive Science*, vol. 6, pp: 205–254
23. Fuhr, N. and Buckley, C. (1991). A probabilistic learning approach for document indexing. *ACM Transactions on Information Systems*, 9(3), 223–248.
24. Garfield, E. (1955). Citation indexes for science. *Science*, p. 108
25. Grossberg, S. (1976). Adaptive pattern classification and universal recoding: I. Parallel development and coding of neural feature detectors. *Biological Cybertnetics*, vol. 23, pp: 121–134
26. Haveliwala, T.H. (1999). *Efficient Computation of PageRank.* Stanford University, http://dbpubs. stanford.edu:8090/pub/1998-31 (visited: 27 Febr 2004)
27. Hopfield, J.J. (1984). Neurons with graded response have collective computational properties like those of two-states neurons. *Proceedings of the National Academy of Sciences*, vol. 81, pp: 3088–3092
28. Huang, Z., Chen, H., Zeng, D. (2004). Applying Associative Retrieval Techniques to Alleviate the Sparsity Problem in Collaborative Filtering. *ACM Transactions on Information Systems*, vol. 22, no. 1, pp: 116–142.
29. James, W. (1890). *Psychology (Briefer Course).* New York: Holt, Chapter XVI, "Association", pp: 253–279
30. Johnson, A., and Fotouhi, F. (1996). Adaptive clustering of hypermedia documents. *Information Systems*, 21, 549–473.
31. Johnson, A., Fotouhi, F., and Goel, N. (1994). Adaptive clustering of scientific data. *Proceedings of the 13th IEEE International Phoenix Conference on Computers and Communication* (pp. 241–247). Tempe, Arizona.
32. Kim, S.J., and Lee, S.H. (2002). An Improved Computation of the PageRank Algorithm. In: Crestani, F., Girolamo, M., and van Rijsbergen, C.J. (eds.) *Proceedings of the European Colloquium on Information Retrieval.* Springer LNCS 2291, pp: 73–85
33. Kleiberg, J. M. (1999). Authoritative Sources in a Hyperlinked Environment. *Journal of the ACM*, vol. 46, no. 5, pp: 604–632.
34. Kohonen, T. (1988). *Self-Organization and Associative Memory.* New York: Springer Verlag.
35. Kraft, D.H., Bordogna, P. and Pasi, G. (1998). Fuzzy Set Techniques in Information Retrieval. *In*: Didier, D. and Prade, H. (Eds.) *Handbook of Fuzzy Sets and Possibility Theory. Approximate Reasoning and Fuzzy Infomation Systems*, (Chp. 8). Kluwer Academic Publishers, AA Dordrecht, The Netherlands.
36. Kwok, K.L. (1989). A Neural Network for the Probabilistic Information Retrieval. In Belkin, N.J. and van Rijsbergen, C.J. (Eds.) *Proceedings of the 12*th *Annual International ACM SIGIR Conference on Research and Development in Information Retrieval.* ACM Press, Cambridge, MA, USA, pp: 21–29.

37. Kwok, K.L. (1990). Application of Neural Networks to Information Retrieval. In Caudill, M. (Ed.) *Proceedings of the International Joint Conference on Neural Networks*, Vol. II (pp. 623–626). Hilldale, NJ, Lawrance Erlbaum Associates, Inc.
38. Kwok, K.L. (1995). A network approach to probabilistic information retrieval. *ACM Transactions on Information Systems*, 13(3), 243–253.
39. Layaida, R., Boughanem, M. and Caron, A. (1994). Constructing an Information Retrieval System with Neural Networks. *Lecture Notes in Computer Science*, 856, Springer, pp: 561–570.
40. Lempel, R., Moran, S. (2001). SALSA: the stochastic approach for link-structure analysis. *ACM Transactions on Information Systems*, vol. 19, no. 2, pp: 131–160.
41. Martin, W. T., Reissner, E. (1961). *Elementary Differential Equations*. Addison-Wesley, Reading-Massachusetts, U.S.A.
42. Niki, K. (1997). Sel-organizing Information Retrieval System on the Web: Sir-Web. In Kasabov, N. et al. (Eds.) *Progress in Connectionist-based Information Systems. Proceedings of the 1997 International Conference on Neural Information Processing and Intelligent Information Systems*, vol. 2, Springer Verlag, Singapore, pp: 881–884.
43. Orponen, P. (1995). Computational Complexity of Neural Networks: A Survey. *Nordic Journal of Computing*, vol. 1, pp: 94–110.
44. Rose, D. E. (1994). *A symbolic and connectionist approach to legal information retrieval.* Hillsdale, NJ, Erlbaum.
45. Rose, D.E. and Belew, R.K. (1991). A connectionist and symbolic hybrid for improving legal research. *International Journal of Man-Machine Studies*, 35(1), 1–33.
46. Roussinov, D.G., Chen, H. (2001). Information navigation on the Web by clustering and summarizing query results. *Information Processing and Management*, vol. 37, pp: 789–816.
47. Ruiz, M.E., Srinivasan, P. (1999). Hierarchical Neural Networks for Text Categorization. *Proceedings of the 22*nd *ACM SIGIR International Conference on Research and Development in Information Retrieval*, Berkeley, California, USA, pp: 281–282.
48. Schlieder, T. (2002). Schema–Driven Evaluation of ApproXQL Queries. *Technical Report B02-01*, Freie Universität Berlin, January 2002. http://www.inf.fu-berlin.de/inst/ag-db/publications/2002/report-B-02-01.pdf (visited: 10 May 2004)
49. Sheikholeslami, G., Chang, W. and Zhang, A. (2002).SemQuery: Semantic Clustering and Querying on Heterogeneous Features for Visual Data. *IEEE Transactions on Knowledge and Data Engineering*, 14(5), pp: 988–1003.
50. Sima, J., Orponen, P. (2003). General-Purpose Computation with Neural Networks: A Survey of Complexity Theoretic Results. *Neural Computation*, vol. 15, pp: 2727–2778.
51. Van Rijsbergen, C.J. (2004). *The Geometry of IR*. Cambridge University Press.
52. Weiss, M.A. (1995). *Data Structures and Algorithm Analysis*. The Benjamin/Cummings Publishing Company, Inc., New York, Amsterdam.
53. Wermter S. (2000). Neural Network Agents for Learning Semantic Text Classification. *Information Retrieval*, 3(2), 87–103.

54. Wong, S.K.M., Cai, Y.J. (1993). Computation of Term Association by Neural Networks. *Proceedings of the 16*th *ACM SIGIR International Conference on Research and Development in Information Retrieval*, Pittsburgh, PA, USA, pp: 107–115.
55. Yang, C.C., Yen, J., Chen, H. (2000). Intelligent internet searching agent based on hybrid simulated annealing. *Decision Support Systems*, vol. 28, pp: 269–277.

Semi-fuzzy Quantifiers for Information Retrieval

David E. Losada[1], Félix Díaz-Hermida[2], and Alberto Bugarín[1]

[1] Grupo de Sistemas Inteligentes, Departamento de Electrónica y Computación. Universidad de Santiago de Compostela
{dlosada,alberto}@dec.usc.es
[2] Departamento de Informática, Universidad de Oviedo
diazfelix@uniovi.es

Summary. Recent research on fuzzy quantification for information retrieval has proposed the application of semi-fuzzy quantifiers for improving query languages. Fuzzy quantified sentences are useful as they allow additional restrictions to be imposed on the retrieval process unlike more popular retrieval approaches, which lack the facility to accurately express information needs. For instance, fuzzy quantification supplies a variety of methods for combining query terms whereas extended boolean models can only handle extended boolean-like operators to connect query terms. Although some experiments validating these advantages have been reported in recent works, a comparison against state-of-the-art techniques has not been addressed. In this work we provide empirical evidence on the adequacy of fuzzy quantifiers to enhance information retrieval systems. We show that our fuzzy approach is competitive with respect to models such as the vector-space model with pivoted document-length normalization, which is at the heart of some high-performance web search systems. These empirical results strengthen previous theoretical works that suggested fuzzy quantification as an appropriate technique for modeling information needs. In this respect, we demonstrate here the connection between the retrieval framework based on the concept of semi-fuzzy quantifier and the seminal proposals for modeling linguistic statements through Ordered Weighted Averaging operators (OWA).

1 Introduction

Classical retrieval approaches are mainly guided by efficiency rather than expressiveness. This yields to Information Retrieval (IR) systems which retrieve documents very efficiently but their internal representations of documents and queries is simplistic. This is especially true for web retrieval engines, which deal with huge amounts of data and their response time is critical. Nevertheless, it is well known that users have often a vague idea of what they are looking for and, hence, the query language should supply adequate means to express her/his information need.

D.E. Losada et al.: *Semi-fuzzy Quantifiers for Information Retrieval*, StudFuzz **197**, 195–220 (2006)
www.springerlink.com

Boolean query languages were traditionally used in most early commercial systems but there exists much evidence to show that ordinary users are unable to master the complications of boolean expressions to construct consistently effective search statements [24]. This provoked that a number of researchers have explored ways to incorporate some elements of the natural language into the query language. To this aim, fuzzy set theory and fuzzy quantifiers have been found useful [2,3]. In particular, fuzzy quantifiers permit to implement a diversity of methods for combining query terms whereas the classic extended boolean methods [24] for softening the basic Boolean connectives are rather inflexible [2]. This is especially valuable for web search as it is well known that users are reluctant to supply many search terms and, thus, it is interesting to support different combinations of the query terms. Indeed, fuzzy linguistic modelling has been identified as a promising research topic for improving the query language of search engines [14]. Nevertheless, the benefits from fuzzy quantification have been traditionally shown through motivating examples in IR whose actual retrieval performance remained unclear. The absence of a proper evaluation, using large-scale data collections and following the well-established experimental methodology for IR, is an important weakness for these proposals.

A first step to augment the availabilty of quantitative empirical data for fuzzy quantification in IR was done in [19], where a query language expanded with quantified expressions was defined and evaluated empirically. This work stands on the concept of semi-fuzzy quantifier (SFQ) and quantifier fuzzification mechanism (QFM). To evaluate a given quantified statement, an appropriate SFQ is defined and a QFM is subsequently applied, yielding the final evaluation score.

In this paper, we extend the research on SFQ for IR in two different ways. First, we show that the framework based on SFQ is general and it handles seminal proposals [30] for applying Ordered Weighted Averaging operators (OWA) as particular cases. Second, the experimentation has been expanded. In particular, we compare here the retrieval performance of the fuzzy model with state-of-the-art IR matching functions. We show that the model is competitive with respect to high-performance extensions of the vector space model based on document length corrections (pivoted document length normalization [28]), which have recurrently appeared among the top performance systems in TREC Web track competitions [13, 27, 29]. This is a promising result which advances the adequacy of fuzzy linguistic quantifiers for enhancing search engines.

The remainder of the paper is organized as follows. Section 2 describes some related work and Section 3 explains the fuzzy model for IR defined in [19]. Section 4 shows that the framework based on SFQ handles the OWA-based quantification as a particular case. The main experimental findings are reported in Section 5. The paper ends with some conclusions and future lines of research.

2 Related Work

Fuzzy set theory has been applied to model flexible IR systems which can represent and interpret the vagueness typical of human communication and reasoning. Many fuzzy proposals have been proposed facing one or more of the different aspects around the retrieval activity. Exhaustive surveys on fuzzy techniques in different IR subareas can be found in [3,6].

In seminal fuzzy approaches for IR, retrieval was naturally modeled in terms of fuzzy sets [15, 16, 21, 23]. The intrisic limitations of the Boolean Model motivated the development of a series of studies aiming at extending the Boolean Model by means of fuzzy set theory. The Boolean Model was naturally extended by implementing boolean connectives through operations between fuzzy sets. Given a boolean expression, each individual query term can be interpreted as a fuzzy set in which each document has a degree of membership. Formally, each individual term, t_i, defines a fuzzy set whose universe of discourse is the set of all documents in the document base, D, and the membership function has the form: $\mu_{t_i} : D \rightarrow [0,1]$. The larger this degree is, the more important the term is for characterizing the document's content. For instance, these values can be easily computed from popular IR heuristics, such as *tf/idf* [25]. Given a Boolean query involving terms and Boolean connectors AND, OR, NOT (e.g. t_1 AND t_2 OR NOT t_3) a fuzzy set of documents representing the query as a whole can be obtained by operations between fuzzy sets. The Boolean connective AND is implemented by an intersection between fuzzy sets, the Boolean OR is implemented by a fuzzy union and so forth. Finally, a rank of documents can be straightforwardly obtained from the fuzzy set of documents representing the query.

These seminal proposals are in one way or another on the basis of many subsequent fuzzy approaches for IR. In particular, those works focused on extending query expressiveness further on boolean expressions are especially related to our research. In [2] an extended query language containing linguistic quantifiers was designed. The boolean connectives AND and OR were replaced by soft operators for aggregating the selection criteria. The linguistic quantifiers used as aggregation operators were defined by Ordered Weighted Averaging (OWA) operators [31]. The requirements of an information need are more easily and intuitively formulated using linguistic quantifiers, such as *all*, *at least k*, *about k* and *most of*. Moreover, the operator *and possibly* was defined to allow for a hierarchical aggregation of the selection criteria in order to express their priorities. This original proposal is very valuable as it anticipated the adequacy of fuzzy linguistic quantifiers for enhancing IR query languages. Nevertheless, the practical advantages obtained from such quantified statements remained unclear because of the lack of reported experiments.

In [19], a fuzzy IR model was proposed to handle queries as boolean combinations of atomic search units. These basic units can be either search terms or quantified expressions. Linguistic quantified expressions were implemented by means of semi-fuzzy quantifiers. Some experiments were reported showing that the approach based on SFQ is operative under realistic circumstances.

In this paper we extend the work developed in [19] at both the theoretical and experimental level. On one hand, we show explicitly the connection between the pioneering proposals on fuzzy quantification for IR [2] and the framework based on SFQ. On the other hand, we compare here the retrieval performance of the SFQ fuzzy model with high performance IR matching functions. This will show whether or not the SFQ approach is comparable to state-of-the-art IR methods.

3 Semi-fuzzy Quantifiers for Information Retrieval

Before proceeding, we briefly review some basic concepts of fuzzy set theory. Next, the approach based on semi-fuzzy quantifiers proposed in [19] is reviewed.

Fuzzy set theory allows us to define sets whose boundaries are not well defined. Given a universe of discourse U, a fuzzy set A can be characterized by a membership function with the form: $\mu_A : U \rightarrow [0,1]$. For every element $u \in U$, $\mu_A(u)$ represents its degree of membership to the fuzzy set A, with 0 corresponding to no membership in the fuzzy set and 1 corresponding to full membership. Operations on fuzzy sets can be implemented in several ways. For instance, the complement of a fuzzy set A and the intersection and union of two fuzzy sets A and B are typically defined by the following membership functions: $\mu_{\overline{A}}(u) = 1 - \mu_A(u)$, $\mu_{A \cup B}(u) = max(\mu_A(u), \mu_B(u))$ and $\mu_{A \cap B}(u) = min(\mu_A(u), \mu_B(u))$.

Some additional notation will be also of help in the rest of this paper. By $\wp(U)$ we refer to the crisp powerset of U and $\widetilde{\wp}(U)$ stands for the fuzzy powerset of U, i.e. the set containing all the fuzzy sets that can be defined over U. Given the universe of discourse $U = \{u_1, u_2, \ldots, u_n\}$, a discrete fuzzy set A constructed over U is usually denoted as: $A = \{\mu_A(u_1)/u_1, \mu_A(u_2)/u_2, \ldots, \mu_A(u_n)/u_n\}$.

Fuzzy quantification is usually applied for relaxing the definition of crisp quantifiers. The evaluation of *unary* expressions such as "approximately 80% of people are tall" or "most cars are fast" is naturally handled through the concept of fuzzy quantifier[1]. Formally,

Definition 1 (fuzzy quantifier). *A unary fuzzy quantifier* $\widetilde{Q}$ *on a base set* $U \neq \varnothing$ *is a mapping* $\widetilde{Q} : \widetilde{\wp}(U) \longrightarrow [0,1]$.

For example, given the fuzzy set $X = \{0.2/u_1, 0.1/u_2, 0.3/u_3, 0.1/u_4\}$, modelling the degree of technical skill of four football players in a team, we

[1] These expressions are called unary because each sentence involves a single vague property (tall in the first example and fast in the second one).

can apply a quantifier of the kind most to determine whether or not most footballers are skillful. Of course, given the membership degrees of the elements in X, any coherent implementation of the most quantifier applied on X would lead to a low evaluation score.

The definition of fuzzy quantifiers for handling linguistic expressions has been widely dealt with in the literature [5, 7, 8, 11, 31, 32, 34]. Unfortunately, given a certain linguistic expression, it is often difficult to achieve consensus on a) the most appropriate mathematical definition for a given quantifier and b) the adequacy of a particular numerical value as the evaluation result for a fuzzy quantified sentence. This is especially problematic when linguistic expressions involve several fuzzy properties. To overcome this problem, some authors have proposed indirect definitions of fuzzy quantifiers through semi-fuzzy quantifiers [9–11]. A fuzzy quantifier can be defined from a semi-fuzzy quantifier through a so-called quantifier fuzzification mechanism (QFM). The motivation of this class of indirect definitions is that semi-fuzzy quantifiers (SFQ) are closer to the well-known crisp quantifiers and can be defined in a more natural and intuitive way. Formally,

Definition 2 (semi-fuzzy quantifier). *A unary semi-fuzzy quantifier Q on a base set $U \neq \varnothing$ is a mapping $Q : \wp(U) \longrightarrow [0,1]$.*

In the next example we show a definition and graphical description of a *relative* semi-fuzzy quantifier *about_half*[2].

Example 1. about_half semi-fuzzy quantifier.
about_half : $\wp(U) \rightarrow [0,1]$

$$about\_half(X) = \begin{cases} 0 & \text{if } \frac{|X|}{|U|} < 0.3 \\ 2\left(\frac{\left(\frac{|X|}{|U|} - 0.3\right)}{0.2}\right)^2 & \text{if } \frac{|X|}{|U|} \geq 0.3 \wedge \frac{|X|}{|U|} < 0.4 \\ 1 - 2\left(\frac{\left(\frac{|X|}{|U|} - 0.5\right)}{0.2}\right)^2 & \text{if } \frac{|X|}{|U|} \geq 0.4 \wedge \frac{|X|}{|U|} < 0.6 \\ 2\left(\frac{\left(\frac{|X|}{|U|} - 0.7\right)}{0.2}\right)^2 & \text{if } \frac{|X|}{|U|} \geq 0.6 \wedge \frac{|X|}{|U|} < 0.7 \\ 0 & \text{otherwise} \end{cases}$$

[2] This is a relative quantifier because it is defined as a proportion over the base set U

Graphically,

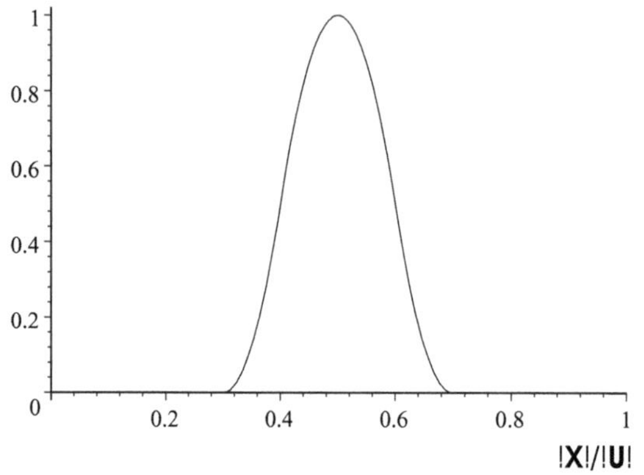

Example of use:
Consider a universe of discourse composed of 10 individuals, $U = \{u_1, u_2, \ldots, u_{10}\}$. Imagine that X is a subset of U containing those individuals which are taller than 1.70m: $X = \{u_1, u_4, u_8, u_{10}\}$
The evaluation of the expression "about half of people are taller than 1.70 m" produces the value: $about\_half(X) = 1 - 2((0.4 - 0.5)/0.2)^2 = 0.5$

Definition 3 (quantifier fuzzification mechanism). *A QFM is a mapping with domain in the universe of semi-fuzzy quantifiers and range in the universe of fuzzy quantifiers*[3]*:*

$$F : (Q : \wp(U) \mapsto [0,1]) \mapsto \left(\widetilde{Q} : \widetilde{\wp}(U) \mapsto [0,1]\right) \tag{1}$$

Different QFMs have been proposed in the literature [9,10]. In the following we will focus on the QFM tested for IR in [19]. Further details on the properties of this QFM and a thorough analysis of its behaviour can be found in [8].

Since this QFM is based on the notion of α-cut, we first introduce the α-cut operation and, next, we depict the definition of the QFM.

The α-cut operation on a fuzzy set produces a crisp set containing certain elements of the original fuzzy set. Formally,

Definition 4 (α-cut). *Given a fuzzy set $X \in \widetilde{\wp}(U)$ and $\alpha \in [0,1]$, the α-cut of level α of X is the crisp set $X_{\geq\alpha}$ defined as $X_{\geq\alpha} = \{u \in U : \mu_X(u) \geq \alpha\}$.*

Example 2. Let $X \in \widetilde{\wp}(U)$ be the fuzzy set $X = \{0.6/u_1, 0.2/u_2, 0.3/u_3, 0/u_4, 1/u_5\}$, then $X_{\geq 0.4} = \{u_1, u_5\}$.

[3] Note that we use the unary version of the fuzzification mechanisms.

In [19], the following quantifier fuzzification mechanism was applied for the basic IR retrieval task:

$$(F(Q))(X) = \int_0^1 Q\left((X)_{\geq\alpha}\right) d\alpha \tag{2}$$

where $Q : \wp(U) \to [0,1]$ is a unary semi-fuzzy quantifier, $X \in \widetilde{\wp}(U)$ is a fuzzy set and $(X)_{\geq\alpha}$ is the α-cut of level α of X.

The crisp sets $(X)_{\geq\alpha}$ can be regarded as crisp representatives for the fuzzy set X. Roughly speaking, 2 averages out the values obtained after applying the semi-fuzzy quantifier to these crisp representatives of X. The original definition of this QFM can be found in [8].

If U is finite, expression (2) can be discretized as follows:

$$(F(Q))(X) = \sum_{i=0}^{m} Q\left((X)_{\geq\alpha_i}\right) \cdot (\alpha_i - \alpha_{i+1}) \tag{3}$$

where $\alpha_0 = 1, \alpha_{m+1} = 0$ and $\alpha_1 \geq \ldots \geq \alpha_m$ denote the membership values in descending order of the elements in U to the fuzzy set X.

Example 3. Imagine a quantified expression such as "about half of people are tall". Let $about\_half : \wp(U) \to [0,1]$ be the semi-fuzzy quantifier depicted in example 1 and let X be the fuzzy set: $X = \{0.9/u_1, 0.8/u_2, 0.1/u_3, 0/u_4\}$. The next table shows the values produced by the semi-fuzzy quantifier $about\_half$ at all α_i cut levels:

	$(X)_{\geq\alpha_i}$	$about\_half\left((X)_{\geq\alpha_i}\right)$
$\alpha_0 = 1$	$\varnothing$	$about\_half(\varnothing) = 0$
$\alpha_1 = 0.9$	$\{u_1\}$	$about\_half(\{u_1\}) = 0$
$\alpha_2 = 0.8$	$\{u_1, u_2\}$	$about\_half(\{u_1, u_2\}) = 1$
$\alpha_3 = 0.1$	$\{u_1, u_2, u_3\}$	$about\_half(\{u_1, u_2, u_3\}) = 0$
$\alpha_4 = 0$	$\{u_1, u_2, u_3, u_4\}$	$about\_half(\{u_1, u_2, u_3, u_4\}) = 0$

Applying (3):

$$\begin{aligned}(F(about\_half))(X) &= about\_half((X)_{\geq 1}) \cdot (1 - 0.9) + \\ &\quad about\_half((X)_{\geq 0.9}) \cdot (0.9 - 0.8) + \\ &\quad about\_half((X)_{\geq 0.8}) \cdot (0.8 - 0.1) + \\ &\quad about\_half((X)_{\geq 0.1}) \cdot (0.1 - 0) + \\ &\quad about\_half((X)_{\geq 0}) \cdot (0 - 0) \\ &= 0.7\end{aligned}$$

This is a coherent result taking into account the definition of the fuzzy set X, where the degree of membership for the elements u_1 and u_2 is very high (0.9 and 0.8 respectively) whereas the degree of membership for u_3 and u_4 is very low (0.1 and 0 respectively). As a consequence, it is likely that about half ot the individuals are actually tall.

3.1 Query Language

Given a set of indexing terms $\{t_1, \ldots, t_m\}$ and a set of quantification symbols $\{Q_1, \ldots, Q_k\}$, query expressions are built as follows: a) any indexing term t_i belongs to the language, b) if e_1 belongs to the language then, NOT e_1 and (e_1) also belong to the language, c) if e_1 and e_2 belong to the language then, e_1 AND e_2 and e_1 OR e_2 also belong to the language and d) if $e_1, e_2, \ldots, e_n$ belong to the language then, $Q_i(e_1, e_2, \ldots, e_n)$ also belongs to the language, where Q_i is a quantification symbol.

Example 4. Given an alphabet of terms $\{a, b, c, d\}$ and the set of quantification symbols $\{most\}$ the expression b AND $most(a, c,$ NOT $c)$ is a syntactically valid query expression.

The range of linguistic quantifiers available determines how flexible the query language is.

3.2 Semantics

Given a query expression q, its associated fuzzy set of documents is denoted by $Sm(q)$. Every indexing term t_i is interpreted by a fuzzy set of documents, $Sm(t_i)$, whose membership function can be computed following classical IR weighting formulas, such as the popular tf/idf method [25]. Given the fuzzy set defined by every individual query term, the fuzzy set representing a Boolean query can be directly obtained applying operations between fuzzy sets.

Given a quantified sentence with the form $Q(e_1, \ldots, e_r)$, where Q is a quantification symbol and each e_i is an expression of the query language, we have to articulate a method for combining the fuzzy sets $Sm(e_1), \ldots, Sm(e_r)$ into a single fuzzy set of documents, $Sm(Q(e_1, \ldots, e_r))$, representing the quantified sentence as a whole.

First, we associate a semi-fuzzy quantifier with every quantification symbol in the query language. For instance, we might include the quantification symbol $about\_half$ in the query language which associated to a semi-fuzzy quantifier similar to the one depicted in example 1[4]. Given a quantification syntactic symbol Q, by Q_s we refer to its associated semi-fuzzy quantifier. Given a QFM, F, $F(Q_s)$ denotes the fuzzy quantifier obtained from Q_s by fuzzification.

Let d_j be a document and $Sm(e_i)$ the fuzzy sets induced by the components of the quantified expression, we can define the fuzzy set C_{d_j}, which represents how much d_j satisfies the individual components of the quantified statement:

$$C_{d_j} = \{\mu_{Sm(e_1)}(d_j)/1, \mu_{Sm(e_2)}(d_j)/2, \ldots, \mu_{Sm(e_r)}(d_j)/r\} \tag{4}$$

[4] Although many times the name of the quantification symbol is the same as the name of the semi-fuzzy quantifier used to handle the linguistic expression, both concepts should not be confused.

From these individual degrees of fulfilment, the expression $Q(e_1, \ldots, e_r)$ can be evaluated by means of the fuzzy quantifier $F(Q_s)$:

$$\mu_{Sm(Q(e_1,\ldots,e_r))}(d_j) = (F(Q_s))(C_{d_j}) \tag{5}$$

For instance, if Q_s is a semi-fuzzy quantifier *about_half* then a document will be assigned a high evaluation score if it has a high degree of membership for about half of the quantifier components and low degrees of membership for the rest of the components.

3.3 Example

Consider the query expression $at\_least\_3(a, b, c, d, e)$ and a document d_j whose degrees of membership in the fuzzy sets defined by each indexing term are: $\mu_{Sm_a(d_j)} = 0$, $\mu_{Sm_b(d_j)} = 0.15$, $\mu_{Sm_c(d_j)} = 0.2$, $\mu_{Sm_d(d_j)} = 0.3$ and $\mu_{Sm_e(d_j)} = 0.4$.

The fuzzy set induced by d_j from the components of the query expression is: $C_{d_j} = \{0/1, 0.15/2, 0.2/3, 0.3/4, 0.4/5\}$.

Consider that we use the following crisp semi-fuzzy quantifier for implementing the quantification symbol *at_least_3*.

$$at\_least\_3 : \wp(U) \rightarrow [0, 1]$$

$$at\_least\_3(X) = \begin{cases} 0 & \text{if } |X| < 3 \\ 1 & \text{otherwise} \end{cases}$$

Now, several crisp representatives of C_{d_j} are obtained from subsequent α-cuts and the semi-fuzzy quantifier *at_least_3* is applied on every crisp representative:

	$(C_{d_j})_{\geq\alpha_i}$	$at\_least\_3\left((C_{d_j})_{\geq\alpha_i}\right)$
$\alpha_0 = 1$	$\varnothing$	$at\_least\_3(\varnothing) = 0$
$\alpha_1 = 0.4$	$\{e\}$	$at\_least\_3(\{e\}) = 0$
$\alpha_2 = 0.3$	$\{d, e\}$	$at\_least\_3(\{d, e\}) = 0$
$\alpha_3 = 0.2$	$\{c, d, e\}$	$at\_least\_3(\{c, d, e\}) = 1$
$\alpha_4 = 0.15$	$\{b, c, d, e\}$	$at\_least\_3(\{b, c, d, e\}) = 1$
$\alpha_5 = 0$	$\{a, b, c, d, e\}$	$at\_least\_3(\{a, b, c, d, e\}) = 1$

And it follows that

$$(F(at\_least\_3))(C_{d_j}) = 0 \cdot 0.6 + 0 \cdot 0.1 + 0 \cdot 0.1 + 1 \cdot 0.05 + \\ +1 \cdot 0.15 + 1 \cdot 0 = 0.2$$

$$\mu_{Sm(at\_least\_3(a,b,c,d,e))}(d_j) = 0.2$$

Indeed, it is unlikely that at least three out of the five query terms are actually related to document d_j because all query terms have low degrees of membership in C_{d_j}.

4 Semi-fuzzy Quantifiers and OWA Quantification

In [19], the modeling of linguistic quantifiers was approached by semi-fuzzy quantifiers and quantifier fuzzification mechanisms (equation (3)) because: 1) this approach subsumes the fuzzy quantification model based on OWA (the OWA method is equivalent to the mechanism defined in equation (3) for increasing unary quantifiers [5,7]) and 2) it has been shown that OWA models [31, 32] do not comply with fundamental properties [1, 9] when dealing with n-ary quantifiers. These problems are not present in the SFQ-based approach defined in [8].

In this section, we enter into details about these issues and, in particular, we show how the implementation of linguistic quantifiers through OWA operators is equivalent to a particular case of the SFQ-based framework. This is a good property of the SFQ approach because seminal models of fuzzy quantification for IR [2], which are based on OWA operators, can be implemented and tested under the SFQ framework. Note that we refer here to the OWA-based unary quantification approach [33]. Although alternative OWA formulations have been proposed in the literature, a thorough study of the role of these alternatives for quantification is out of the scope of this work.

4.1 Linguistic Quantification using OWA Operators

OWA operators [30] are mean fuzzy operators whose results lie between those produced by a fuzzy MIN operator and those yielded by a fuzzy MAX operator.

An ordered weighted averaging (OWA) operator of dimension n is a non linear aggregation operator: OWA: $[0,1]^n \to [0,1]$ with a weighting vector $W = [w_1, w_2, \ldots, w_n]$ such that:

$$\sum_{i=1}^{n} w_i = 1, w_i \in [0,1]$$

and

$$\text{OWA}(x_1, x_2, \ldots, x_n) = \sum_{i=1}^{n} w_i \cdot Max_i(x_1, x_2, \ldots, x_n)$$

where $Max_i(x_1, x_2, \ldots, x_n)$ is the i-th largest element across all the x_k, e.g. $Max_2(0.9, 0.6, 0.8)$ is 0.8.

The selection of particular weighting vectors W allows the modeling of different linguistic quantifiers (e.g. *at least*, *most of*, etc.).

Given a quantified expression $Q(e_1, \ldots, e_r)$ and a document d_j, we can apply OWA quantification for aggregating the importance weights for the selection conditions e_i. Without loss of generality, these weights will be denoted here as $\mu_{Sm(e_i)}(d_j)$. Formally, the evaluation score produced would be:

$$\text{OWA}_{op}(\mu_{Sm(e_1)}(d_j), \ldots, \mu_{Sm(e_r)}(d_j)) = \sum_{i=1}^{r} w_i \cdot Max_i(\mu_{Sm(e_1)}(d_j), \ldots, \mu_{Sm(e_r)}(d_j)) \tag{6}$$

where OWA_{op} is an OWA operator associated with the quantification symbol Q.

Following the modelling of linguistic quantifiers via OWA operators [2,4], the vector weights w_i associated to the $\text{OWA}_{operator}$ operator are defined from a monotone non-decreasing relative fuzzy number $FN : [0,1] \to [0,1]$ as follows:

$$w_i = FN(i/r) - FN((i-1)/r), i : 1, \ldots, r \tag{7}$$

The fuzzy numbers used in the context of OWA quantification are *coherent*. This means that it is guaranteed that $FN(0) = 0$ and $FN(1) = 1$.

Without loss of generality, we can denote $Max_1(\mu_{Sm(e_1)}(d_j), \mu_{Sm(e_2)}(d_j), \ldots, \mu_{Sm(e_r)}(d_j))$ as α_1, $Max_2(\mu_{Sm(e_1)}(d_j), \mu_{Sm(e_2)}(d_j), \ldots, \mu_{Sm(e_r)}(d_j))$ as α_2, etc. and the evaluation value equals:

$$\sum_{i=1}^{r} w_i \cdot \alpha_i = \sum_{i=1}^{r} (FN(i/r) - FN((i-1)/r)) \cdot \alpha_i \tag{8}$$

This equation depicts the evaluation score produced by an OWA operator. In the next section we show that an equivalent result can be obtained within the SFQ framewok if particular semi-fuzzy quantifiers are selected.

4.2 Linguistic Quantification using SFQ

Recall that, given a quantified expression $Q(e_1, \ldots, e_r)$ and a document d_j, the evaluation scored computed following the SFQ approach is:

$$\mu_{Sm(Q(e_1,\ldots,e_r))}(d_j) = (F(Q_s))(C_{d_j})$$

A key component of this approach is the quantifier fuzzification mechanism F, whose discrete definition (equation 3) is repeated here for the sake of clarity:

$$(F(Q))(X) = \sum_{i=0}^{m} Q\left((X)_{\geq \alpha_i}\right) \cdot (\alpha_i - \alpha_{i+1})$$

where $\alpha_0 = 1, \alpha_{m+1} = 0$ and $\alpha_1 \geq \ldots \geq \alpha_m$ denote the membership values in descending order of the elements in the fuzzy set X.

Putting all together:

$$\mu_{Sm(Q(e_1,\ldots,e_r))}(d_j) = \sum_{i=0}^{r} Q_s\left(\left(C_{d_j}\right)_{\geq \alpha_i}\right) \cdot (\alpha_i - \alpha_{i+1}) \tag{9}$$

Without loss of generality, we will assume that the e_i terms are ordered in decreasing order of membership degrees in C_{d_j}, i.e. $\mu_{C_{d_j}}(e_1) = \alpha_1 \geq \mu_{C_{d_j}}(e_2) = \alpha_2 \ldots \geq \mu_{C_{d_j}}(e_r) = \alpha_r$. Note also that equation 9 stands on a sequence of successive α-cuts on the fuzzy set C_{d_j}. The first cut (α_0) is done at the membership level 1 and the last cut (α_r) is performed at the level 0. This means that the equation can be rewritten as:

$$\mu_{Sm(Q(e_1,\ldots,e_r))}(d_j) = \sum_{i=0}^{r} Q_s\left(CS_i\right) \cdot (\alpha_i - \alpha_{i+1}) \tag{10}$$

where $CS_0 = \varnothing$ and $CS_i = \{e_1, \ldots, e_i\}$, $i = 1, \ldots, r$.

The equation can be developed as:

$$\begin{aligned}
\mu_{Sm(Q(e_1,\ldots,e_r))}(d_j) &= \sum_{i=0}^{r} Q_s\left(CS_i\right) \cdot (\alpha_i - \alpha_{i+1}) \\
&= Q_s(\varnothing) \cdot (1 - \alpha_1) + \\
&\quad Q_s(\{e_1\}) \cdot (\alpha_1 - \alpha_2) + \ldots + \\
&\quad Q_s(\{e_1, e_2, \ldots, e_r\}) \cdot \alpha_r \\
&= Q_s(\varnothing) + \alpha_1 \cdot (Q_s(\{e_1\}) - Q_s(\varnothing)) + \\
&\quad \alpha_2 \cdot (Q_s(\{e_1, e_2\}) - Q_s(\{e_1\})) + \ldots + \\
&\quad \alpha_r \cdot (Q_s(\{e_1, e_2, \ldots, e_r\}) - Q_s(\{e_1, e_2, \ldots, e_{r-1}\}))
\end{aligned} \tag{11}$$

The unary semi-fuzzy quantifier Q_s can be implemented by means of a fuzzy number as follows: $Q_s(CS_i) = FN(|CS_i|/r)$. Hence, the previous equation can be rewritten in the following way:

$$\begin{aligned}
\mu_{Sm(Q(e_1,\ldots,e_r))}(d_j) &= \sum_{i=0}^{r} Q_s\left(CS_i\right) \cdot (\alpha_i - \alpha_{i+1}) \\
&= FN(0) + \alpha_1 \cdot (FN(1/r) - FN(0)) + \\
&\quad \alpha_2 \cdot (FN(2/r) - FN(1/r)) + \ldots + \\
&\quad \alpha_r \cdot (FN(1) - FN((r-1)/r))
\end{aligned} \tag{12}$$

It is straightforward that we can replicate the OWA-based evaluation (equation 8) if we select a SFQ whose associated fuzzy number is the same as the one used in the OWA equation. Note that, $FN(0) = 0$ provided that the fuzzy number is coherent.

4.3 Remarks

Given a query and a document d_j, the application of SFQ for IR proposed in [19] involves a single fuzzy set, C_{d_j}. In these cases, as shown in the last section, equivalent evaluation results can be obtained by an alternative OWA formulation[5]. This means that the advantages shown empirically for the SFQ framework can be directly extrapolated to OWA-based approaches such as the one designed in [2]. This is good because the evaluation results apply not only for a particular scenario but for other well-known proposals whose practical behaviour for large document collections was unclear.

Nevertheless, some counterintuitive problems have been described for OWA operators when handling expressions involving several fuzzy sets. We offer now additional details about these problems and we sketch their implications in the context of IR. A thorough comparative between different fuzzy operators can be found in [1,9].

One of the major drawbacks of OWA's method is its nonmonotonic behaviour for propositions involving two properties [1]. This means that, given two quantifiers Q_1, Q_2 such that Q_1 is more specific than Q_2[6], it is not assured that the application of the quantifiers for handling a quantified proposition maintains specificity. This is due to the assumption that any quantifier is a specific case of OWA interpolation between two extreme cases: the existential quantifier and the universal quantifier. Let us illustrate this with an example. Consider two quantifiers *at_least_60%* and *at_least_80%* and two fuzzy sets of individuals representing the properties of being blonde and tall, respectively. Obviously, *at_least_80%* should produce an evaluation score which is less than or equal than the score produced by *at_least_60%*. Unfortunately, the evaluation of a expression such as *at_least_80% blondes are tall* does not necessarily produce a value which is less or equal than the value obtained from *at_least_60% blondes are tall*. This means that, given two fuzzy sets *blondes* and *tall*, it is possible that these sets are better at satisfying the expression *at_least_80% blondes are tall* than satisfying the expression *at_least_60% blondes are tall*. This is clearly unacceptable.

This is also problematic for the application in IR. Imagine two quantifiers such that Q_1 is more specific than Q_2. This means that Q_1 is more restrictive than Q_2 (e.g. a crisp *at_least_5* vs a crisp *at_least_3*). The application of these quantifiers for handling expressions with the form Q_i $A's$ are $B's$ cannot be faced using OWA operators. This is an important limitation because it prevents the extension of the fuzzy approach in a number of ways. For instance, expressions such as *most* t_i *are* t_k, where t_i and t_k are terms, can be used to determine whether or not most documents dealing with t_i are also related

[5] That is, the SFQ formulation is equivalent to the OWA formulation for monotonic unary expressions.

[6] Roughly speaking, if Q_1 is more specific than Q_2 then for all the elements of the domain of the quantifier the value produced by Q_1 is less or equal than the value produced by Q_2.

to t_k. In general, statements with this form involving several fuzzy sets are promising for enhancing the expressiveness of IR systems in different tasks.

These problems are not present for the fuzzification mechanisms defined in [8], which stand on the basis of the SFQ-based framework. This fact and the intrinsic generality of the SFQ-based approach are convenient for the purpose of IR.

5 Experiments

The behaviour of the extended fuzzy query language has been evaluated empirically. This experimental research task is fundamental in determining the actual benefits that retrieval engines might obtain from linguistic quantifiers. The empirical evaluation presented in this section expands the experimentation carried out in [19]. In particular, only a basic tf/idf weighting scheme was tested in [19]. We report here performance results for evolved weighting approaches. Our hypothesis is that these weighting methods, which have traditionally performed very well in the context of popular IR models, might increase the absolute performance attainable by the fuzzy approach. The results of the experimentation conducted in [19] are also shown here because we want to check whether or not the same trends hold when different weighting schemes are applied.

The experimental benchmark involved the Wall Street journal (WSJ) corpora from the TREC collection, which contains about 173000 news articles spread over six years (total size: 524 Mb), and 50 topics from TREC-3 [12] (topics #151-#200). Common words were removed from documents and topics[7] and Porter's stemmer [22] was applied to reduce words to their syntactical roots. The inverted file was built with the aid of GNU mifluz [20], which supplies a C++ library to build and query a full text inverted index.

As argued in Section 3.2, every indexing term t_i is interpreted as a fuzzy set of documents, $Sm(t_i)$, whose membership function can be computed following classical IR weighting formulas. In [19], a normalized version of the popular tf/idf weighting scheme was applied as follows. Given a document d_j, its degree of membership in the fuzzy set defined by a term t_i is defined as:

$$\mu_{Sm(t_i)}(d_j) = \frac{f_{i,j}}{max_k f_{k,j}} * \frac{idf(t_i)}{max_l idf(t_l)} \qquad (13)$$

In the equation, $f_{i,j}$ is the raw frequency of term t_i in the document d_j and $max_k f_{k,j}$ is the maximum raw frequency computed over all terms which are mentioned by the document d_j. By $idf(t_i)$ we refer to a function computing an inverse document frequency factor[8]. The value $idf(t_i)$ is divided

[7] The stoplist was composed of 571 common words.

[8] The function used in [19] was $idf(t_i) = log(max_l n_l / n_i)$, where n_i is the number of documents in which the term t_i appears and the maximum $max_l n_l$ is computed

by $max_l idf(t_l)$, which is the maximum value of the function *idf* computed over all terms in the alphabet. Note that $\mu_{Sm(t_i)}(d_j) \in [0,1]$ because both the *tf* and the *idf* factors are divided by its maximum possible value.

Although the basic *tf/idf* weighting was very effective on early IR collections, it is now accepted that this classic weighiting method is non-optimal [26]. The characteristics of present datasets required the development of methods to factor document length into term weights. In this line, pivoted normalization weighting [28] is a high-performance method which has demonstrated its merits in exhaustive TREC experimentations [26]. It is also especially remarkable that pivot-based approaches are also competitive for web retrieval purposes [13, 27, 29]. As a consequence, it is important to check how this effective weighting scheme works in the context of the SFQ-based method. Furthermore, a comparison between the fuzzy model powered by pivoted weigths and a high performance pivot-based IR retrieval method will also help to shed light on the adequacy of the fuzzy approach to enhance retrieval engines. More specifically, we will compare the fuzzy model against the inner product matching function of the vector-space model with document term weights computed using pivoted normalization.

The fuzzy set of documents induced by every individual query term can be defined using pivoted document length as follows:

$$\mu_{Sm(t_i)}(d_j) = \frac{\frac{1+ln(1+ln(f_{i,j}))}{(1-s)+s\frac{dl_j}{avg_{dl}}}}{norm\_1} * \frac{qtf_i}{max_l qtf_l} * \frac{ln(\frac{N+1}{n_i})}{norm\_2} \tag{14}$$

where $f_{i,j}$ is the raw frequency of term t_i in the document d_j, s is a constant (the pivot) in the interval $[0,1]$, dl_j is the length of document d_j, avg_{dl} is the average document length, qtf_i is the frequency of term t_i in the query and $max_l qtf_l$ is the maximum term frequency in the query. The value N is the total number of documents in the collection and n_i is the number of documents which contain the term t_i. The normalizing factors $norm\_1$ and $norm\_2$ are included to maintain $\mu_{Sm(t_i)}(d_j)$ between 0 and 1. In the experiments reported here $norm\_1$ is equal to $\frac{1+\ln(1+\ln(maxdl)))}{(1-s)}$ ($\max dl$ is the size of the largest document) and $norm\_2$ is equal to $\ln(N+1)$. This formula arises straightforwardly from the pivot-based expression detailed in [26].

The rationale behind both equations (13) and (14), is that t_i will be a good representative for documents with high degree of membership in $Sm(t_i)$ whereas t_i poorly represents the documents with low degree of membership in $Sm(t_i)$. Note that, it is not guaranteed that there exists a document d_j such that $\mu_{Sm(t_i)}(d_j) = 1$. Indeed, the distribution of the values $\mu_{Sm(t_i)}(d_j)$ depends largely on the characteristics of the document collection[9]. Anyway,

over all terms in the indexing vocabulary. The same function has been used in the new experiments reported here.

[9] For instance, in eq. (13) this will only happen if the term(s) that appear(s) the largest number of times within the document is/are also the most infrequent one(s) across the whole collection.

for medium/large collections, such as WSJ, most $\mu_{Sm(t_i)}(d_j)$ values tend to be small. We feel that the large success of tf/idf weighting schemes and their evolved variations in the context of IR is a solid warranty for its application in the context of the SFQ framework. Other mathematical shapes could have been taken into account for defining a membership function. Nevertheless, these membership definitions are convenient because, as sketched in the next paragraphs, the SFQ framework can thus handle popular IR methods as particular cases.

For both weighing methods (equations 13 and 14) we implemented a baseline experiment by means of a linear fuzzy quantified sentence Q_{lin}, whose associated semi-fuzzy quantifier is:

$$Q_{lin} : \wp(U) \rightarrow [0,1]$$
$$Q_{lin}(X) = \frac{|X|}{|U|}$$

Terms are collected from the TREC topic and, after stopword and stemming, a fuzzy query with the form $Q_{lin}(t_1, \ldots, t_n)$ is built. It can be easily proved that the ranking produced from such a query is equivalent to the one generated from the inner product matching function in the vector-space model [25]. The details can be found in appendix A. This is a good property of the fuzzy approach because it can handle popular IR retrieval methods as particular cases.

5.1 Experiments: *tf/idf*

The first pool of experiments considered only terms from the topic title. In order to check whether *non-linear* quantifiers are good in terms of retrieval performance, relaxed versions of *at least* quantifiers were implemented. For example, a usual crisp implementation of an *at least 6* quantifier (left-hand side) and its proposed relaxation (right-hand side) can be defined as:

$$at\_least\_6 : \wp(U) \rightarrow [0,1]$$
$$at\_least\_6(X) = \begin{cases} 0 & \text{if } |X| < 6 \\ 1 & \text{otherwise} \end{cases}$$

$$at\_least\_6 : \wp(U) \rightarrow [0,1]$$
$$at\_least\_6(X) = \begin{cases} (10/6) * (|X|/10)^2 & \text{if } |X| < 6 \\ |X|/10 & \text{otherwise} \end{cases}$$

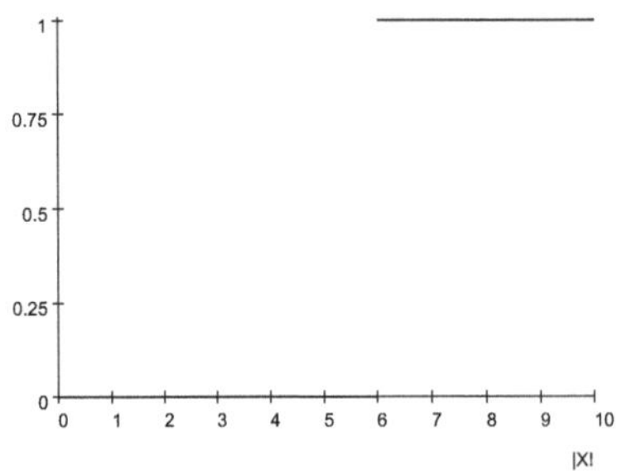

a) crisp definition

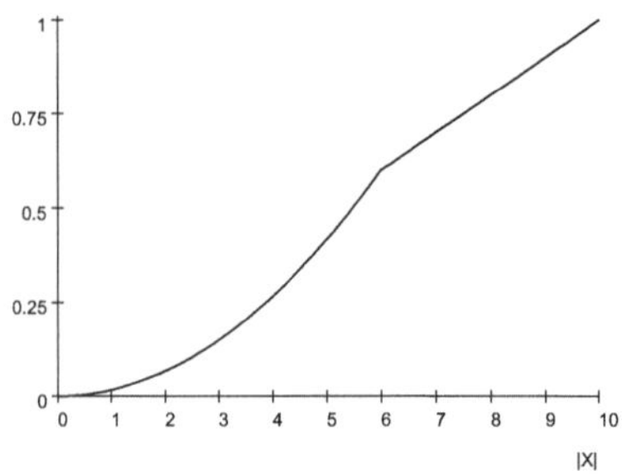

b) relaxed definition

The crisp *at least* implementation is too rigid to be applied in IR. It is not fair to consider that a document matching 9 query terms is as good as one matching only 6 terms. On the other hand, it is too rigid to consider that a document matching 0 query terms is as bad as one matching 5 query terms. The intuitions behind *at least* quantifiers can be good for retrieval purposes if implemented in a relaxed form. In particular, intermediate implementations, between a classical *at least* and a linear implementation (which is typical in popular IR matching functions, as shown above), were proposed and tested in [19]. Non-relevant documents might match a few query terms simply by chance. To minimize this problem the relaxed formulation makes that documents matching few terms (less than 6 for the example depicted above) receive a lower score compared to an alternative linear implementation. On the other hand, unlike the rigid *at least* implementation, documents matching many terms (more than 6 for the example) receive a score that grows linearly with the number of those terms.

The first set of results, involving the baseline experiment ($Q_{lin}(t_1, t_2, \ldots, t_n)$) and several *at least* formulations, are shown in Table 1. The *at least* quantifiers were relaxed in the form shown in the example above. Although topic titles consist typically of very few terms, the outcome of these experiments clearly shows that flexible query formulations can lead to significant improvements in retrieval performance. There is a steady increment of performance across all recall levels and for *at_least_x* with $x \geq 8$ the performance values became stabilized.

The next pool of experiments used all topic subfields (Title, Description & Narrative). Different strategies were tested in order to produce fuzzy queries from topics. For all experiments, every subfield is used for generating a single fuzzy quantifier and the fuzzy query is the conjunction of these quantifiers.

Table 1. Effect of simple *at least* queries on retrieval performance

		at_least_k						
Recall	Q_{lin}	$k = 2$	$k = 3$	$k = 4$	$k = 5$	$k = 6$	$k = 7$	$k = 8$
0.00	0.5979	0.6165	0.6329	0.6436	0.6568	0.6580	0.6582	0.6597
0.10	0.4600	0.4776	0.4905	0.4968	0.5019	0.5036	0.5035	0.5037
0.20	0.3777	0.3997	0.4203	0.4208	0.4243	0.4251	0.4253	0.4253
0.30	0.3092	0.3336	0.3454	0.3479	0.3486	0.3483	0.3483	0.3483
0.40	0.2430	0.2680	0.2751	0.2805	0.2792	0.2786	0.2784	0.2784
0.50	0.1689	0.2121	0.2191	0.2234	0.2228	0.2226	0.2226	0.2226
0.60	0.1302	0.1592	0.1704	0.1774	0.1772	0.1768	0.1770	0.1770
0.70	0.0853	0.1100	0.1215	0.1261	0.1267	0.1269	0.1273	0.1273
0.80	0.0520	0.0734	0.0855	0.0888	0.0892	0.0892	0.0892	0.0892
0.90	0.0248	0.0428	0.0467	0.0497	0.0496	0.0495	0.0495	0.0495
1.00	0.0034	0.0070	0.0107	0.0106	0.0105	0.0105	0.0105	0.0105
Avg.prec. (non-interpolated)	0.2035	0.2241	0.2362	0.2403	0.2409	0.2410	0.2410	0.2411
% change		+10.12%	+16.07%	+18.08%	+18.4%	+18.4%	+18.4%	+18.5%

TREC topic:
⟨title⟩ Topic: Vitamins – The Cure for or Cause of Human Ailments
⟨desc⟩ Description:
Document will identify vitamins that have contributed to the cure for human diseases or ailments or documents will identify vitamins that have caused health problems in humans.
⟨narr⟩ Narrative:
A relevant document will provide information indicating that vitamins may help to prevent or cure human ailments. Information indicating that vitamins may cause health problems in humans is also relevant. A document that makes a general reference to vitamins such as "good for your health" or "having nutritional value" is not relevant. Information about research being conducted without results would not be relevant. References to derivatives of vitamins are to be treated as the vitamin.
Fuzzy query:
at_least_4(vitamin,cure,caus,human,ailment) ∧ at_least_4(document,identifi, vitamin,contribut,cure,human,diseas,ailment,caus,health,problem) ∧ at_least_3(relevant,document,provid,inform,indic,vitamin,prevent,cure, human,ailment,caus,health,problem,make,gener,refer,good, nutrit,research,conduct,result,deriv,treat)

Fig. 1. Fuzzy query from a TREC topic.

Figure 1 exemplifies the articulation of fuzzy queries from a TREC topic[10]. This simple method allows to obtain fuzzy representations from TREC topics in an automatic way. This advances that fuzzy query languages might be adequate not only to assist users when formulating their information needs but also to transform textual queries into fuzzy expressions.

We tested several combinations of *at least* and linear quantifiers. For implementing the conjunction connective both the fuzzy MIN operator and the product operator were applied. Performance results are summarized in Tables 2 (MIN operator) and 3 (product operator). In terms of average precision, the product operator is clearly better than the MIN operator to implement the boolean AND connective. Indeed, all the columns in Table 3 depict better performance ratios than their respective columns in Table 2.

On the other hand, the combination of linear quantifiers is clearly inferior to the combination of *at_least_x* quantifiers. There is a progressive improvement in retrieval performance as the value of x grows from 2 to 8. This happens independently of the operator applied for implementing the conjunction. Performance becomes stabilized for values of x around 8. It is important to emphasize that a combination of linear quantifiers is not a common characteristic of popular IR approaches, where a single linear operation is usually applied over all topic terms. As a consequence, the comparison presented in

[10] We use the symbol ∧ to refer to the Boolean AND connective.

Table 2. Conjunctions between quantifiers - MIN operator

	Q_{lin}(**title terms**) $\wedge$ Q_{lin}(**desc terms**) $\wedge$ Q_{lin}(**narr terms**)	**at_least_x(title terms)** $\wedge$ **at_least_x(desc terms)** $\wedge$ **at_least_x(narr terms)**			
Recall		$x = 2$	$x = 3$	$x = 4$	$x = 8$
0.00	0.6822	0.6465	0.6642	0.6917	0.7577
0.10	0.4713	0.4787	0.4739	0.4804	0.5290
0.20	0.3839	0.3803	0.4011	0.4080	0.4408
0.30	0.3071	0.3132	0.3236	0.3283	0.3371
0.40	0.2550	0.2621	0.2671	0.2720	0.2722
0.50	0.2053	0.2127	0.2190	0.2221	0.2256
0.60	0.1557	0.1457	0.1578	0.1613	0.1709
0.70	0.1053	0.1117	0.1146	0.1192	0.1311
0.80	0.0641	0.0685	0.0744	0.0788	0.0849
0.90	0.0397	0.0440	0.0436	0.0403	0.0430
1.00	0.0060	0.0097	0.0140	0.0142	0.0171
Avg.prec. (non-interpolated)	0.2225	0.2232	0.2282	0.2321	0.2481
% change		+0.3%	+2.6%	+4.3%	+11.5%

Tables 2 and 3 aims at checking the effect of *at least* quantifiers vs linear quantifiers within the SFQ fuzzy approach and, later on, we will compare the best SFQ results with a classic approach in which a linear quantifier is applied over all topic terms.

Experiments using averaging-like operators (such as the ones tested by Lee and others in [17,18]) for implementing the boolean conjunction were also run but further improvements in performance were not obtained. This might indicate that, although T-norm operators (e.g. MIN and product) worked bad

Table 3. Conjunctions between quantifiers - product operator

	Q_{lin}(**title terms**) $\wedge$ Q_{lin}(**desc terms**) $\wedge$ Q_{lin}(**narr terms**)	**at_least_x(title terms)** $\wedge$ **at_least_x(desc terms)** $\wedge$ **at_least_x(narr terms)**			
Recall		$x = 2$	$x = 3$	$x = 4$	$x = 8$
0.00	0.7277	0.7473	0.7311	0.7375	0.7664
0.10	0.5513	0.5524	0.5576	0.5542	0.5991
0.20	0.4610	0.4665	0.4711	0.4671	0.4769
0.30	0.3608	0.3802	0.3869	0.3830	0.3983
0.40	0.2915	0.3142	0.3133	0.3154	0.3172
0.50	0.2428	0.2684	0.2638	0.2643	0.2660
0.60	0.1857	0.2069	0.2111	0.2077	0.2136
0.70	0.1160	0.1407	0.1496	0.1531	0.1561
0.80	0.0720	0.0882	0.0932	0.0972	0.1014
0.90	0.0431	0.0522	0.0559	0.0609	0.0634
1.00	0.0067	0.0089	0.0126	0.0136	0.0157
Avg.prec. (non interpolated)	0.2572	0.2722	0.2760	0.2750	0.2849
% change		+5.8%	+7.3%	+6.9%	+10.8%

Table 4. Linear quantifier query vs more evolved query

Recall	Q_{lin}(**title, desc & narr terms**)	**at_least_8(title terms)** $\wedge$ **at_least_8(desc terms)** $\wedge$ **at_least_8(narr terms)**
0.00	0.6354	0.7664
0.10	0.4059	0.5991
0.20	0.3188	0.4769
0.30	0.2382	0.3983
0.40	0.1907	0.3172
0.50	0.1383	0.2660
0.60	0.0885	0.2136
0.70	0.0530	0.1561
0.80	0.0320	0.1014
0.90	0.0158	0.0634
1.00	0.0019	0.0157
Avg.prec.	0.1697	0.2849
% change		+67.9%

to combine terms within conjunctive boolean representations [17, 18], they could play an important role to combine more expressive query components, such as quantifiers.

In order to conduct a proper comparison against popular IR methods, an additional baseline experiment was carried out. In this test, all terms from all topic subfields were collected into a single linear quantifier. Recall that this is equivalent to the popular vector-space model with the inner product matching function (appendix A). The results obtained are compared to the previous best results in Table 4. The application of relaxed non-linear quantifiers leads to very significant improvements in retrieval performance. Clearly, a linear strategy involving all topic terms is not the most appropriate way to retrieve documents. On the other hand, expressive query languages provide us with tools to capture topic's contents in a better way. In particular, our evaluation shows clearly that non-linear fuzzy quantifiers are appropriate for enhancing search effectiveness. For example, *at least* quantifiers appear as powerful tools to establish additional requirements for a document to be retrieved. Although the combination of linear quantifiers (e.g. Table 3, col. 2) outperforms significantly the single linear quantifier approach (Table 4, col. 2), it is still clear that a Boolean query language with linear quantifiers is not enough because further benefits are obtained when *at least* quantifiers are applied (e.g. Table 3, cols. 3–5).

5.2 Experiments: Pivoted Document Length Normalization

It is well known that the classic tf/idf weighting approach is nowadays overcome by weighting schemes based on document length corrections [26]. Thus, the actual impact of the SFQ fuzzy approach can only be clarified after a proper comparison against state-of-the-art matching functions. Moreover, there is practical evidence on the adequacy of pivoted weights for web retrieval purposes [13, 27, 29] and, hence, the comparison presented in this section will help to shed light on the role of fuzzy quantifiers to enhance web retrieval engines.

We have run additional experiments for evaluating the SFQ approach with pivot-based weighting methods (equation 14). For the sake of brevity, we will not report here every individual experiment but we will summarize the main experimental findings. Our discussion will be focused on tests using all topic subfields because the inner product matching function (baseline experiment with linear quantifier) yields its top performance when applied to all topic subfields. Indeed, as expected, the performance of the baseline experiment is substantially better than the tf/idf baseline (Table 5, column 2 vs Table 4, column 2).

The following enumeration sketches the main conclusions from the new pool of tests:

1. Again, the product operator is better than the MIN operator to implement the boolean AND connective.
2. The fuzzy approach with relaxed *at least* statements was not able to produce better performance results than the inner product matching function (baseline).
3. The fuzzy approach with a linear quantifier applied on every individual topic subfield (whose results are combined with the product operator) is able to produce modest improvements with respect to the baseline.

The pivot constant s was fixed to the value of 0.2[11]. The main performance results are shown in Table 5.

Further research is needed to determine the actual role of *at least* statements in the context of a high performance weighting technique such as pivoted document length normalization. At this point, the aplication of relaxed *at least* expressions produced performance results which are worse than those obtained for the baseline. As depicted in Table 5, the overall performance gets worse as the *at least* statement becomes stricter. An *at_least_2* statement is slightly worse than the baseline (1.5% worse) but the *at_least_8* formulation yields significantly worse performance ratios (average precision decreases by 15%). In the near future we plan to make extensive testing on different relaxations of *at least* formulations in order to shed light on this issue.

On the contrary, the fuzzy model with linear quantifiers was able to overcome the baseline. Although the baseline experiment follows a high performance state-of-the-art IR retrieval technique (inner product matching function of the vector-space model with pivoted document length normalized

[11] Some tests with varying values of s were run for the fuzzy model (with both linear & *at least* statements) but no improvements were found. The baseline performance is also optimal for the value of 0.2. Indeed, the ideal value of the pivot s has also been considered very stable in previous experimentations on pivoted document length normalization schemes [26].

Table 5. Experimental results - Pivoted document length normalization

	Q_{lin}(**title, desc & narr terms**) (baseline)	Q_{lin}(**title terms**) $\wedge$ Q_{lin}(**desc terms**) $\wedge$ Q_{lin}(**narr terms**)	**at_least_x(title terms)** $\wedge$ **at_least_x(desc terms)** $\wedge$ **at_least_x(narr terms)**			
Recall			$x = 2$	$x = 3$	$x = 4$	$x = 8$
0.00	0.8741	0.8637	0.8702	0.8470	0.8065	0.8030
0.10	0.7211	0.7276	0.7114	0.6984	0.6733	0.6563
0.20	0.6159	0.6467	0.6326	0.6160	0.5787	0.5544
0.30	0.5074	0.5405	0.5253	0.5114	0.4869	0.4487
0.40	0.4380	0.4584	0.4265	0.4128	0.3965	0.3697
0.50	0.3673	0.3722	0.3509	0.3430	0.3292	0.3024
0.60	0.3186	0.3200	0.2910	0.2778	0.2652	0.2434
0.70	0.2461	0.2511	0.2276	0.2138	0.2052	0.1864
0.80	0.1761	0.1876	0.1737	0.1567	0.1502	0.1340
0.90	0.1122	0.1239	0.1082	0.1027	0.0982	0.0854
1.00	0.0374	0.0350	0.0380	0.0375	0.0377	0.0365
Avg.prec. (non-interpolated)	0.3858	0.3977	0.3799	0.3666	0.3488	0.3278
% change		+3.1%	-1.5%	-5.0%	-9.6%	-15%

weights), the fuzzy approach was still able to construct slightly better rankings. This is an important circumstance as it anticipates that fuzzy methods can say a word in future retrieval engines.

6 Conclusions and Further Work

Classical IR approaches tend to oversimplify the content of user information needs whereas flexible query languages allow to articulate more evolved queries. For instance, the inclusion of quantified statements in the query language permits to express additional constraints for the retrieved documents. IR matching functions can be relaxed in different ways by means of quantified statements whose implementation is handled efficiently by semi-fuzzy quantifiers and quantified fuzzification mechanisms.

In this work we showed that our proposal based on the concept of semi-fuzzy quantifier handles pioneering fuzzy quantification proposals for IR as particular cases. On the other hand, we conducted large-scale experiments showing that this fuzzy approach is competitive with state-of-the-art IR techniques. These popular IR methods have recurrently appeared among the best retrieval methods for both adhoc and web retrieval tasks and, hence, it is very remarkable that our SFQ approach performs at the same level.

It is also important to observe that the benefits shown here empirically are not restricted to our particular fuzzy apparatus, but also hold in the framework of the seminal proposals of fuzzy quantification for IR. This is guaranteed by the subsumption proved in this work.

We applied very simple methods for building automatically fuzzy queries from TREC topics. In the near future we plan to study other means for obtaining fuzzy statements from user queries. It is particularly interesting to design methods for building n-ary statements involving several fuzzy sets. On the other hand, future research efforts will also be dedicated to analyze the practical behaviour of alternative models of fuzzy quantification. In this respect, besides *at least* expressions, we plan to extend the evaluation to other kind of quantifiers. For the basic retrieval task we have only found benefits in retrieval performance when this sort of quantifiers were applied. Nevertheless, we will study the adequacy of other sort of linguistic quantifiers in the context of other IR tasks.

Acknowledgements

Authors wish to acknowledge support from the Spanish Ministry of Education and Culture (project ref. TIC2003-09400-C04-03) and Xunta de Galicia (project ref. PGIDIT04SIN206003PR). D. E. Losada is supported by the "Ramón y Cajal" R&D program, which is funded in part by "Ministerio de Ciencia y Tecnología" and in part by FEDER funds.

References

1. S. Barro, A. Bugarín, P. Cariñena, and F. Díaz-Hermida. A framework for fuzzy quantification models analysis. *IEEE Transactions on Fuzzy Systems*, 11:89–99, 2003.
2. G. Bordogna and G. Pasi. Linguistic aggregation operators of selection criteria in fuzzy information retrieval. *International Journal of Intelligent Systems*, 10(2):233–248, 1995.
3. G. Bordogna and G. Pasi. Modeling vagueness in information retrieval. In F. Crestani M. Agosti and G. Pasi, editors, *Lectures on Information Retrieval (LNCS 1980)*. Springer Verlag, 2000.
4. G. Bordogna and G. Pasi. Modeling vagueness in information retrieval. In M. Agosti, F. Crestani, and G. Pasi, editors, *ESSIR 2000, LNCS 1980*, pages 207–241. Springer-Verlag Berlin Heidelberg, 2000.
5. P. Bosc, L. Lietard, and O. Pivert. Quantified statements and database fuzzy querying. In P. Bosc and J. Kacprzyk, editors, *Fuzziness in Database Management Systems*, volume 5 of *Studies in Fuzziness*, pages 275–308. Physica-Verlag, 1995.
6. F. Crestani and G. Pasi (eds). *Soft Computing in Information Retrieval: techniques and applications.* Studies in fuzziness and soft computing. Springer-Verlag, 2000.
7. M. Delgado, D. Sánchez, and M. A. Vila. Fuzzy cardinality based evaluation of quantified sentences. *International Journal of Approximate Reasoning*, 23(1):23–66, 2000.
8. F. Díaz-Hermida, A. Bugarín, P. Cariñena, and S. Barro. Voting model based evaluation of fuzzy quantified sentences: a general framework. *Fuzzy Sets and Systems*, 146:97–120, 2004.

9. I. Glöckner. A framework for evaluating approaches to fuzzy quantification. Technical Report TR99-03, Universität Bielefeld, May 1999.
10. I. Glöckner. *Fuzzy Quantifiers in Natural Language: Semantics and Computational Models.* PhD thesis, Universität Bielefeld, 2003.
11. I. Glöckner and A. Knoll. A formal theory of fuzzy natural language quantification and its role in granular computing. In W. Pedrycz, editor, *Granular computing: An emerging paradigm*, volume 70 of *Studies in Fuzziness and Soft Computing*, pages 215–256. Physica-Verlag, 2001.
12. D. Harman. Overview of the third text retrieval conference. In *Proc. TREC-3, the 3rd text retrieval conference*, 1994.
13. D. Hawking, E. Voorhees, N. Craswell, and P. Bailey. Overview of the trec-8 web track. In *Proc. TREC-8, the 8th Text Retrieval Conference*, pages 131–150, Gaithersburg, United States, November 1999.
14. E. Herrera-Viedma and G. Pasi. Fuzzy approaches to access information on the web: recent developments and research trends. In *Proc. International Conference on Fuzzy Logic and Technology (EUSFLAT 2003)*, pages 25–31, Zittau (Germany), 2003.
15. D.H. Kraft and D.A. Buell. A model for a weighted retrieval system. *Journal of the american society for information science*, 32(3):211–216, 1981.
16. D.H. Kraft and D.A. Buell. Fuzzy sets and generalized boolean retrieval systems. *International journal of man-machine studies*, 19:45–56, 1983.
17. J. H. Lee. Properties of extended boolean models in information retrieval. In *Proc. of SIGIR-94, the 17th ACM Conference on Research and Development in Information Retrieval*, Dublin, Ireland, July 1994.
18. J. H. Lee, W. Y. Kim, and Y. J. Lee. On the evaluation of boolean operators in the extended boolean framework. In *Proc. of SIGIR-93, the 16th ACM Conference on Research and Development in Information Retrieval*, Pittsburgh, USA, 1993.
19. D. E. Losada, F. Díaz-Hermida, A. Bugarín, and S. Barro. Experiments on using fuzzy quantified sentences in adhoc retrieval. In *Proc. SAC-04, the 19th ACM Symposium on Applied Computing - Special Track on Information Access and Retrieval*, Nicosia, Cyprus, March 2004.
20. GNU mifluz. http://www.gnu.org/software/mifluz. 2001.
21. Y. Ogawa, T. Morita, and K. Kobayashi. A fuzzy document retrieval system using the keyword connection matrix and a learning method. *Fuzzy sets and systems*, 39:163–179, 1991.
22. M.F. Porter. An algorithm for suffix stripping. In K.Sparck Jones and P.Willet, editors, *Readings in Information Retrieval*, pages 313–316. Morgan Kaufmann Publishers, 1997.
23. T. Radecki. Outline of a fuzzy logic approach to information retrieval. *International Journal of Man-Machine studies*, 14:169–178, 1981.
24. G. Salton, E. A. Fox, and H. Wu. Extended boolean information retrieval. *Communications of the ACM*, 26(12):1022–1036, 1983.
25. G. Salton and M.J. McGill. *Introduction to modern information retrieval.* McGraw-Hill, New York, 1983.
26. A. Singhal. Modern information retrieval: a brief overview. *IEEE Data Engineering Bulletin*, 24(4):35–43, 2001.
27. A. Singhal, S. Abney, M. Bacchiani, M. Collins, D. Hindle, and F. Pereira. At&t at trec-8. In *Proc. TREC-8, the 8th Text Retrieval Conference*, pages 317–330, Gaithersburg, United States, November 1999.

28. A. Singhal, C. Buckley, and M Mitra. Pivoted document length normalization. In *Proc. SIGIR-96, the 19th ACM Conference on Research and Development in Information Retrieval*, pages 21–29, Zurich, Switzerland, July 1996.
29. A Singhal and M. Kaszkiel. At&t at trec-9. In *Proc. TREC-9, the 9th Text Retrieval Conference*, pages 103–116, Gaithersburg, United States, November 2000.
30. R.R. Yager. On ordered weighted averaging aggregation operators in multi criteria decision making. *IEEE Transactions on Systems, Man and Cybernetics*, 18(1):183–191, 1988.
31. R.R. Yager. Connectives and quantifiers in fuzzy sets. *Fuzzy Sets and Systems*, 40:39–75, 1991.
32. R.R. Yager. A general approach to rule aggregation in fuzzy logic control. *Applied Intelligence*, 2:333–351, 1992.
33. R.R. Yager. Families of owa operators. *Fuzzy Sets and Systems*, 59(2):125–244, 1993.
34. L.A. Zadeh. A computational approach to fuzzy quantifiers in natural languages. *Comp. and Machs. with Appls.*, 8:149–184, 1983.

Appendix A

Given a query expression such as $Q_{lin}(t_1, \ldots, t_n)$, where each t_i is an atomic term, and a document d_j, the fuzzy set induced by the document can be expressed as: $C_{d_j} = \{w_{1,j}/1, \ldots, w_{n,j}/n\}$.

Without any loss of generality, we will assume that query terms are sorted in descending order of its membership degree in C_{d_j}.

The linear semi-fuzzy quantifier Q_{lin} operates on C_{d_j} as follows:

$$\begin{aligned}(F(Q_{lin}))(C_{d_j}) &= (1 - w_{1,j}) * Q_{lin}((C_{d_j})_{\geq 1}) + (w_{1,j} - w_{2,j}) * Q_{lin}((C_{d_j})_{\geq w_{1,j}}) + \\ &\quad (w_{2,j} - w_{3,j}) * Q_{lin}((C_{d_j})_{\geq w_{2,j}}) + \ldots + \\ &\quad (w_{n-1,j} - w_{n,j}) * Q_{lin}((C_{d_j})_{\geq w_{n-1,j}}) + w_{nj} * Q_{lin}((C_{d_j})_{\geq w_{n,j}}) = \\ &= (1 - w_{1,j}) * 0 + (w_{1,j} - w_{2,j}) * (1/n) + \\ &\quad (w_{2,j} - w_{3,j}) * (2/n) + \ldots + (w_{n-1,j} - w_{n,j}) * ((n-1)/n) + w_{nj} * 1 = \\ &= (1/n) * ((w_{1,j} - w_{2,j}) + (w_{2,j} - w_{3,j}) * 2 + \\ &\quad + \ldots + (w_{n-1,j} - w_{n,j}) * (n-1) + w_{nj} * n) = \\ &= (1/n) * \sum_{t_i \in q} w_{ij}\end{aligned}$$

leading to:

$$\mu_{Sm(Q_{lin}(t_1,\ldots,t_n))}(d_j) = (1/n) * \sum_{t_i \in q} w_{ij}$$

Let us now analyze the two weighting schemes (equations 13 and 14) independently:

- ***tf/idf* weights** (equation 13). Consider now a vector-space approach in which document vectors are weighted as in equation (13) and query vectors are binary. The inner product equation, $\sum w_{i,j} * q_i$, where $w_{i,j}$ (q_i) is the weight for term t_i in document d_j (query), can be reduced to $\sum_{t_i \in q} w_{i,j}$ when query weights are binary. It follows that both approaches result in the same ranking of documents because the value $1/n$ does not affect the ranking of every query.
- **pivoted weights** (equation 14). Consider now a vector-space approach in which document vectors are weighted as:

$$\frac{\frac{1+ln(1+ln(f_{i,j}))}{(1-s)+s\frac{dl_j}{avgdl}}}{norm\_1} * \frac{ln(\frac{N+1}{n_i})}{norm\_2}$$

and query vector weights are:

$$\frac{qtf_i}{max_l qtf_l}$$

Again, it follows that the inner product matching yields the same ranking that the one constructed from the fuzzy model.

Helping Users in Web Information Retrieval Via Fuzzy Association Rules

M.J. Martín-Bautista[1], D. Sánchez[1], J.M. Serrano[2], and M.A. Vila[1]

[1] University of Granada. Department of Computer Science and Artificial Intelligence. C/ Periodista Daniel Saucedo Aranda s/n, 18071 Granada, Spain
mbautis@decsai.ugr.es, daniel@decsai.ugr.es, vila@decsai.ugr.es
[2] University of Jaén. Department of Computer Science. Campus Las Lagunillas, 23071 Jaén, Spain
jschica@ujaen.es

Summary. We present an application of fuzzy association rules to find new terms that help the user to search in the web. Once the user has made an initial query, a set of documents is retrieved from the web. Representing these documents as text transactions, each item in the transaction means the presence of the term in the document. From the set of transactions, fuzzy association rules are extracted. Based on the thresholds of support and certainty factor, a selection of rules is carried out and the terms in those rules are offered to the user to be added to the query and to improve the retrieval.

1 Introduction

Finding information in the web is not so easy as users expect. Most of the documents retrieved as a result of a web search meet the search criteria but do not satisfy the user's preferences. Generally, this can be due to a not suitable formulation of the query, either because the query terms of the user does not match the indexed terms of the collection, or because the user does not know more vocabulary related to the search topic at the query moment.

To solve this problem, the query can be modified by adding or removing terms to discard uninteresting retrieved documents and/or to retrieve interesting documents that were not retrieved by the query. This problem has been named as *query refinement* or *query expansion* in the field of Information Retrieval [14].

In this work, we propose the use of mining techniques to solve this problem. For this purpose, we use fuzzy association rules to find dependence relations among the presence of terms in an initial set of retrieved documents. A group of selected terms from the extracted rules generates a vocabulary related to the search topic that helps the user to refine the query with the aim of improving the retrieval effectiveness. Data mining techniques have been applied

M.J. Martín-Bautista et al.: *Helping Users in Web Information Retrieval Via Fuzzy Association Rules*, StudFuzz **197**, 221–237 (2006)
www.springerlink.com

successfully in the last decade in the field of Databases, but also to solve some classical Information Retrieval problems such as document classification [26] and query refinement [35].

This paper is organized as follows: a survey of query refinement solutions found in the literature is given in Section 2. The concepts of association rules, fuzzy association rules and fuzzy transactions are presented briefly in Section 3, while the process to refine queries via fuzzy association rules is explained in Section 4. The obtention of the document representation and the extraction of fuzzy association rules are given in Section 5 and Section 6, respectively. Finally, some experimental examples are shown in Section 7 and the conclusions and future work are presented in 8.

2 Query Refinement

The query refinement process, also called query expansion, is a possible solution to the problem of dissatisfaction of the user with the answer of an information retrieval system, given a certain query. This problem is due, most of the times, to the terms used to query, which meet the search criteria, but do not reflects exactly what the user is really searching. This occurs, most of the times because the user does not know the vocabulary of the topic of the query, or the query terms do not come to user's mind at the query moment, or just because the vocabulary of the user does not match with the indexing words of the collection. This problem is even strong when the user is searching in the web, due to the amount of available information which makes that the user feels overwhelmed with the retrieved set of documents.

The process of query refinement solves this problem by modifying the search terms so the system results are more adequate to user's needs. There are mainly two different approaches in query refinement regarding how the terms are added to the query. The first one is called *automatic query expansion* [7], [18] and consist of the augmentation of query terms to improve the retrieval process without the intervention of the user. The second one is called *semi-automatic query-expansion* [30,37], where new terms are suggested to the user to be added to the original query in order to guide the search towards a more specific document space.

We can also distinguish different cases based on the source from which the terms are selected. By this way, terms can be obtained from the collection of documents [2,39], from user profiles [23], from user behavior [21] or from other users' experience [16], among others. If a document collection is considered as a whole from which the terms are extracted to be added to the query, the technique is called *global analysis*, as in [39]. However, if the expansion of the query is performed based on the documents retrieved from the first query, the technique is denominated *local analysis*, and the set of documents is called *local set*.

Local analysis can also be classified into two types. On the one hand, *local feedback* adds common words from the top-ranked documents of the local set. These words are identified sometimes by clustering the document collection [2]. In this group we can include the relevance feedback process, since the user has to evaluate the top ranked documents from which the terms to be added to the query are selected. On the other hand, *local context analysis* [39], which combines global analysis and context local feedback to add words based on relationships of the top-ranked documents. The calculus of co-occurrences of terms is based on passages (text windows of fixed size), as in global analysis, instead of complete documents. The authors show that, in general, local analysis performs better than global one.

There are several approaches using different techniques to identify terms that should be added to the original query. The first group is based on their association relation by co-occurrence to query terms [36]. Instead of simply terms, in [39] the authors find co-occurrences of concepts given by noun groups with the query terms. Another approach based on the concept space approach is [8]. The statistical information can be extracted from a clustering process and a ranking of documents from the local set, as it is shown in [9] or by similarity of the top-ranked documents [28]. All these approaches where a co-occurrence calculus is performed has been said to be suitable to construct specific knowledge base domains, since the terms are related, but they can not be distinguished how [4].

In the second group of techniques, search terms are selected on the basis of their similarity to the query terms, by constructing a similarity term thesaurus [31]. Other approaches in this same group use techniques to find out the most discriminatory terms, which are the candidates to be added to the query. These two characteristics can be combined by first calculating the nearest neighbors and second, by measuring the discriminatory ability of the terms [30]. The last group is formed by approaches based on lexical variants of query terms extracted from a lexical knowledge base such as Wordnet [27]. Some approaches in this group are [38], and [4] where a semantic network with term hierarchies is constructed. The authors reveal the adequacy of this approach for general knowledge bases, which can be identified in general terms with global analysis, since the set of documents from which the hierarchies are constructed is the corpus, and not the local set of a first query. Previous approaches with the idea of hierarchical thesaurus can be also found in the literature, where an expert system of rules interprets the user's queries and controls the search process [18].

3 Association Rules and Fuzzy Association Rules

We use association rules and fuzzy association rules to find the terms to be added to the original query. In this section, we briefly review association rules and some useful extensions able to deal with weighted sets of items in a fuzzy framework.

3.1 Association Rules

Given a database of transactions, where each transaction is an itemset, we can extract association rules [1]. Formally, let T be a set of transactions containing items of a set of items I. Let us consider two itemsets (sets of items) $I_1, I_2 \subset I$, where $I_1, I_2 \neq \varnothing$ and $I_1 \cap I_2 = \varnothing$. An association rule [1] $I_1 \Rightarrow I_2$ is an implication rule meaning that the apparition of itemset I_1 in a transaction implies the apparition of itemset I_2 in the same transaction. The reciprocal does not have to happen necessarily [22]. I_1 and I_2 are called antecedent and consequent of the rule, respectively. The rules obtained with this process are called boolean association rules or, in general, association rules since they are generated from a set of boolean or crisp transactions.

3.2 Fuzzy Association Rules

Fuzzy association rules are defined as those rules extracted from a set of fuzzy transactions FT where the presence of an item in a transaction is given by a fuzzy value of membership [3,10,19,24,25]. Though most of these approaches have been introduced in the setting of relational databases, we think that most of the measures and algorithms proposed can be employed in a more general framework. A broad review, including references to papers on extensions to the case of quantitative attributes and hierarchies of items, can be found in [11].

In this paper we shall employ the model proposed in [10]. This model considers a general framework where data is in the form of fuzzy transactions, i.e., fuzzy subsets of items. A (crisp) set of fuzzy transactions is called a FT-set, and fuzzy association rules are defined as those rules extracted from a FT-set. Fuzzy relational databases can be seen as a particular case of FT-set. Other datasets, such as the description of a set of documents by means of fuzzy subsets of terms, are also particular cases of FT-sets but fall out of the relational database framework.

Given a FT-set $\widetilde{T}$ on a set of items I and a fuzzy transaction $\tilde{\tau} \in \widetilde{T}$, we note $\tilde{\tau}(i)$ the membership degree of i in $\tilde{\tau}$ $\forall i \in I$. We also define $\tilde{\tau}(I_0) = \min_{i \in I_0} \tilde{\tau}(i)$ for every itemset $I_0 \subseteq I$.

With this scheme, we have a degree in $[0,1]$ associated to each pair $\langle \tilde{\tau}, I_0 \rangle$. Sometimes it is useful to see this information in a different way by means of what we call the *representation* of an itemset. The idea is to see an itemset as a fuzzy subset of transactions. The representation of an itemset $I_0 \subseteq I$ in a FT-set $\widetilde{T}$ is the fuzzy subset $\widetilde{\Gamma}_{I_0} \subseteq \widetilde{T}$ defined as

$$\widetilde{\Gamma}_{I_0} = \sum_{\tilde{\tau} \in \widetilde{T}} \tilde{\tau}\,(I_0)/\tilde{\tau} \tag{1}$$

On this basis, a fuzzy association rule is an expression of the form $I_1 \Rightarrow I_2$ that holds in a FT-set $\widetilde{T}$ iff $\widetilde{\Gamma}_{I_1} \subseteq \widetilde{\Gamma}_{I_2}$. The only difference with the definition of crisp association rule is that the set of transactions is a FT-set, and the inclusion above is the usual between fuzzy sets.

3.3 Measures for Association and Fuzzy Association Rules

There are two relevant aspects of association rules that we need to measure. On the one hand, an association rule can be interesting even if there are some exceptions to the rule in the set T, so we are interested in assessing the *accuracy* of the rule and to decide on its basis whether the rule is accurate or not. On the other hand, an accurate rule that holds in few transactions is not interesting since it is not representative of the whole data and its possible application is limited. Hence, we need to measure the amount of transactions supporting the rule and to decide on that basis whether the rule is important or not.

The assessment of association rules is usually based on the values of *support* and *confidence*. We shall note $supp(I_k)$ the support of the itemset I_k. The support and the confidence of the rule $I_1 \Rightarrow I_2$ noted by $Supp\,(I_1 \Rightarrow I_2)$ and $Conf\,(I_1 \Rightarrow I_2)$, respectively. Support is the percentage of transactions containing an itemset, calculated by its probability, while confidence measures the strength of the rule calculated by the conditional probability of the consequent with respect to the antecedent of the rule. Only itemsets with a support greater than a threshold *minsupp* are considered, and from the resulting association rules, those ones with a confidence less than a threshold *minconf* are discarded. Both thresholds must be fixed by the user before starting the process.

To deal with the imprecision of fuzzy transactions, we need to obtain the support and the confidence values with alternative methods which can be found mainly in the framework of approximate reasoning. We have selected the the evaluation of quantified sentences presented in [40], calculated by means of method GD presented in [13].

Moreover, as an alternative to confidence, we propose the use of *certainty factors* to measure the accuracy of association rules, since they have been revealed as a good measure in knowledge discovery too [17]. Basically, the problem with confidence is that it does not take into account the support of I_2, hence it is unable to detect statistical independence or negative dependence, i.e., a high value of confidence can be obtained in those cases. This problem is specially important when there are some items with very high support. In the worst case, given an itemset I_C such that $supp(I_C) = 1$, every rule of the form $I_A \Rightarrow I_C$ will be strong provided that $supp(I_A) > minsupp$. It has been shown that in practice, a large amount of rules with high confidence are misleading because of the aforementioned problems.

The certainty factor (CF) of an association rule is defined as $I_1 \Rightarrow I_2$ based on the value of the confidence of the rule. If $Conf\,(I_1 \Rightarrow I_2) > supp\,(I_2)$ the value of the factor is given by expression (2); otherwise, is given by expression (3), considering that if $supp(I_2) = 1$, then $CF(I_1 \Rightarrow I_2) = 1$ and if $supp(I_2) = 0$, then $CF(I_1 \Rightarrow I_2) = -1$

$$CF(I_1 \Rightarrow I_2) = \frac{Conf(I_1 \Rightarrow I_2) - supp(I_2)}{1 - supp(I_2)} \tag{2}$$

$$CF(I_1 \Rightarrow I_2) = \frac{Conf(I_1 \Rightarrow I_2) - supp(I_2)}{supp(I_2)} \tag{3}$$

4 Query Refinement via Fuzzy Association Rules

Besides the techniques explained in Section 2, we also consider fuzzy association rules (which generalize the crisp ones) as a way to find presence dependence relations among the terms of a document set. A group of selected terms from the extracted rules generate a vocabulary related to the search topic that helps the user to refine the query.

In a text framework, association rules can be seen as rules with a semantic of presence of terms in a group of documents (we explain it with detail in the following section). This way, we can obtain rules such as $t_1 \Rightarrow t_2$ meaning that the presence of t_1 in a document imply the presence of term t_2, but the opposite do not have to occur necessarily. This concept is different from the co-occurrences where, given an occurrence between t_1 and t_2, the presence of both terms is reciprocal, that is, if one occurs, the other also does [22]. Only when the association rule $t_1 \Rightarrow t_2$ and its opposite $t_2 \Rightarrow t_1$ are extracted, we can say there is a co-occurrence between t_1 and t_2. In query refinement association rules extend the use of co-occurrences since it allows not only to substitute one term by other, but also to modify the query making it more specific or more general.

The process occurs as follows: before query refinement can be applied, we assume that a retrieval process is performed. The user's initial query generates a set of ranked documents. If the top-ranked documents do not satisfy user's needs, the query improvement process starts. Since we start from the initial set of documents retrieved from a first query, we are dealing with a *local analysis* technique. And, since we just considered the top-ranked documents, we can classify our technique as a *local feedback* one.

From the initial retrieved set of documents, called *local set*, association rules are found and additional terms are suggested to the user in order to refine the query. As we have explained in Section 2, there are two general approaches to query refinement: automatic and semi-automatic. In our case, as we offer to the user a list of terms to add to the query, the system performs a semi-automatic process. Finally, the user selects from that list the terms to add to the query so the query process starts again. The whole process is summarized in the following:

Semi-automatic query refinement process using association rules

1. The user queries the system
2. A first set of documents is retrieved
3. From this set, the representation of documents is extracted and association rules are generated
4. Terms that appear in certain rules are shown to the user (Subsection 6.1)
5. The user selects those terms more related to her/his needs
6. The selected terms are added to the query, which is used to query the system again

Once the first query is constructed, and the association rules are extracted, we make a selection of rules where the terms of the original query appear. However, the terms of the query can appear in the antecedent or in the consequent of the rule. If a query term appears in the antecedent of a rule, and we consider the terms appearing in the consequent of the rule to expand the query, a generalization of the query will be carried out. Therefore, a generalization of a query gives us a query on the same topic as the original one, but looking for more general information. However, if query terms appear in the consequent of the rule, and we reformulate the query by adding the terms appearing in the antecedent of the rule, then a specialization of the query will be performed, and the precision of the system should increase. The specialization of a query looks for more specific information than the original query but in the same topic. In order to obtain as much documents as possible, terms appearing in both sides of the rules can also be considered.

5 Document Representation for Association Rule Extraction

From that initial retrieved set of documents, a valid representation for extracting the rules is needed. Different representations of text for association rules extraction can be found in the literature: bag of words, indexing keywords, term taxonomy and multi-term text phrases [12]. In our case, we use automatic indexing techniques coming from Information Retrieval [34] to obtain *word items*, that is, single words appearing in a document where stop-list and/or stemming processes can be applied. Therefore, we represent each document by a set of terms where a weight meaning the presence of the term in the document can be calculated. There are several term weighting schemes to consider [33]. In this work, we study three different weighting schemes [22]:

Boolean weighting scheme: It takes values {0,1} indicating the absence or presence of the word in the document, respectively.

Frequency weighting scheme: It associates to each term a weight meaning the relative frequency of the term in the document. In a fuzzy framework, the normalization of this frequency can be carried out by dividing the number of occurrences of a term in a document by the number of occurrences of the most frequent term in that document [6].

TFIDF weighting scheme: It is a combination of the within-document word frequency (*TF*) and the inverse document frequency (*IDF*). The expressions of these schemes can be found in [33]. We use this scheme in its normalized form in the interval [0, 1] according to [5]. In this scheme, a term that occurs frequently in a document but infrequently in the collection is assigned a high weight.

5.1 Text Transactions

Once we have the representation of the documents in a classical information retrieval way, a transformation of this representation into a transactional one is carried out. In a text framework, we identify each transaction with the representation of a document. Therefore, from a collection of documents $D = \{d_1, \ldots, d_n\}$ we can obtain a set of terms $I = \{t_1, \ldots, t_m\}$ which is the union of the keywords for all the documents in the collection. The weights associated to these terms in a document d_i are represented by $W_i = (w_{i1}, \ldots, w_{im})$. For each document d_i, we consider an extended representation where a weight of 0 will be assigned to every term appearing in some of the documents of the collection but not in d_i.

Considering these elements, we can define a *text transaction* $\tau_i \in T$ as the extended representation of document d_i. Without loosing generalization, we can write the set of transactions associated to the collection of document D as $T_D = \{d_1, \ldots, d_n\}$.

When the weights $W_i = (w_{i1}, \ldots, w_{im})$ associated to the transactions take values in {0,1}, that is, following the boolean weighting scheme of the former section, the transactions can be called boolean or crisp transactions, since the values of the tuples are 1 or 0 meaning that the attribute is present in the transaction or not, respectively.

Fuzzy Text Transactions

As we have explained above, we can consider a weighted representation of the presence of the terms in the documents. In the fuzzy framework, a normalized weighting scheme in the unit interval is employed. We call them *fuzzy weighting schemes.* Concretely, we consider two fuzzy weighting schemes, namely the frequency weighting scheme and the TFIDF weighting scheme, both normalized. Therefore, analogously to the former definition of text transactions, we can define a set of *fuzzy text transactions* $FT_D = \{d_1, \ldots, d_n\}$, where each document d_i corresponds to a fuzzy transaction $\tilde{\tau}_i \in FT$, and where the weights $W = \{w_{i1}, \ldots, w_{im}\}$ of the keyword set $I = \{t_1, \ldots, t_m\}$ are fuzzy values from a fuzzy weighting scheme.

6 Extraction of Fuzzy Association Rules

As described in the previous subsection, we consider each document as a transaction. Let us consider $T_D = \{d_1, \ldots d_n\}$ as the set of transactions from the collection of documents D, and $I = \{t_1, \ldots, t_m\}$ as the text items obtained from all the representation documents $d_i \in D$ with their membership to the transaction expressed by $W_i = (w_{i1}, \ldots, w_{im})$. On this set of transactions we apply Algorithm 1 to extract the association rules. We must note that we do not distinguish in this algorithm the crisp and the fuzzy case, but we give general steps to extract association rules from text transactions. The specific cases will be given by the item weighting scheme that we consider in each case.

Algorithm 1 Basic algorithm to obtain the association rules from text

Input: a set of transactions $T_D = \{d_1, \ldots d_n\}$
a set of term items $I = \{t_1, \ldots, t_m\}$ with their associated weights $W_i = (w_{i1}, \ldots, w_{im})$ for each document d_i.
Output: a set of association rules.

1. Construct the itemsets from the set of transactions T.
2. Establish the threshold values of minimum support *minsupp* and minimum confidence *minconf*
3. Find all the itemsets that have a support above threshold *minsupp*, that is, the *frequent itemsets*
4. Generate the rules, discarding those rules below threshold minconf

We must point out that, as it has been explained in [15], [32], in the applications of mining techniques to text, documents are usually categorized, in the sense of documents which representation is a set of keywords, that is, terms that really describe the content of the document. This means that usually a full text is not considered and its description is not formed by all the words in the document, even without stop words, but also by keywords. The authors justify the use of keywords because of the appearing of useless rules. Some additional commentaries about this problem regarding the poor discriminatory power of frequent terms can be found in [30], where the authors comment the fact that the expanded query may have worst performance than the original one due to the poor discriminatory ability of the added terms.

Therefore, the problem of selecting good terms to be added to the query has two faces. On the one hand, if the terms are not good discriminators, the expansion of the query may not improve the result. But, on the other hand, in dynamic environments or systems where the response-time is important, the application of a pre-processing stage to select good discriminatory terms may not be suitable. In our case, since we are dealing with a problem of query refinement in Internet, information must be shown on-line to the user, so a time constraint is present.

Solutions for both problems can be given. In the first case, discriminatory schemes almost automatic can be used alternatively to a preprocessing stage for selecting the most discriminatory terms. This is the case of the *TFIDF* weighting scheme (see Section 5). In the second case, when we work in a dynamic environment, we have to remind that to calculate the term weights following the *TFIDF* scheme, we need to know the presence of a term in the whole collection, which limits in some way its use in dynamic collections, as usually occurs in Internet. Therefore, instead of improving document representation in this situation, we can improve the rule obtaining process. The use of alternative measures of importance and accuracy such as the ones presented in Section 3 is considered in this work in order to avoid the problem of non appropriate rule generation.

Additionally to the representation of the documents by terms, an initial categorization of the documents can be available. In that case, the categories can appear as items to be included in the transactions with value $[0, 1]$ based on the membership of the document to that category. This way, the extracted rules not only provide additional terms to the query, but also information about the relation between terms and categories.

6.1 The Selection of Terms for Query Refinement

The extraction of rules is usually guided by several parameters such as the minimum support *(minsupp)*, the minimum value of certainty factor *(mincf)*, and the number of terms in the antecedent and consequent of the rule. Rules with support and certainty factor over the respective thresholds are called *strong rules.*

Strong rules identify dependence in the sense of nontrivial inclusion of the set of transactions where each itemset (set of terms in this case) appears. This information is very useful for us in order to refine the query. First, the minimum support restriction ensures that the rules apply to a significant set of documents. Second, the minimum accuracy restriction, though allowing for some exceptions, ensures that the inclusion holds to an important degree.

Once the strong association rules are extracted, the selection of useful terms for query refinement depends on the appearance in antecedent and/or consequent of the terms. Let us suppose that *qterm* is a term that appears in the query and let $term \in S$, $S_0 \subseteq S$. Some possibilities are the following:

- Rules of the form $term \Rightarrow qterm$ such that $qterm \Rightarrow term$ has low accuracy. This means that the appearance of *term* in a document "implies" the appearance of *qterm*, but the reciprocal does not hold significantly, i.e.,

$\Gamma_{term} \subseteq \Gamma_{qterm}$ to some extent. Hence, we could suggest the word *term* to the user as a way to restrict the set of documents obtained with the new query.

- Rules of the form $S_0 \Rightarrow qterm$ with $S_0 \subseteq S$. We could suggest the set of terms S_0 to the user as a whole, i.e., to add S_0 to the query. This is again uninteresting if the reciprocal is a strong rule.
- Rules of the form $qterm \Rightarrow term$ with $term \in S$ and $term \Rightarrow qterm$ a not strong rule. We could suggest the user to replace *qterm* with *term* in order to obtain a set of documents that include the actual set (this is interesting if we are going to perform the query again in the web, since perhaps *qterm* is more specific that the user intended).
- Strong rules of the form $S_0 \Rightarrow qterm$ or $term \Rightarrow qterm$ such that the reciprocal is also strong. This means co-occurrence of terms in documents. Replacing *qterm* with S_0 (or *term*) can be useful in order to search for similar documents where *qterm* does not appear. These rules can be interesting if we are going to perform the query again in Internet, since new documents not previously retrieved and interesting for the user can be obtained by replacing *qterm* with *term*.

7 Experimental Examples

The experiments have been carried out in the web with the search engine *Google (http://www.google.es)*. Three different queries have been submitted to the engine, with the search and results in English, namely: *networks*, *learning* and *genetic*. The purpose of our system is to find additional terms that can modify the query but narrow the set of retrieved documents in most of the cases, and/or improve the retrieval effectiveness. Therefore, if the user has the intention of searching for documents about *genetic* with a Computer Science and an Artificial Intelligence meaning, but she/he does not know more vocabulary related to that concept, the resulting rules can suggest her/him some terms to add to the query. This new query can discard the documents related to other meanings (always that the additional terms are not in the vocabulary of the other meanings).

Once the queries have been submitted to the search engine for the first time, an initial set of documents is retrieved, from which we take the first 100 top-ranked documents. Since we start from the initial set of documents retrieved from a first query, we are dealing with a *local analysis* technique. And, since we just considered the top-ranked documents, we can classify our technique as a *local feedback* one. From this *local set*, a document representation is obtained as in classical information retrieval, and a transformation of this representation into a transactional one is carried out. These transactions are mined for each query to obtain a set of association rules so additional terms can be offered to the user to refine the query. The number of results in each query, the number of text transactions and the number of terms (items)

Table 1. Queries with number of results, transactions and terms

Query	N. Results	N. Transactions	N. Terms
networks	94.200.000	100	839
learning	158.000.000	100	832
genetic	17.500.000	100	756

can be seen in Table 1. It must be remarked the difference in the length of the dimensions of the set of transactions obtained. In traditional data mining, the number of transactions is usually greater while the number of items is lower. In our case it is the opposite, although the goodness of the rules has not to be affected.

The weighted schemes considered are those proposed in Section 5, that is, the boolean, the frequency and the TFIDF weighting scheme. We must point out that the first one is crisp, while the other two are fuzzy values. The threshold of support is established to 2% for the crisp and the frequency case, while for the TFIDF we decide to remove the threshold, since no rules appear with more than a 2% for all the queries. For the obtention of the rules, we have established a level of the rule of 5, which implies that the number of components appearing in the rule (antecedent and consequent) can not be more than 5 adding both sides of the rule). The number of rules obtained for each weighting scheme with these thresholds can be seen in Table 2. In this table, we can observe the main advantages of the fuzzy weighting schemes against the crisp case. We must remember that the boolean scheme assigns 0 if the term does not appear in the document, and 1 if the terms appears, no matter how many times. This implies that the importance of a term will be 1 either if the term appears 1 or 20 times in the same document, which does not reflect the real presence of a term in a document. From the point of view of rules, this generates a huge number of them which give not very realistic presence relations among the terms, so they are not very useful for the user.

In the case of the TFIDF case, this scheme assigns a low weight to those items appearing very frequently in the whole collection. When the TFIDF scheme is used, the term query, for instance, *networks* is assigned a weight of 0, since it appears in all the documents of the collection. This means that no rule with the term *networks* will appear in the set of extracted rules in this case. This effect is the same that is obtained with the selection of rules, where

Table 2. Number of rules for each query with different weighting schemes

Query	Boolean	Norm. Freq.	TFIDF
networks	1118	95	56
learning	296	73	10
genetic	233	77	10

high frequent terms are not considered since they do not give new information. However, this lack of new information does not mean that the terms appearing in the same rule as the query term do not help to refine the query to decrease the number of retrieved documents and increase the satisfaction of the user.

The best scheme to analyze cases is the normalized frequency scheme. This scheme assigns a weight to a term meaning the normalized relative frequency of the term in the document, which is more realistic than the boolean scheme but less discriminatory than the TFIDF one. For instance, in the document set retrieved as the answer of query *genetic*, there are documents related to Biology and to Computer Science. If a novel user does not know the vocabulary of the topic, and the intention of the search is looking for *genetic* in the field of Computer Science, rules such as $programming \Rightarrow genetic$, can suggest to the user a new term, *programming*, in order to add it to the query so the results of the refined query are more suitable to user's needs. This case is of type $term \Rightarrow qterm$, where the rule $programming \Rightarrow genetic$ holds with a certainty factor of 1 while the opposite rule $genetic \Rightarrow programming$ holds with a certainty factor of 0.013.

Other example in this case is related to the query *learning*. Let us suppose that the user has the intention of searching about learning and the new technologies, but only use the query term *learning* so millions of documents are retrieved by the search engine. Some interesting rules obtained in this case related learning and new technologies are shown in Table 3, where the terms appearing in the antecedent of the rules are shown in the left column and the terms appearing in the consequent of the rules are shown in the first row of the table.

Table 3. Confidence/Certainty Factor values of some rules with the normalized frequency weighting scheme for the query *learning*

	learning	technology	web	online
learning	–	0.04/0.01	0.06/0.01	0.15/0.03
technology	0.94/0.94	–	–	–
web	0.8/0.79	–	–	–
online	0.79/0.76	–	–	–

We can also observe a case of substitution of terms when both $term \Rightarrow qterm$ and its reciprocal are strong rules. For instance, with the query of *networks*, the rule $rights \Rightarrow reserved$ and its reciprocal $reserved \Rightarrow rights$, appears with a support of 2.3% and a certainty factor of 1. This means that these two terms are equivalent to be used as additional terms to refine the query.

Regarding the information retrieval effectiveness values, as we add terms to the query, in our experiments the precision increases while the recall decreases. For instance, let us suppose again the example of the user looking for documents related to *genetic* in the field of Computer Science. If the user submit the query with only the term *genetic*, the recall value is 1 while the precision value is of 0.16 in the top-ranked first 100 documents. As the rule *programming* $\Rightarrow$ *genetic* has a support of 6% and a certainty factor of 1, it will be selected to show to the user the term *programming* to be added to the query. With the refined query, the recall decreases to 0.375, but the precision increases to 1.

8 Conclusions and Future Work

We have presented a possible solution to the Information Retrieval problem of query refinement in the web by means of fuzzy association rules. The fuzzy framework allows to represent documents by terms with an associated weight of presence. This representation improves the traditional ones based on binary presence/ausence of terms in the document, since it allows to distinguish between terms appearing in a document with different frequencies. This representation of documents by weighted terms is transformed into a transactional one, so text rules can be extracted following a mining process. From all the extracted rules, a selection process is carried out, so only the rules with a high support and certainty factor are chosen. The terms appearing in these rules are shown to the user, so a semi-automatic query refinement process is carried out. As it has been shown in the experimental examples, the refined queries reflect better the user's needs and the retrieval process is improved. The selection of rules and the chance to make the query more general, specific or to change the terms with the same meaning in order to improve the results lead us to consider this approach an useful tool for query refinement.

In the future, we will study the extraction of rules over a collection of documents as in the global analysis techniques, so we could compare with the actual system.

Acknowledgements

This work is supported by the research project Fuzzy-KIM, CICYT TIC2002-04021-C02-02.

References

1. Agrawal, R., Imielinski, T. & Swami, A. "Mining Association Rules between Set of Items in Large Databases". In *Proc. of the 1993 ACM SIGMOD Conference*, 207–216, 1993.
2. Attar, R. & Fraenkel, A.S. "Local Feedback in Full-Text Retrieval Systems". *Journal of the Association for Computing Machinery 24(3)*:397–417, 1977.

3. Au, W.H. & Chan, K.C.C. "An effective algorithm for discovering fuzzy rules in relational databases". In *Proc. Of IEEE International Conference on Fuzzy Systems*, vol II, 1314–1319, 1998.
4. Bodner, R.C. & Song, F. "Knowledge-based approaches to query expansion in Information Retrieval". In McCalla, G. (Ed.) *Advances in Artificial Intelligence*:146–158. New-York, USA: Springer Verlag, 1996.
5. Bordogna, G., Carrara, P. & Pasi, G. "Fuzzy Approaches to Extend Boolean Information Retrieval". In Bosc., Kacprzyk, J. *Fuzziness in Database Management Systems*, 231-274. Germany: Physica Verlag, 1995.
6. Bordogna, G. & Pasi, G. "A Fuzzy Linguistic Approach Generalizing Boolean Information Retrieval: A Model and Its Evaluation". *Journal of the American Society for Information Science 44(2)*:70–82, 1993.
7. Buckley, C., Salton. G., Allan, J. & Singhal, A. "Automatic Query Expansion using SMART: TREC 3". *Proc. of the 3rd Text Retrieval Conference*, Gaithersburg, Maryland, 1994.
8. Chen, H., Ng, T., Martinez, J. & Schatz, B.R. "A Concept Space Approach to Addressing the Vocabulary Problem in Scientific Information Retrieval: An Experiment on the Worm Community System". *Journal of the American Society for Information Science 48(1)*:17–31, 1997.
9. Croft, W.B. & Thompson, R.H. "I^3R: A new approach to the design of Document Retrieval Systems". *Journal of the American Society for Information Science 38(6)*, 389–404, 1987.
10. Delgado, M., Marín, N., Sánchez, D. & Vila, M.A. "Fuzzy Association Rules: General Model and Applications". *IEEE Transactions on Fuzzy Systems 11*:214–225, 2003a.
11. Delgado, M., Marín, N., Martín-Bautista, M.J., Sánchez, D. & Vila, M.A. "Mining Fuzzy Association Rules: An Overview". *2003 BISC International Workshop on Soft Computing for Internet and Bioinformatics"*, 2003b.
12. Delgado, M., Martín-Bautista, M.J., Sánchez, D. & Vila, M.A. "Mining Text Data: Special Features and Patterns". In *Proc. of EPS Exploratory Workshop on Pattern Detection and Discovery in Data Mining*, London, September 2002a.
13. Delgado, M., Sánchez, D. & Vila, M.A. "Fuzzy cardinality based evaluation of quantified sentences". *International Journal of Approximate Reasoning 23*:23–66, 2000c.
14. Efthimiadis, E. "Query Expansion". *Annual Review of Information Systems and Technology 31*:121–187, 1996.
15. Feldman, R., Fresko, M., Kinar, Y., Lindell, Y., Liphstat, O., Rajman, M., Schler, Y. & Zamir, O. "Text Mining at the Term Level". In *Proc. of the 2nd European Symposium of Principles of Data Mining and Knowledge Discovery*, 65–73, 1998.
16. Freyne J, Smyth B. (2005) Communities, collaboration and cooperation in personalized web search. In Proc. of the 3rd Workshop on Intelligent Techniques for Web Personalization (ITWP'05). Edinburgh, Scotland, UK
17. Fu, L.M. & Shortliffe, E.H. "The application of certainty factors to neural computing for rule discovery". *IEEE Transactions on Neural Networks 11(3)*:647–657, 2000.
18. Gauch, S. & Smith, J.B. "An Expert System for Automatic Query Reformulation". *Journal of the American Society for Information Science 44(3)*:124–136, 1993.

19. Hong, T.P., Kuo, C.S. & Chi, S.C. "Mining association rules from quantitative data." *Intelligent Data Analysis 3*:363–376, 1999.
20. Jiang, M.M., Tseng, S.S. & Tsai, C.J. "Intelligent query agent for structural document databases." *Expert Systems with Applications 17*:105–133, 1999.
21. Kanawati R., Jaczynski M., Trousse B., Andreoli J.M. (1999) Applying the Broadway recommendation computation approach for implementing a query refinement service in the CBKB meta search engine. In Proc. of the French Conference of CBR (RaPC99), Palaiseau, France
22. Kraft, D.H., Martín-Bautista, M.J., Chen, J. & Sánchez, D. "Rules and fuzzy rules in text: concept, extraction and usage". *International Journal of Approximate Reasoning 34*, 145–161, 2003.
23. Korfhage R.R. (1997) Information Storage and Retrieval. John Wiley & Sons, New York
24. Kuok, C.-M., Fu, A. & Wong, M.H. "Mining fuzzy association rules in databases," *SIGMOD Record 27(1)*:41–46, 1998.
25. Lee, J.H. & Kwang, H.L. "An extension of association rules using fuzzy sets". In *Proc. of IFSA'97*, Prague, Czech Republic, 1997.
26. Lin, S.H., Shih, C.S., Chen, M.C., Ho, J.M., Ko, M.T., Huang, Y.M. "Extracting Classification Knowledge of Internet Documents with Mining Term Associations: A Semantic Approach". In *Proc. of ACM/SIGIR'98*, 241–249. Melbourne, Australia, 1998.
27. Miller, G. "WordNet: An on-line lexical database". *International Journal of Lexicography 3(4)*:235–312, 1990.
28. Mitra, M., Singhal, A. & Buckley, C. "Improving Automatic Query Expansion". In *Proc. Of ACM SIGIR*, 206–214. Melbourne, Australia, 1998.
29. Moliniari, A. & Pasi, G. "A fuzzy representation of HTML documents for information retrieval system." *Proceedings of the fifth IEEE International Conference on Fuzzy Systems, vol. I*, pp. 107–112. New Orleans, EEUU, 1996.
30. Peat, H.P. & Willet, P. "The limitations of term co-occurrence data for query expansion in document retrieval systems". *Journal of the American Society for Information Science 42(5)*, 378–383, 1991.
31. Qui, Y. & Frei, H.P. "Concept Based Query Expansion". In *Proc. Of the Sixteenth Annual International ACM-SIGIR'93 Conference on Research and Development in Information Retrieval*, 160–169, 1993.
32. Rajman, M. & Besançon, R. "Text Mining: Natural Language Techniques and Text Mining Applications". In *Proc. of the 3rd International Conference on Database Semantics (DS-7)*. Chapam & Hall IFIP Proceedings serie, 1997.
33. Salton, G. & Buckley, C. "Term weighting approaches in automatic text retrieval". *Information Processing and Management 24(5)*, 513–523, 1988.
34. Salton, G. & McGill, M.J. *Introduction to Modern Information Retrieval.* McGraw-Hill, 1983.
35. Srinivasan, P., Ruiz, M.E., Kraft, D.H. & Chen, J. "Vocabulary mining for information retrieval: rough sets and fuzzy sets". *Information Processing and Management 37*:15–38, 2001.
36. Van Rijsbergen, C.J., Harper, D.J. & Porter, M.F. "The selection of good search terms". *Information Processing and Management 17*:77–91, 1981.
37. Vélez, B., Weiss, R., Sheldon, M.A. & Gifford, D.K. "Fast and Effective Query Refinement". In *Proc. Of the 20th ACM Conference on Research and Development in Information Retrieval (SIGIR'97)*. Philadelphia, Pennsylvania, 1997.

38. Voorhees, E. "Query expansion using lexical-semantic relations. *Proc. of the 17th International Conference on Research and Development in Information Retrieval (SIGIR)*. Dublin, Ireland, July, 1994.
39. Xu, J. & Croft, W.B. "Query Expansion Using Local and Global Document Analysis". In *Proc. of the Nineteenth Annual International ACM SIGIR Conference on Research and Development in Information Retrieval*, 4–11, 1996.
40. Zadeh, L.A. "A computational approach to fuzzy quantifiers in natural languages". *Computing and Mathematics with Applications 9(1)*:149–184, 1983.

Combining Soft and Hard Techniques for the Analysis of Batch Retrieval Tasks

Francisco J. Valverde-Albacete

Dpto. de Teoría de la Señal y de las Comunicaciones, Universidad Carlos III de Madrid, Avda. de la Universidad, 30. Leganés 28911. Spain
fva@tsc.uc3m.es

Summary. In this chapter we present a formal model that provides an analysis of the batch retrieval phase of a Web retrieval interaction or any other batch retrieval task by decomposing it into three subproblems: first, given a perfect relevance relation in representation space find a system that implements it; second, transform query and document representations into descriptions while maintaining a prescribed relevance relation; third and final, given a partially specified relevance relation in description space, build a system that implements it. We give solutions to problems 1 and 2 based in rough-set theory and Formal Concept Analysis and claim that to understand how such a task behaves neither hard nor soft techniques are sufficient of their own but rather, their combinations should be brought to bear.

1 Introduction: a Prototypical Batch Retrieval Task

This chapter deals with the batch phase of a typical web retrieval interaction, be it the querying of a portal or a more proprietary system. We believe such a task, sending some sort of query to a repository of documents and expecting a document-bearing structure in return, will remain a central phase in Web retrieval interactions, hence the understanding of its underpinnings remains crucial.

To put in context this research, we adapt the formal model put forward by Fuhr [1] reproduced in (Fig. 1), although we interpret the signs there differently, and we let $\overline{Q}$, $\overline{D}$, and $\overline{R}$ respectively stand for a set of information needs for a querying user, a set of information-bearing percepts and psychological capability whereby a particular user is going to judge the relevance of the information percepts for her information needs. We may call them *queries*, *documents* and *relevance assignments of documents to queries.*

Likewise, let Q, D and R be the outcome of as many instantiation processes of the above-mentioned information needs, information supplies and relevance judgments, respectively. We will call them *query representations*, *documents representations* and *relevance judgments* and assume that the relevance judgment representations may adopt the form of a relevance relation, $R \subseteq D \times Q$.

F.J. Valverde-Albacete: *Combining Soft and Hard Techniques for the Analysis of Batch Retrieval Tasks*, StudFuzz **197**, 239–256 (2006)
www.springerlink.com

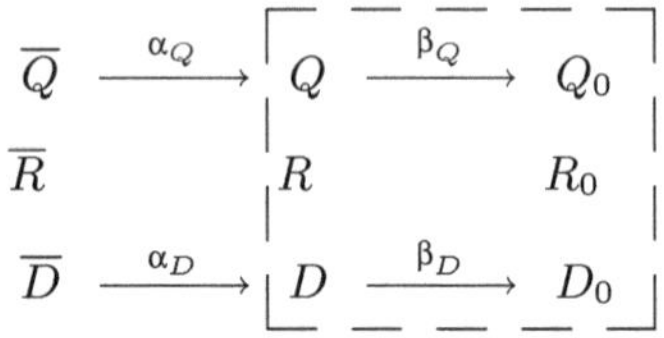

Fig. 1. An adaptation of the conceptual model of Fuhr [1] with the concepts dealt with in this paper highlighted.

Finally, let Q_0, D_0, R_0 be the *query descriptions*, *document descriptions* and the *relevance judgments in description space* respectively, so that $R_0 \subseteq D_0 \times Q_0$. These arise when we try to approximate the information content of query representations and document representations for surrogates in a system capable of estimating an approximate relevance relation $\hat{R}_0$ to a certain degree of competence, that is, an information retrieval system. Hence descriptions will sometimes also be called *surrogates* for their representations.

The model also posited four maps between the above-introduced domains, of which we concentrate on two and change their meaning to represent:

A query description process, $\beta_Q : Q \to Q_0$, mapping from query representations to query descriptions suitable for processing in a particular information retrieval system, that is, a (query) representation adaptation process.

A document description process, $\beta_D : D \to D_0$, mapping from document representations to document descriptions, that is, a (document) representation adaptation process.

Therefore, we limit ourselves to the domains, mappings and sets enclosed by the square in Fig. 1, the representation and description-related domains.

The Statement of the Task

A batch retrieval task is an elaboration of the so-called "Cranfield model of Information Retrieval system evaluation" [2] as a set of document records, or *collection*, D_T, a set of sampled query records, or *questions*, Q_T, and a set of relevance judgments involving documents and query records, $R_T \subseteq D_T \times Q_T$. Curiously, the existence of a generalisation of the set of queries $Q \supseteq Q_T$ is implicit in this definition and therefore the existence of a bigger collection of documents, $D \supseteq D_T$, and an ideal relation $R \supseteq D \times Q$ capturing the perfect relevance of all possible documents to all possible queries, of which $R_T \subseteq R$ is but a part.

An ideal system $S_{D,Q}(R) = \langle \varrho_R \rangle$ would consist in a (query) retrieval function with respect to the perfect relevance, ϱ_R, returning only relevant documents where $\varrho_R(q_i)$ is the set of documents *relevant to query* q_i as dictated by the ideal relevance relation, R.

$$\varrho_R : Q \to 2^D \qquad (1)$$
$$q_i \mapsto \varrho_R(q_i) = \{d_j \in D | d_j R q_i\}$$

Because we may incur in errors, approximations and the like in the process of building an information retrieval system, $S_{D,Q}$, we accept that the actual relevance relation implemented will be some $\hat{R} \neq R$. Hence the actual system implemented $S_{D,Q}(\hat{R}) = \langle \varrho_{\hat{R}} \rangle$ may return $\varrho_{\hat{R}}(q_i)$ the set of documents *retrieved for the same query as dictated by the approximate relevance* $\hat{R}$. Assessing the quality of $S_{D,Q}(\hat{R})$ means, essentially, comparing R and $\hat{R}$.

Making Some Hypotheses Explicit

So far, we have gathered results that could be readily obtained from the information given in the relevance relation. However, the information retrieval paradigm seems to use implicitly two notions of order between queries and between documents based in the "accumulation of content":

1. (focus narrowing) the more content a query or document has, the more specific it is.
2. (scope widening) the more content a query or document has has, the more generic it is.

We refer to these orders as *specificity orders* and will try to capture them through the statement of two formal search hypotheses: *conjunctive querying* considers the increase in content to narrow the search (as with a Boolean AND operator) whereas *disjunctive querying* considers the increase in content to broaden the search (as per a Boolean OR operator). For these hypotheses on sets we lift the definition of the retrieval function in 1 to sets of queries,

$$\varrho_R^1 : 2^Q \to 2^D \qquad \varrho_R^2 : 2^Q \to 2^D \qquad (2)$$
$$\varrho_R^1(B) = \{d_i \in D | \forall q \in B, d_i R q\} \qquad \varrho_R^2(B) = \{d_i \in D | \exists q \in B, d_i R q\}$$

where ϱ_R^1 implies the conjunctive querying hypothesis and ϱ_R^2 the disjunctive querying hypothesis above. Note that on a single query both definitions coincide.

Hypothesis 1 (Conjunctive querying).

1. *For queries, $q_i \leq q_j$ means query q_j is more specific than query q_i, that is, all documents relevant to query q_j are also relevant to query q_i. In other words, the more specific a query, the less relevant documents will be retrieved for it. Therefore, ϱ_R^1 is contra-variant in the amount of content*

$$q_i \leq q_j \Rightarrow \varrho_R^1(q_i) \supseteq \varrho_R^1(q_j) \qquad (3)$$

Consequently, the answer to a set of queries is less extensive or as extensive as the answer to any of the queries alone,

$$\forall B \in Q, \forall q \in B \qquad \varrho_R^1(B) \subseteq \varrho_R^1(q)$$

2. *For sets, the set of documents relevant to a subset is more extensive than the set of documents relevant to its superset, hence a conjunctive retrieval function on sets is also* contra-variant,

$$\forall B_1, B_2 \in Q \qquad B_1 \subseteq B_2 \Rightarrow \varrho^1_R(B_1) \supseteq \varrho^1_R(B_2) \tag{4}$$

Hypothesis 2 (Disjunctive querying).

1. *For queries, $q_i \leq q_j$ means that query q_j is more general than query q_i, that is, all documents relevant to query q_i are also relevant to query q_j. In other words, the more generic a query, the more relevant documents will be retrieved for it, that is, ϱ_R is covariant in the amount of content*

$$q_i \leq q_j \Rightarrow \varrho^2_R(q_i) \subseteq \varrho^2_R(q_j) \tag{5}$$

   Consequently, the answer to a set of queries is more extensive or as extensive as the answer to any of the queries alone,

$$\forall B \in Q, \forall q \in B \qquad \varrho^2_R(B) \supseteq \varrho^2_R(q)$$

2. *For sets, the set of documents relevant to a subset is less extensive than the set of documents relevant to its superset, hence a disjunctive retrieval function on sets is* covariant,

$$\forall B_1, B_2 \in Q \qquad B_1 \subseteq B_2 \Rightarrow \varrho^2_R(B_1) \subseteq \varrho^2_R(B_2) \tag{6}$$

Therefore, choosing any of this hypothesis implies choosing whether the retrieval function is covariant or contra-variant; likewise, prescribing that the retrieval function is order-preserving or order-reversing amounts to choosing between these two hypotheses. In any case, under any of them, each $(Q, \leq)$, $(D, \leq)$ is at least a partial order, a constraint that is usually nowhere made explicit in retrieval tasks. We will consider only the conjunctive hypothesis in this chapter from now on.

A Decomposition of the Problem

We have found convenient to solve a batch retrieval task by decomposing it in the following subproblems:

Problem 1. Given domains of documents D and queries Q (whether they be descriptions or representations), a querying hypothesis and a prescribed (given) relevance relation R[1], build an information retrieval system that faithfully implements the prescribed relevance.

[1] We use here R as a variable ranging over possible relation values, not necessarily the optimal one.

Problem 2. Given different spaces of representations, Q, and descriptions, Q_0, for queries, find a mapping, β_Q, between them. Do likewise for document representations, D, descriptions, D_0, and mapping, β_D, between them.

Problem 3. Given local information about R, a relevance relation in the form of a training subset $R_T \subseteq R$, extend/generalise such information to the whole of $D \times Q$ in problem 1 above.

In the rest of the chapter, we first look in Sect. 2 for a solution to problem 1 in terms of Formal Concept Analysis notions, to analyse it later in Sect. 3 in terms of some approximation spaces introduced by the solution, for which we provide an example; in Sect. 4 we look into the implications of using description mappings over the sets of queries and documents and we try to give a solution to problem 2 that takes into consideration that given for problem 1. Finally, we review previous work related to the use of Rough Set analysis and Formal Concept Analysis for information retrieval and discuss on possible avenues of research regarding our solution to problem 2. A (very) preliminary treatment of solutions for problem 3 may be found in [3].

2 A Perfect Information Retrieval System

A solution to problem 1 in Sect. 1 above has been presented in [3] based in the concept of a *polarity* or Galois connection between two power-sets [4].

Galois Connections and Adjunctions

Let $\mathcal{D} = \langle D, \leq \rangle$ be a partially ordered set or *poset* and denote its order-theoretic dual by $\mathcal{D}^{op} = \langle D, \geq \rangle$.

Definition 1. *[5] A Galois Connection between posets $\mathcal{D} = \langle D, \leq \rangle$ and $\mathcal{Q} = \langle Q, \leq \rangle$ is a quadruple $\pi = \langle \mathcal{D}, \pi_*, \pi^*, \mathcal{Q} \rangle$ with $\pi_* : D \rightarrow Q$ and $\pi^* : Q \rightarrow D$ a pair of functions such that*

$$\forall d \in D, q \in Q, \quad d \leq \pi^*(q) \Leftrightarrow \pi_*(d) \leq q \tag{7}$$

and we also write $\pi : \mathcal{D} \multimap Q$ for the whole connection, or just $\langle \pi_*, \pi^* \rangle$, where π_* and π^* are the left and right adjoint parts *of π, respectively.*

Of particular interest to us are the Galois connections between the power-sets of two sets D and Q, because all such connections are induced by binary relations $R \in 2^{D \times Q}$ between the underlying sets [6, 5, 7, 4]:

Proposition 1 (Polarity). *Every relation $R \in 2^{D \times Q}$ induces a Galois connection $R_*^* : 2^D \multimap (2^Q)^{op}$, between the power-set of D (as an order), $\langle D, \subseteq \rangle$ and the (order-theoretic) dual power-set of Q, $\langle Q, \supseteq \rangle = \langle Q, \subseteq^{op} \rangle$ whose components, $\langle R_*, R^* \rangle$,*

$$\begin{aligned} R_* &: 2^D \to 2^Q & R^* &: 2^Q \to 2^D \\ R_*(A) &= \{q \in Q \mid \forall d \in A, dRq\} & R^*(B) &= \{d \in D \mid \forall q \in B, dRq\} \end{aligned} \tag{8}$$

are dually adjoint *to each other, because they both commute with indexed intersections, and the dual adjunction condition is:*

$$B \subseteq R_*(A) \Leftrightarrow A \subseteq R^*(B). \tag{9}$$

Galois connections go hand in hand with a complementary notion, that of *adjunction* or order-preserving (instead of order-reversing) homomorphisms between partially ordered sets. We will consider in particular, *axialities*, which are the complementary notion to polarities, that is, adjunctions between power-sets:

Proposition 2 (Axiality). *Every relation* $\beta \in 2^{D \times D_0}$ *induces an adjunction* ${\beta_\exists}^\forall : 2^D \leftrightharpoons 2^{D_0}$*, between the power-set of* D *(as an order),* $\langle 2^D, \subseteq \rangle$ *and the power-set of* D_0*,* $\langle 2^{D_0}, \subseteq \rangle$*, known as an* axiality *whose components,* $\langle \beta_\exists, \beta^\forall \rangle$*,*

$$\begin{aligned} \beta_\exists &: 2^D \to 2^{D_0} & \beta^\forall &: 2^{D_0} \to 2^D \\ \beta_\exists(A) &= \{d_0 \in D_0 \mid \exists d \in A, d\beta d_0\} & \beta^\forall(A_0) &= \{d \in D \mid d\beta \subseteq A_0\} \end{aligned} \tag{10}$$

are said adjoint *to each other, because* $\beta_\exists$ *commutes with indexed unions,* $\beta^\forall$ *commutes with indexed intersections, and the adjunction conditions is:*

$$A \subseteq \beta^\forall(A_0) \Leftrightarrow \beta_\exists(A) \subseteq A_0 \tag{11}$$

The Polarity of Conjunctive Querying

A cursory inspection shows that the conjunctive retrieval function ϱ^1_R of a particular relevance relation and the *first* adjoint of the polarity it induces are the same function.

This suggests, that we hypothesise the existence of a *document retrieval function* ι^1_R equivalent to the dual adjoint R_*:

$$\iota^1_R(A) = R_*(A) = \{q \in Q \mid \forall d \in A, dRq\} \tag{12}$$

returning all the queries which are pertinent to a set of documents. This function completes the pair $\langle \iota^1_R, \varrho^1_R \rangle$ which we will call the *polarity of conjunctive querying*. To improve readability we will often use an operator-like notation:

$$\varrho^1_R[\iota^1_R(A)] = {R_*}^*(A) = {A_*}^* \qquad \iota^1_R[\varrho^1_R(B)] = {R^*}_*(B) = {B^*}_* \tag{13}$$

the latter when the relevance relation is understood from context.

The case of polarities has been extensively studied in the domain of Formal Concept Analysis [7, 8], where the triple $\mathbb{K} = (D, Q, R)$ is called a *formal context*. Pairs (A, B) such that $A = R^*(B) \Leftrightarrow B = R_*(A)$ are called *formal* *concepts (of context* $\mathbb{K}$ *)*, the set of formal concepts of $\mathbb{K}$, notated as $\mathfrak{B}(\mathbb{K})$ and their first component is the *extent* of the concept while the second component is the *intent* of the concept.

$$\begin{array}{ll} \text{ext} : \mathfrak{B}(\mathbb{K}) \to 2^D & \text{int} : \mathfrak{B}(\mathbb{K}) \to 2^Q \\ \text{ext}\,(A, B) = A & \text{int}\,(A, B) = B \end{array} \tag{14}$$

$\mathfrak{B}(\mathbb{K})$, is naturally ordered by the partial inclusion order of extents and the reverse inclusion order of intents

$$(A_1, B_1) \leq (A_2, B_2) \Leftrightarrow A_1 \subseteq A_2 \Leftrightarrow B_1 \supseteq B_2 \tag{15}$$

In fact we have the following theorem asserting that it is a complete lattice:

Theorem 1 (Basic theorem on Concept Lattices ([8], p. 20)).

1. *The concept lattice* $\underline{\mathfrak{B}}(D, Q, R)$ *is a complete lattice in which infimum and supremum are given by:*

$$\bigwedge_{i\in I}(A_i, B_i) = \left(\bigcap_{i\in I} A_i, \left[\bigcup_{i\in I} B_i\right]^*_{\;*}\right) \tag{16}$$

$$\bigvee_{i\in I}(A_i, B_i) = \left(\left[\bigcup_{i\in I} A_i\right]^*_{\;*}, \bigcap_{i\in I} B_i,\right) \tag{17}$$

2. *A complete lattice* $\mathcal{V} = \langle V, \leq\rangle$ *is isomorphic to* $\underline{\mathfrak{B}}(D, Q, R)$ *if and only if there are mappings* $\tilde{\gamma} : D \to V$ *and* $\tilde{\mu} : Q \to V$ *such that* $\tilde{\gamma}(D)$ *is supremum-dense in* $\mathcal{V}$*,* $\tilde{\mu}(Q)$ *is infimum-dense in* V *and* dRq *is equivalent to* $\tilde{\gamma}(d) \leq \tilde{\mu}(q)$ *for all* $d \in D$ *and* $q \in Q$*. In particular* $\mathcal{V} \cong \underline{\mathfrak{B}}(V, V, \leq)$*.*

Gathering Results: the Solution to Problem 1

With the construction sketched in the previous section, essentially every formal concept analysis intuition and result translates into information retrieval ones. In a nutshell, building a perfect information retrieval system amounts to building the concept lattice of the relevance relation. The extrema of the lattice, are especially interesting elements:

- The *top*, $\top = ({\varnothing_Q}^*, {{\varnothing_Q}^*}_*)$, represents the whole set of documents (maximal extent) and the queries for which all documents are relevant (minimal intent), hence it exactly detects those queries which are useless for distinguishing between documents.
- The *bottom*, $\bot = ({\varnothing_{D*}}^*, \varnothing_{D*})$, represents the whole set of queries (maximal intent) and the documents relevant to all queries (minimal extent), hence those documents which are useless for distinguishing between queries.

For single documents or queries we may obtain their relevance concepts by two special functions reminiscent of those in the theorem:

$$\begin{aligned} \gamma_R &: D \to \underline{\mathfrak{B}}(D, Q, R) & \mu_R &: Q \to \underline{\mathfrak{B}}(D, Q, R) \\ \gamma_R(d) &= (\{d\}_*{}^*, \{d\}_*) & \mu_R(q) &= (\{q\}^*, \{q\}^*{}_*) \end{aligned} \tag{18}$$

and we will refer to them as a *document* and *query concepts*, respectively.

For more complicated document sets $A \in D$ and queries $B \in Q$ we obtain the important fact that any relevance concept, $(A, B) \in \underline{\mathfrak{B}}(D, Q, R)$, may be obtained through conjunctive querying or conjunctive grouping of documents and the concept building functions:

$$\bigvee_{d_j \in A} \gamma_R(d_j) = (A, B) = \bigwedge_{q_i \in B} \mu_R(q_i) \tag{19}$$

The above and the definition of meets and joins allow to write succinctly the two interaction constructs with the system as a conjunction of queries or documents:

$$\begin{aligned} \tilde{\gamma}_R &: 2^D \to \underline{\mathfrak{B}}(D, Q, R) & \tilde{\mu}_R &: 2^Q \to \underline{\mathfrak{B}}(D, Q, R) \\ \tilde{\gamma}_R(A) &= \bigvee_{d_j \in A} \gamma_R(d_j) = (A_*{}^*, A_*)) & \tilde{\mu}_R(B) &= \bigwedge_{q_i \in B} \mu_R(q_i) = (B^*, B^*{}_*) \end{aligned} \tag{20}$$

Notice that for all $B \subseteq Q$, $\tilde{\mu}_R(B)$ encloses in its extent the documents relevant to B, and in its intent, the list of queries which are pertinent to the same documents as B for the relevance context, $\mathbb{K}$, and likewise, mutatis mutandis for $\tilde{\gamma}_R(A), A \subseteq D$. Therefore an implementation of a retrieval function in terms of Formal Concept Analysis is, $\varrho_R^1(q) = \mathrm{ext}(\mu_R(q))$, and any lattice construction algorithm is essentially a solution to problem 1 above that may be used to solve two instances of the system-building problem:

- Building the perfect information retrieval system in representations space, $S_{(D,Q)}(R)$, given R and
- Building the surrogate information retrieval system in description space, $S_{(D_0,Q_0)}(R_0)$, given R_0.

Building such lattices is not out main purpose in this paper. Rather, the remarkable fact that the sets of queries and documents, as related by a relevance relation, exhibit the extraordinary structure of a concept lattice.

3 Relevance-Induced Analysis of Retrieval Systems

Although the description achieved at in the previous section uses "hard" techniques we want to state in the present section that the interpretation of their results can only be accomplished by the use of "soft" techniques, in particular, rough set analysis.

Equivalences and Partitions Defined by Relevance Relations

For that purpose, we now build two relations over queries θ_R and documents φ_R based in the basic functions above:

$$\begin{aligned} \forall q_i, q_j \in Q, q_i \equiv q_j \,(\text{mod}\; \theta_R) \Leftrightarrow \mu_R(q_i) = \mu_R(q_j) \Leftrightarrow \varrho^1_R(q_i) = \varrho^1_R(q_j) \\ \forall d_i, d_j \in D, d_i \equiv d_j \,(\text{mod}\; \varphi_R) \Leftrightarrow \gamma_R(d_i) = \gamma_R(d_j) \Leftrightarrow \iota_R(d_i) = \iota_R(d_j) \end{aligned} \quad (21)$$

These are clearly equivalences: θ_R amounts to partitioning the set of queries Q into classes $[q]_{\theta_R}$ of queries pertinent to the same set of documents, or *iso-pertinence classes*, and φ_R amounts to partitioning the set of documents into *iso-relevance* classes, $[d]_{\varphi_R}$, such that:

$$[q]_{\theta_R} = \{q_j \in Q | q \equiv q_j \,(\text{mod}\; \theta_R)\} \quad [d]_{\varphi_R} = \{d_i \in D | d \equiv d_i \,(\text{mod}\; \varphi_R)\} \quad (22)$$

Such equivalences induce partitions π_{θ_R} and π_{φ_R} of the original sets of queries and documents, respectively:

$$\pi_{\theta_R} = \{[q]_{\theta_R} \mid q \in Q\} \qquad \pi_{\varphi_R} = \{[d]_{\varphi_R} \mid d \in D\} \quad (23)$$

Rough Set Analysis of Relevance

Borrowing a construction from rough set analysis [9, 10] already present in information retrieval [11, 12], we call an *approximation space for queries* the tuple $\langle Q, \theta \rangle$ of a domain and an equivalence relation $\theta \in 2^{Q \times Q}$. Likewise, we can define an *approximation space for documents*, $\langle D, \varphi \rangle$.

In such approximation spaces we may define an *upper approximation* $\overline{}^{\theta}$ and a *lower approximation* $\underline{}_{\theta}$ operators based in the equivalence classes, $c_i \in \pi_\theta$:

$$\overline{B}^{\theta} = \bigcup \{c_i \in \pi_\theta | c_i \cap B \neq \varnothing\} \qquad \underline{B}_{\theta} = \bigcup \{c_i \in \pi_\theta | c_i \subseteq B\} \quad (24)$$

where the upper approximation is a closure operator and the lower approximation an interior operator [7], hence $\underline{B}_\theta \subseteq B \subseteq \overline{B}^{\theta}$. An *exact* set (in the approximation space induced by θ) is a set composed of equivalence classes (of θ). All other sets are *approximate*. One can easily prove that B is exact if and only if $\underline{B}_\theta = \overline{B}^{\theta}$.

Note that, given a relevance relation R, for a particular query $q \in Q$ and a document $d \in D$:

$$\overline{\{q\}}^{\theta_R} = [q]_{\theta_R} \supseteq \{q\} \qquad \overline{\{d\}}^{\varphi_R} = [d]_{\varphi_R} \supseteq \{d\} \quad (25)$$

$$\varnothing \neq \underline{\{q\}}_{\theta_R} \Leftrightarrow \underline{\{q\}}_{\theta_R} = \{q\} \qquad \varnothing \neq \underline{\{d\}}_{\varphi_R} \Leftrightarrow \underline{\{d\}}_{\varphi_R} = \{d\} \quad (26)$$

that is, singleton query (document) sets are only exact when they are iso-pertinence (iso-relevance) classes. Therefore:

- any rational agent[2] operating as an information retrieval system will return the same set of documents for all queries iso-pertinent to q, that is, *queries are submitted as iso-pertinence classes*:

$$\forall q_i \in \overline{q}^{\theta_R} \Rightarrow \varrho^1_R(q_i) = \varrho^1_R(\overline{q}^{\theta_R}) \tag{27}$$

- if the set of relevant documents for a query q is non-void, for each relevant document, d, with dRq, all its iso-relevance documents will also have to be judged relevant, that is, *the retrieved relevant set is an aggregate of classes of iso-relevance*

$$\varrho^1_R(q) \neq \varnothing \Rightarrow \varrho^1_R(q) = \overline{\varrho^1_R(q)}^{\varphi_R} = \overline{\varrho^1_R(\overline{q}^{\theta_R})}^{\varphi_R} \tag{28}$$

Given a perfect retrieval system for a prescribed relevance R, we may conclude that the retrieval set provided by any implemented system is *doubly approximated*. To see this, suppose we managed to approximate the relevance relation by $\hat{R}$ in an implemented information retrieval system. For a given query q, the system would retrieve documents $\varrho_{\hat{R}}(q)$ whereas the relevant documents would still be given by the prescribed relevance as $\varrho^1_R(q)$. Therefore the retrieved relevant documents for each query $q \in Q$ would be $\varrho^1_R(q) \cap \varrho_{\hat{R}}(q)$, and we would have precision and recall as:

$$P_{\hat{R}}(q) = \frac{|\varrho^1_R(q) \cap \varrho_{\hat{R}}(q)|}{|\varrho_{\hat{R}}(q)|} \qquad R_{\hat{R}}(q) = \frac{|\varrho^1_R(q) \cap \varrho_{\hat{R}}(q)|}{|\varrho^1_R(q)|} \tag{29}$$

Notice that R and $\hat{R}$ being different, they generate different partitions, and classes of iso-pertinence and iso-relevance:

$$\overline{q}^{\theta_R} \neq \overline{q}^{\theta_{\hat{R}}} \qquad \overline{d}^{\theta_R} \neq \overline{d}^{\theta_{\hat{R}}} \tag{30}$$

Hence, due to the approximation effect described in 27, the prescribed system would request $\overline{q}^{\theta_R}$, but the implemented system actually requests $\overline{q}^{\theta_{\hat{R}}}$. Similarly, due to the effect made evident in 28, the prescribed system would retrieve the "granular" approximation $\overline{\varrho^1_R(q)}^{\varphi_R}$ but the implemented system actually returns $\overline{\varrho_{\hat{R}}(q)}^{\varphi_{\hat{R}}}$ based in other granules.

Notice that the different granularities are responsible for the ordering between partitions: if all the granules in one partition π_i appear in a second one, π_j, we say that the partition is *finer* and write $\pi_i \leq \pi_j$, but it may well be the case that none is finer than the other. Given that for perfect precision and recall we would like:

$$P_{\hat{R}}(q) = 1 \Leftrightarrow \varrho^1_R(q) \supseteq \varrho_{\hat{R}}(q) \qquad R_{\hat{R}}(q) = 1 \Leftrightarrow \varrho^1_R(q) \subseteq \varrho_{\hat{R}}(q) \tag{31}$$

In terms of partitions this would entail:

[2] We refer to "rational agents" in this context in the sense of [13] as programs embodying decision rules driven by utility functions.

$$P_{\hat{R}}(q) = 1 \Leftrightarrow \pi_{\theta_R} \geq \pi_{\theta_{\hat{R}}} \qquad R_{\hat{R}}(q) = 1 \Leftrightarrow \pi_{\theta_R} \leq \pi_{\theta_{\hat{R}}} \tag{32}$$

that is, to obtain perfect precision and recall the equivalences induced by R and $\hat{R}$ must generate the same partition. Although, this is a less stringent result than demanding that both relations be *equal*, it is clearly a very strong requisite, failing which precision and recall will not be the optimal.

An Example: the Relevance Lattice of a TREC Task

In order to clarify the concepts in the last sections we present below the concept lattice associated to a task developed over the Wall Street Journal corpus used in old TREC evaluations. Note that, because lattices can be used as indexing, searching and browsing devices [14], this amounts to building an information retrieval system for the (scarce) data provided by the task as training information. However, we do not intend this example to imply that this is the correct way to use concept lattices in building information retrieval systems.

For this particular visualisation we chose topics 201 to 250 as queries and their related documents, with the relevance judgments included in the task taken to mean that each document is relevant for the adequate topic. We used the ConExp tool [15] to upload the data and generate the concept lattice, whose cursory inspection shows that:

- There are initially no non-pertinent queries or irrelevant documents, nor always pertinent or relevant ones. This can be gleaned from the top and bottom concepts.
- Almost half the queries are only pertinent to a single document in its turn only relevant to that particular query. These are single-document, single-query concepts, essentially *incomparable*, that is, with no relation to each other or the rest of the lattice, which clearly form an anti-chain in the order of specificity. In Fig. 2 they have been excised to prevent visual cluttering although they are an important clue that the data do not capture adequately the structure of the underlying domain.
- A more interesting and structured sublattice is focused on in Fig. 2, comprising concepts with a single query in their intent but several documents in their extent, some of which are shared among different concepts. We have highlighted the concept where query 237 appears as pertinent to half a dozen documents also relevant for other queries. Note, however, that 237 is still unrelated to other queries (there are no links to query concepts).

Overall, almost no order structure can be found in the query set and very little in the document set. This seems to be the case also for other corpora in similar batch tasks, suggesting that these queries, documents and topics are too sparse a sampling of the content space they try to capture.

Besides, all iso-pertinent and most iso-relevant classes are singletons, hence it would be fairly easy for an implemented system to generate a partition

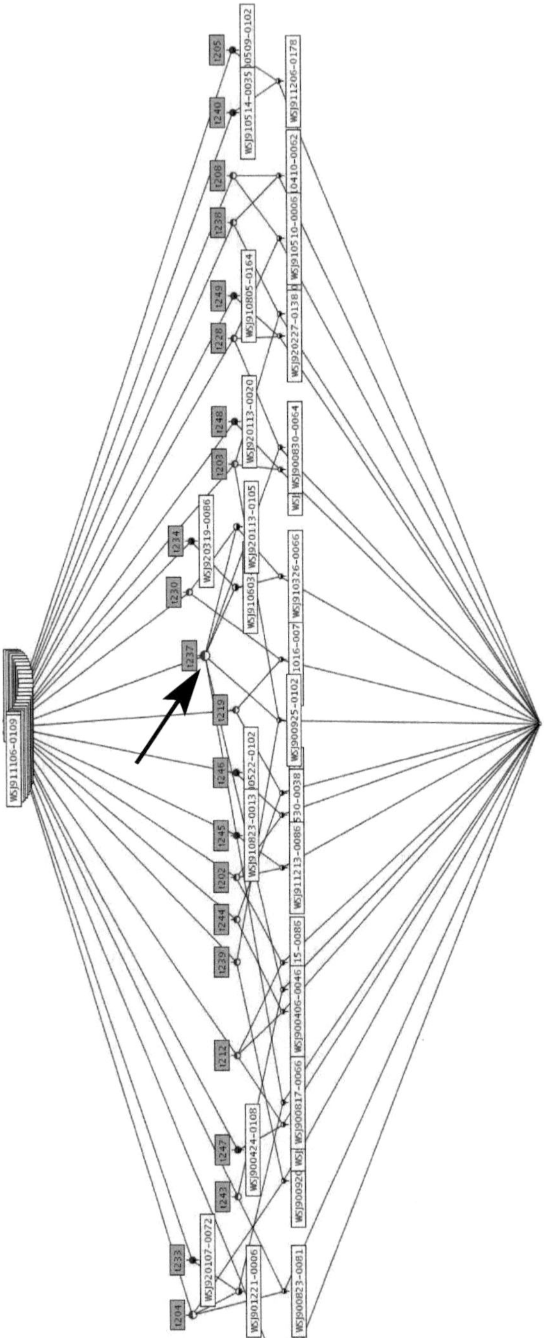

Fig. 2. The prescribed concept lattice for the WSJ topics 201-250 with single-query nodes excised to prevent further visual cluttering. Nodes represent concepts which are annotated with queries and documents above and below respectively. The top concept shows documents irrelevant to all appearing queries, precisely because their queries have been excised. The arrow points at the concept for query 237 mentioned in the text.

similar to this one, so that the performance on this (training) set be quite good while not generalising correctly to unseen data. This is a typical situation for a poorly representative task corpus resulting in a hard learning task.

We believe both these reasons may be at the root of the difficulty in developing information retrieval systems using these data, and their inability to generalise correctly to the whole spectrum of content.

The considerations above also suggest that in developing tasks for broader information spaces, like the Web, not only the quantity of data is important, but also the adequacy of the data to represent the wealth of interactions between pieces of content that we want to capture as queries, documents and the relevance of the latter to the former.

4 Designing the Description Mappings

We now turn to problem 2, that of building the representation-to-description mappings, or simply *description mappings*. First we need to characterise them in order-theoretic terms: it will turn out that the tools of Galois connection theory are of great help for this purpose as well.

Constraints for the Description Mappings

To obtain constraints on the way the functions of Fig. 1 transform representations and descriptions we detail the procedure of retrieving documents for a particular query:

1. First transform a query $q \in Q$ through the query mapping, $\beta_Q(q)$,
2. Then find the retrieval set in the domain of descriptions, $\varrho_{R_0}\left[\beta_Q(q)\right]$,
3. Next retrieve the documents for such descriptions, $\beta_D^{-1}\{\varrho_{R_0}\left[\beta_Q(q)\right]\}$,
4. And finally compare them with the relevant documents, $\varrho_R^1(q)$.

In order to maximise precision and recall, we need:

$$\varrho_R^1(q) = \beta_D^{-1}\{\varrho_{R_0}\left[\beta_Q(q)\right]\} \tag{33}$$

but remembering that retrieval uses iso-relevance classes, this amounts to demanding that iso-relevance classes in representation space map to iso-relevance classes in description space and vice-versa. Also, because of the tight integration of iso-relevance classes (of documents) and iso-pertinence classes (of queries) through the formal concepts of the relevance context, we may expect to find such a bijective mapping between iso-pertinence classes as well[3].

[3] This is more soundly argued in terms of the extensional and intensional continuity of the mappings [16], but reasons of length preclude the discussion in this chapter in such terms.

Hence, in order to simplify formulae, we prefer to compare in the space of document descriptions, thus instead of step 3, we transform the relevant document representations to obtain a necessary constraint:

$$\beta_D\left[\varrho^1_R(q)\right] = \varrho_{R_0}\left[\beta_Q(q)\right] \tag{34}$$

that is, the diagram focused on in Fig. 1 commutes when we interpret the relevance relations R and R_0 as their respective retrieval functions ϱ^1_R and ϱ_{R_0}.

A Solution: Infomorphisms

The notion of an infomorphism applies to the task at hand admirably: given two contexts $\mathbb{K} = (G, M, I)$ and $\mathbb{K}[L] = (H, N, J)$ an informorphism is a pair of mappings $(f^{\rightarrow}, f^{\leftarrow})$ such that $gIf^{\leftarrow}(n) = f^{\rightarrow}(g)Jn$ holds for arbitrary $g \in G$ and $n \in N$ ([17]; [16], def. 7).

The rather cryptic condition on the mappings unfolds in a very convenient theorem ([16], lemma 6): given contexts $\mathbb{K}^{\,d} = (Q, D, R^d)$ and $\mathbb{K}[L]^d = (Q_0, D_0, R^d_0)$, for any pair of informorphisms $(f^{\rightarrow} = \beta_Q, f^{\leftarrow} = \beta_D^{-1})$ the following holds (among other conditions)[4]

$$\forall B \subseteq Q, \beta_D\left[R^{d^*}(B)\right] = R^d_{0*}\left[\beta_Q(B)\right] \tag{35}$$

which, recalling the definition of the retrieval functions in 8, instantiates to 34, that is, gives a much stronger result in which not only singleton queries behave as required but also any set of queries.[5]

In such circumstances, we can further assert ([16], theorem 8) that there exists an axiality between the extent lattices[6] of queries and their representations $\psi_Q : 2^Q \leftrightharpoons 2^{Q_0}$ for which the adjoint maps are defined as:

$$\begin{aligned} \psi_\exists &: 2^Q \to 2^{Q_0} & \psi^\forall &: 2^{D_0} \to 2^D \\ \psi_\exists(B) &= R^{d\;*}_{0*}\left[\beta_Q(B)\right] & \psi^\forall(B_0) &= \beta_Q^{-1}(B_0) \\ &= R^{d^*}_0(\beta_D\left[R^d_*(B)\right]) & &= R^{d^*}(\beta_D^{-1}\left[R^d_{0*}(B_0)\right]) \end{aligned} \tag{36}$$

Given such an adjunction, one may wonder whether there is another adjunction for documents. In fact, theorem 8 and proposition 4 of [16] state that the adjunction above is obtained from a relation $\tilde{\beta}_Q \in 2^{Q\times Q_0}$ such that $\tilde{\beta}_Q(q) = R^{d\;*}_{0*}[\beta_Q(q)]$, and that there is another relation $\tilde{\beta}_D \in 2^{D\times D_0}$

---

[4] The dualisation of contexts is a technical requisite because of our previous treatment of documents as objects and queries as attributes.

[5] We have assumed the involution of inverses, $(\beta_D^{-1})^{-1} = \beta_D$.

[6] We have actually expressed them between the power-sets of queries and representations.

such that $\tilde{\beta}_D^{-1}(d_0) = R^{cd*}{}_*[\beta_D^{-1}(d_0)]$. The latter relation induces an adjunction $\psi_D : 2^D \leftrightharpoons 2^{D_0}$, in a manner compatible with the previous one.

The question of whether the classes of queries or documents transform adequately is in this way completely solved: the consideration of several notions of continuity and closure between the lattices of extents and intents of the relevance relations shows that, with the requirement that $\tilde{\beta}_Q$ and $\tilde{\beta}_D^{-1}$ form an infomorphism, they transform iso-relevance classes of R in representation space into iso-relevance classes of R_0 in description space and likewise for the iso-pertinence classes in representation and description space. But the questions remain open whether less stringent conditions allow the definition of similar, precision- and recall-preserving mappings, and whether efficient algorithmic procedures exist to find such mappings.

5 Related Work and Discussion

Within the Formal Concept Analysis community, techniques like the ones we used in section 2 were applied to *building* information retrieval systems as early as fifteen years ago, starting with the work by Godin and colleagues [18], but the work by Mooers [19] already describes the essential of conceptualising retrieval in a lattice with extents and intents. Ten years ago there were already systems built and evaluated with this technique [20, 21].

Carpineto and Romano later extended this model [21]: their approach was to consider document pre-processing a problem in conceptual clustering including indexing, and browsing the "natural" mode of interfacing the lattice. They also realised that concept intents are related to queries and extents represent all documents that match a query, and found that the use of a thesaurus of terms improves recall while leaving precision unharmed, which was later capitalised on by Priss [22]. An up-to-date review of the state of the art for this approach is contained in [14], where a later development, CREDO, allows to structure the results of a Web interaction into a Concept Lattice. JBraindead [23] is another system aiming at ephemeral clustering of Web retrieval results in terms of Concept Lattices which shows promising performance.

In this paper, unlike previous work on Formal Concept Analysis used for the *design and implementation* of information retrieval systems, we present a technique to *analyse* the performance of information retrieval systems for batch retrieval tasks: by using a Formal Concept Analysis formulation, however, this analysis can easily be transformed into a system building procedure.

In all work reviewed in the previous paragraphs, the explicit consideration of the relevance relation is lacking: the preferred method is to treat relevance indirectly by decomposing texts into terms (the inverted file) with which to scale contexts so as to build the concept lattice, thus considering the mapping between document representations and descriptors the main information-bearing structure [18]. Queries are in this view, sets of terms, so the disparity between queries and documents evident in our framework is lacking there.

Moreover, the analytical framework for Information Retrieval that we present in this paper is missing.

In contrast, our approach tries to understand what the implications for the building of each particular $\hat{R}$ (an approximation of the prescribed relevance relation R) are. We have seen that Formal Concept Analysis can be though of a solution to it given by three choices:

- The choice of the training subdomains of representation space, and the training subrelation, $\mathbb{K}_T = (D_T, Q_T, R_T)$, which implies a bias in the substructure of the ideal relevance relation being focused on.
- The choice of mappings β_D and β_Q from representations into descriptions and vice-versa, both for the documents and queries.
- The choice in the way of extending the training relation in description space, $\mathbb{K}_{0T} = (D_{0T}, Q_{0T}, R_{0T})$, or its lattice $\mathfrak{B}(\mathbb{K}_{0T})$ to cover the whole of description space, $\hat{\mathbb{K}}_{0T} = (D_0, Q_0, \hat{R}_0)$.

These choices notwithstanding, the quality of the system depends crucially on the Rough Set approximations of iso-relevance document sets and iso-pertinence query sets, in the sense that any retrieval action actually refers to such equivalence classes rather than isolated queries or documents. In this sense we claim that the analysis of batch information retrieval tasks both need hard techniques, like Formal Concept Analysis casting of IR concepts in Section 2, and soft techniques, like the Approximation Spaces from Rough Set theory of Section 3.

We have shown in Section 4 that if we want to preserve precision and recall for prescribed R, R_0, it is enough to guarantee that the mappings form an infomorphism. Another matter is whether given a relevance relation R we can find, at the same time a relevance relation in description space R_0, a query mapping β_Q and an inverse document mapping β_D^{-1} so that the performance of the system is effective. In fact the problem reduces, at least, to finding either of the mappings and the adequate R_0, because the second mapping follows from a result we have mentioned: both mappings give raise to an adjunction any of whose adjoint functions uniquely defines the other. In the case where no care is taken to make the mappings and the relevance in description space agree, for instance, all of the Formal Concept Analysis -based systems mentioned above and most (all?) present-day batch information retrieval systems, we cannot expect to obtain perfect precision and recall. In such cases, rough set analysis can help detect those retrieval instances further away from the optimal cases.

Finally, we want to state that our use of Galois connections to formally analyse Information Retrieval concepts has precedents in the interpretation of an inverted file index as a Galois connection and the ensuing claim that the logic of Information Retrieval cannot be Boolean [24], a parallel development which we later found out about.

Acknowledgements

This work has been partially supported by a Spanish Government CICYT grant TIC2002-03713.

References

1. Fuhr, N.: Probabilistic models or information retrieval. The Computer Journal **35** (1992) 243–255
2. Meadow, C.T., Boyce, B.R., Kraft, D.H.: Text Information Retrieval Systems. Second edn. Library and Information Sciences. Academic Press, San Diego and San Francisco (2000)
3. Valverde-Albacete, F.J.: A Formal Concept Analysis look into the analysis of information retrieval tasks. In Ganter, B., Godin, R., Mephu-Nguifo, E., eds.: Formal Concept Analysis. Proceedings of the 3nd International Conference on Formal Concept Analysis, ICFCA 2005, Lens, France. Supplementary volume. IUT de Lens – Universit d'Artois, France, Lens, France (2005) 31–46
4. Denecke, K., Erné, M., Wismath, S., eds.: Galois Connections and Applications. Number 565 in Mathematics and Its Applications. Kluwer Academic, Dordrecht, Boston and London (2004)
5. Erné, M., Koslowski, J., Melton, A., Strecker, G.: A primer on Galois connections. In Todd, A., ed.: Proceedings of the 1991 Summer Conference on General Topology and Applications in Honor of Mary Ellen Rudin and Her Work. Volume 704 of Annals of the New York Academy of Sciences., Madison, WI, USA, New York Academy of Science (1993) 103–125
6. Birkhoff, G.: Lattice theory. 3rd ed. edn. American Mathematical Society (1967)
7. Davey, B., Priestley, H.: Introduction to lattices and order. 2nd edn. Cambridge University Press, Cambridge, UK (2002)
8. Ganter, B., Wille, R.: Formal Concept Analysis: Mathematical Foundations. Springer, Berlin, Heidelberg (1999)
9. Pawlak, Z.: Rough Sets. Theoretical Aspects of Reasoning about Data. Kluwer, Dordrecht (1991)
10. Polkowski, L.: Rough Sets. Mathematical Foundations. Advances in Soft Computing. Physica-Verlag, Heidelberg and New York (2002)
11. Das-Gupta, P.: Rough sets and information retrieval. In: Proceedings of the 11th annual international ACM SIGIR conference on Research and development in information retrieval, ACM Press (1988) 567–581
12. Wong, S.K.M., Ziarko, W.: A machine learning approach in information retrieval. In: Proceedings of the 9th annual international ACM SIGIR conference on Research and development in information retrieval, ACM Press (1986) 228–233
13. Ferber, J.: Multi-Agent Systems: An Introduction to Distributed Artificial Intelligence. Addison-Wesley (1995)
14. Carpineto, C., Romano, G.: Concept Data Analysis: Theory and Applications. Wiley (2004)
15. Yevtushenko, S.A.: System of data analysis "Concept Explorer". In: Proceedings of the 7th national conference on Artificial Intelligence KII-2000, Russia, ACM (2000) 127–134 (In Russian) http://sourceforge.net/projects/conexp. 18 Francisco J. Valverde-Albacete

16. Hitzler, P., Krötzsch, M., Zhang, G.Q.: Morphisms in context. In: Proceedings of the 13th International Conference on Conceptual Structures, ICCS'05. (2005)
17. Barwise, J., Selignan, J.: Information Flow. The Logic of Distributed Systems. Volume 44 of Cambridge Tracts IN Theoretical Computer Science. Cambridge University Press (1997)
18. Godin, R., Gecsel, J., Pichet, C.: Design of a browsing interface for information retrieval. In: Proceedings of the 12th International Conference on Research and Development in Information Retrieval (ACM SIGIR '89), Cambridge, MA, ACM (1989) 32–39
19. Mooers, C.: A mathematical theory of language symbols in retrieval. In: Proceedings of the International Conference on Scientific Information, Washington, D.C. (1958)
20. Godin, R., Missaoui, M., April, A.: Experimental comparison of navigation in a galois lattice with a conventional information retrieval methods. International Jounal of Man-Machine Studies **38** (1993) 747–767
21. Carpineto, C., Romano, G.: A lattice conceptual clustering system and its application to browsing retrieval. Machine Learning **24** (1996) 95–122
22. Priss, U.E.: A graphical interface for document retrieval based on formal concept analysis. In Santos, E., ed.: Proceedings of the 8th Midwest Artificial Intelligence and Cognitive Science Conference and AAAI Technical Report CF-97-01. (1997) 66–70
23. Cigarrán, J.M., Peñas, A., Gonzalo, J., Verdejo, F.: Automatic selection of noun phrases as document descriptors in an FCA-based information retrieval system. In Ganter, B., Godin, R., eds.: Formal Concept Analysis. Proceedings of the 3rd International Conference on Formal Concept Analysis, ICFCA 2005,Lens, France. Number 3403 in LNAI, Berlin, Heidelberg, Springer (2005) 49–63
24. van Rijsbergen, C.: The Geometry of Information Retrieval. Cambridge University Press, Cambridge, New York, Melbourne, Madrid and Cape Town (2004)

Part IV

Web Application

Search Advertising

Marco Cristo[1], Berthier Ribeiro-Neto[12], Paulo B. Golgher[2], and Edleno Silva de Moura[3]

[1] Computer Science Department, Federal University of Minas Gerais, Belo Horizonte, Brazil
{marco,berthier}@dcc.ufmg.br
[2] Google Brazil, Belo Horizonte, Brazil
{berthier,golgher}@google.com
[3] Computer Science Department, Federal University of Amazonas, Manaus, Brazil
edleno@dcc.ufam.edu.br

Summary. The current boom of the Web Marketing is associated with the revenues originated from search advertising, which has become the driving force sustaining monetization of Web services. The search advertising market is forecast to grow from US $3.6 billion in 2004 to US $11.2 billion by 2010. Further, forecasts suggest that its influence will increase in the upcoming years through diversification and the production of new types of search-related advertising. This rapidly consolidating market involves complex business networks and increasingly sophisticated technology. Thus, the exploitation of new forms of search services requires advances in two fronts: the commercial front and the technology front. In this chapter we discuss in somewhat detail the key concepts and variables related to search advertising, both in the commercial and in the technology fronts.

1 Introduction

The Internet's emergence represented a new marketing opportunity to any company – the possibility of global exposure to a large audience at a dramatically low cost. In fact, during the 90's many enterprises were willing to spend great sums on advertising in the Internet with apparently no concerns about their investment return [56]. As a result, the Internet became the media of fastest growth in its first five years, according to the Interactive Advertising Bureau [17].

This situation radically changed in the following decade, when the failure of many Web companies led to a dropping in supply of cheap venture capital. This lead to wide concern over the value of these companies as reliable marketing partners and, as a result, to considerable reduction in on-line advertising investments [55, 56]. Such reduction caused consecutive declines of quarterly company revenues in the US market, beginning with the first quarter of 2001.

M. Cristo et al.: *Search Advertising*, StudFuzz **197**, 259–285 (2006)
www.springerlink.com

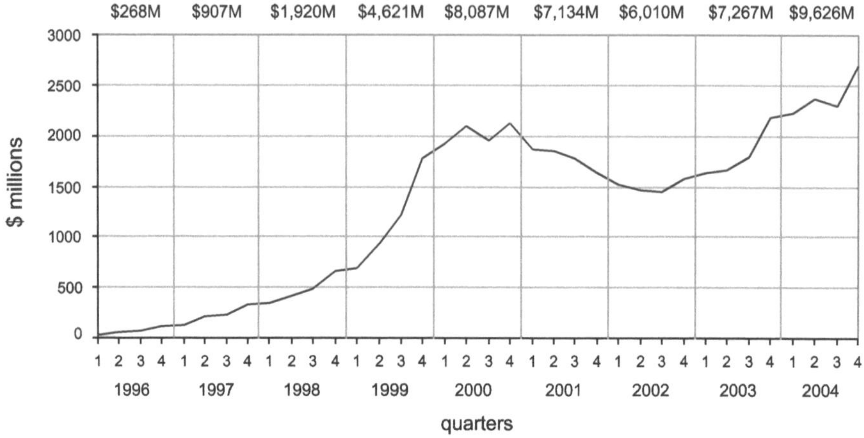

Fig. 1. Quarterly Revenue Growth Comparisons 1996-2004. Source: IAB, 1996-2004.

This loss trend, however, has been reversed by the end of 2002 as seen in Fig. 1. Further, it has been growing steadily since reaching peak values by the end of 2004 [17].

To better understand the reasons for this recover of the online industry, we have to analyze how different Web advertising formats have performed over time. Table 1 shows revenues generated by eight distinct forms of Internet advertising, as measured by IAB[1]: display ads, sponsorships, email, classifieds/auctions, rich media, search, referrals, and slotting fees [1].

As we can see in Table 1, there were important changes in the popularity of the various forms of advertisement (ad). For example, display ads (which include banners) gradually declined from 56 percent in 1998 to 19 percent in 2004. Similar decrease in usage is observed for sponsorships. On the other hand, search advertising rose from 1 percent in 2000 to 40 percent in 2004, becoming the leading form of Internet advertising. Thus, the recovery of Web advertising coincided with the increasing adoption of search advertising. This

[1] Display ads is the format in which advertisers pay on-line companies to display banners or logos on one or more of the company's pages. In Sponsorship advertising, an advertiser sponsors targeted Web site or email areas to build good-will more than traffic to its site. E-mail advertising accounts for ads associated with commercial e-mail communications. In Classifieds and Auctions, advertisers pay on-line companies to list specific products or services. Rich media is a generic term for a variety of interactive ads that integrate video and/or audio. In Referrals, advertisers pay on-line companies for references to qualified leads or purchase inquiries. In Slotting Fees, advertisers pay on-line companies for preference positioning of an ad on the company site. In search advertising, advertisers pay on-line companies to list and/or link the company site to a specific search keyword or page content, as well as to optimize their pages for search engines and ensure their insertion in search indexes.

Table 1. Internet advertising revenues by format in percents – 1998–2004. Source: IAB, 1998–2004

Advertising Formats	1998	1999	2000	2001	2002	2003	2004
Display ads	56	56	48	36	29	21	19
Sponsorships	33	27	28	26	18	10	8
Email	–	2	3	3	4	3	1
Classifieds/auctions	–	–	7	16	15	17	18
Rich media	5	4	6	5	10	10	10
Search	–	–	1	4	15	35	40
Referrals	–	–	4	2	1	1	2
Slotting fees	–	–	–	8	8	3	2
Other	6	11	3	–	–	–	–
Total	100	100	100	100	100	100	100

growth has not been restricted to the USA, since similar gains have been reported in Europe [43]. It is not either a transitory phenomenon, since both advertisers and publishers have announced plans to increase their investments in search advertising [20, 47]. In fact, according to Forrester Research projections, by 2010, search advertising alone will represent a market of US$11.2 billion [29]. As a consequence, an entire new industry offering search advertising related services has emerged, in part by reverse engineering the search engine ranking algorithms [9]. Such services comprehend consultancy on keyword selection, performance analysis, site optimization, etc.

Searching advertising has become the driving force sustaining monetization of Web services through online marketing. Further, forecasts suggest that the influence of search advertising will increase in the upcoming years through diversification and the production of new types of search-related advertising. For instance, in a study conducted by the paid list provider Kanoodle[2], 63% of the advertisers said they would like to target users according to their past behavior and demographics, and 49% would like to look at geographic location of the targets [20]. This clearly points to strong trends such as local search and personalization.

In local search, the user query is answered using data, products, services, and pages that are restricted to a given geographical space such as a state, a city, or a neighborhood. In personal search, the user query is answered using data, products, services, and pages associated with the individual preferences of the user. Both forms of search open interesting new opportunities of targeting online ads with greater precision. This will generate higher interest from the user, increasing ad click-through rate and, consequently, additional revenues for the advertisers. These, on their turn, will tend to increase investments on online markets, closing a cycle of positive reinforcement.

[2] `http://www.kanoodle.com`

The exploitation of new forms of search services requires advances in two fronts: the commercial front and the technology front. In this chapter we discuss in somewhat detail the key concepts and variables related to search advertising, both in the commercial and in the technology fronts. This is important because new matching and ranking algorithms might be dependent on variables affecting the commercial model, and vice-versa.

Given the large success of search advertising as an effective form of monetization and the lack of specialized literature in the academic forefront, we understand that there is a great need of technical biased literature on the subject. Our work here is an attempt of partially address these needs.

2 Basic Concepts

Search advertising is in constant evolution and many different technical approaches have been proposed over the last few years. In this section, we discuss the two main types of search advertising described by IAB: keyword-targeted advertising and content-targeted advertising. We also present the network of actors involved in Search Advertising.

2.1 Keyword-targeted Advertising

Keyword-targeted advertising was introduced by Overture[3] in 1998 [11]. It consists in showing a list of ads at the top or at the right hand side of a Web page search results. The ads displayed have to be "related" to the content of the user query (normally composed of one or more words). Figure 2 shows an example of keyword-targeted advertising.

The list of ads is called a *paid list.* It is composed of a small number of ads, in general, three or five, as shown in the right hand side of Fig. 2. To associate a certain keyword K with one of its ads, the advertiser has to bid on the keyword K in an auction type system. The more the advertiser bids on the keyword K, greater are the chances that its ads will be shown in the paid list associated with that keyword. Notice that the advertisers will only pay for their bids when the users click on their ads. Because of this, keyword-targeted advertising is called a pay-per-performance system. Normally, a minimum click-through rate is required to maintain the association between an ad and a given keyword.

The ads can be presented to the users in different formats. However, they are generally showed as static text, to simplify ad creation and reduce impact on the page download time. This text normally comprises a title, a short description, and a URL address. The content of the title and the description is also called a *creative.* The creative consists of a concise action-oriented text designed to attract the user. As a consequence, hot words such as "free" and

[3] `http://www.overture.com`

Fig. 2. Example of keyword-targeted advertising. The ads are related to the user query "canon camera".

call-to-action phrases like "Click here" or "Enter now to win a..." are very common. Further, ad titles normally include its associated keywords. This is common because ads whose keyword words appear in the title are more likely to be clicked by the users [12]. For the top ad exhibited in Fig. 2, the title is "All the Canon Cameras", the description is "All the Canon 20D Digital Savings! Smart Camera Shoppers Start Here", and the URL address is "Canon.20D.AlltheBrands.com".

By creating a compelling ad, the advertiser expects to convey users to click on it and jump to its *landing page*, that is, the page indicated by the URL address. In the landing page the user will find more information related to the ad or to the company, its products, and services, and possibly will start a transaction. If this is the case, we say that the click was converted into a transaction, an event called *conversion*. Here, a transaction is any action with which an advertiser associates value, such as the purchasing of a product or service, the downloading of a white paper, the requesting of a commercial proposal, etc. We call conversion rate the proportion of received clicks that are converted into transactions.

2.2 Content-targeted Advertising

Content-targeted advertising was introduced by Google[4] in 2002 [20, 24]. It is analogous to keyword-targeted ad in the form advertisers select and pay for keywords. The selection of the ads to be displayed, however, is based on the content of the page p being viewed, instead of on a user query. The page p is called *triggering page*. The mapping process that determines which keywords

[4] `http://www.google.com`

are associated with the content of the triggering page can be partially or totally automated [54]. In some cases, the publisher can inform directly the key topics associated with their pages by means of keywords or categories [24, 33, 48].

Nowadays, content-targeted advertising is the dominant contextual approach in Web marketing. It has raised more interest than other approaches that, previously, were considered more promising, such as behavioral targeting [47]. Once the most relevant and profitable ads are known, they have to be shown to the users. As in the case of keyword-targeted advertising, the ads can be grouped in paid lists and positioned in the triggering page. Figure 3 shows an example of content-targeted advertising in which the ads are exhibited in a paid list in the triggering page.

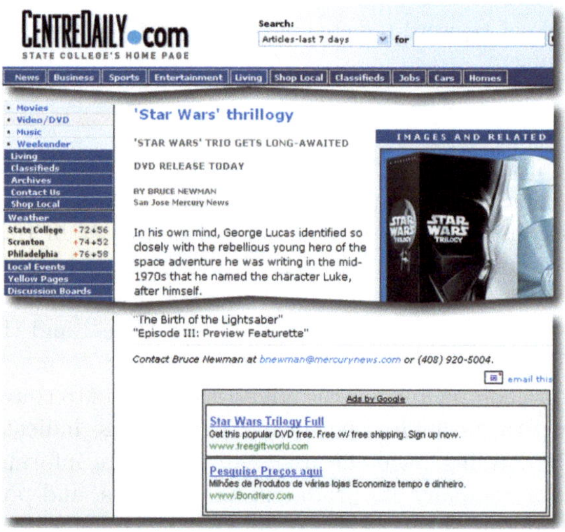

Fig. 3. Example of content-targeted advertising in the page of a newspaper. The middle slice of the page shows the beginning of an article about the launch of a movie on DVD. At the bottom slice, we can see ads picked for this page by Google's content-targeted advertising system, AdSense.

2.3 The Search Advertising Network

Many of the key elements of search advertising are not new. Both contextual placement and payment per performance have been used since the beginning of the Internet. However, these first attempts were crude. Matching systems were very simple. As a result, irrelevant messages were often shown to users, annoying them. Also, questionable practices were very common such as analyzing user behavior without its consent and popping up ads in pages without the permission of their publishers, associating their image with improper companies, products, and services [15, 21, 53]. Such strategies were also not good for most of the advertisers since user attitude towards these approaches was very negative [28, 53]. Consequently, this commonly implied in branding problems for advertisers [15].

It is not surprising that the same elements that work so well in search advertising have failed so consistently in the past. Contrary to these first efforts, search advertising companies have formed networks where all the participant actors are benefited. In fact, the success of search advertising can be credited to the formation of such reliable networks [25]. In general terms, these networks are composed of four main actors: the broker, the advertisers, the publishers, and the users, as depicted in Fig. 4.

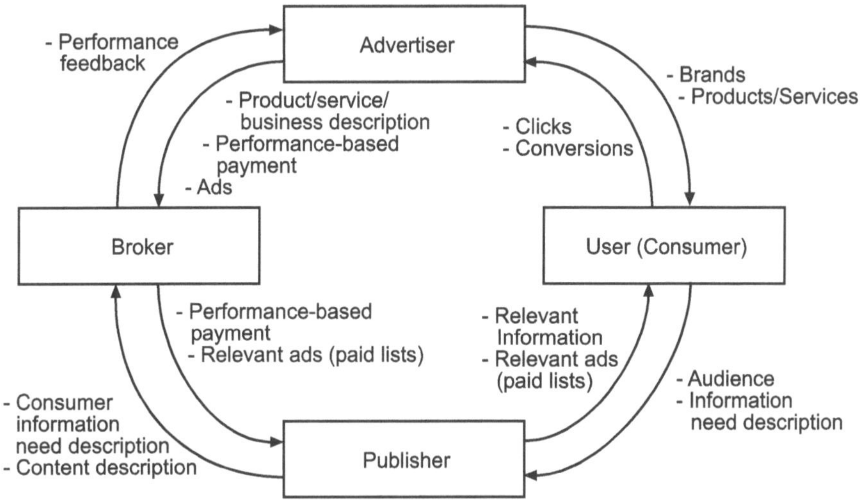

Fig. 4. Search advertising network.

The broker is responsible for the maintenance of the network. It determines which advertisers and publishers can participate of the network as well as the publishing policies to be followed. For example, brokers cannot allow pornographic content, improper language, and copyright violation. They also want to avoid the participation of companies that promote or deal with illegal matters such as drugs or gambling games. Further, they are responsible for making the auction system works by offering tools (interfaces, databases, controlled vocabularies) that the advertisers use to bid on the keywords necessary to describe their products and services. The broker is also responsible for the technology that will be used to match keywords/content and ads [26] and for the measurement systems that will allow evaluating the performance of the publishers and the advertisers [11].

The advertisers participate of the network with the expectation that they will be referred to quality users by the publishers. From the advertisers' point of view, quality users are those which are interested or could become interested in their products or services. This is the case of many users looking for information in directories and search engines or browsing editorial content in the Web. The advertisers compete among themselves for keywords by bidding in an auction system [9,11,18]. They pay to the broker according to the traffic provided to them by the publishers. Based on the performance reports they receive, it is possible to tune their campaigns dynamically, which allows them to maximize their revenues and, by extent, the quality of the overall system [12,45].

The publishers are interested in monetizing their pages through the loyalty of their audience. Thus, the publishers provide the broker with a description of the content of their pages by using keywords and/or categories [33]. In general, however, an automatic system provided by the broker will infer *essencial topics* related to a broker-provided content in an automatic or semi-automatic fashion. Notice that we also consider information gatekeepers, like search engines and directories, as publishers. They provide the broker with the user query or with the directory entries selected by the user. In this case, the publishers payment is based on the traffic they provide.

The last actor in our network is the user or consumer. The users are interested in getting relevant information from the publishers. So, they naturally segment themselves by describing their information needs by means of keywords or by surfing in Web pages whose content is of their interest [43]. Occasionally, they can click on the ads exhibited, jump to the advertisers' pages, and start commercial transactions.

In the next subsections we discuss in more detail the benefits provided by search advertising networks to the users, to the advertisers, and to the publishers.

The Users

An increasing number of people are using the Internet as their main source of information. As a consequence, searching and browsing time are very critical moments in which many users are obtaining information related to their interests or taking direct actions related to them [11]. In fact, an average of four hundred million queries per day was submitted to search engines in 2003. From these, it is estimated that 40% were of commercial nature [28]. The motivation behind search advertising is to catch the user attention in such critical moments. For this, the user should be presented with useful information related to his immediate interest which can be accomplished by algorithms that try to avoid associations in irrelevant contexts or intrusive ways [15,21,28,40,53]. Further, user interests provide a genuine stimulus for the production of higher quality content by the publishers [51]. This occurs because the success of the publishers, in terms of online monetization, depends on the user satisfaction.

Thus, the publishers are stimulated to produce more interesting content to their users in order to ensure their loyalty [53].

In summary, users are more likely to receive content of high quality and relevant to their interests by means of nonintrusive approaches.

The Advertisers

Targeting at users in search and browse time provides the advertisers with the opportunity of popularizing their brands and commercializing their products and services. This leads to a new marketing reality, dominated by a self service of offers [13,35,45,46,53].

Other advantage of Internet as a marketing medium is its flexibility. It allows measuring the performance of publishers and advertisers, providing detailed feedback about the performance of marketing campaigns and strategies. This accountability allows advertisers to control their costs precisely and contributes to the adoption of fair payment strategies, directly related to the acquired benefits [11]. It also makes possible to target audiences by inferring dynamically its past behavior, demographic information, and local information [9].

The Publishers

As mentioned previously, publishers are interested in monetizing their content and increasing user loyalty [38]. However, monetization is a very hard task even for popular sites, which frequently have difficulties in filling all their ad slot spaces [20]. Moreover, experience has shown that users do not want to pay for content and are not willing to pay for traditional Web marketing formats, as display advertising and interstitials have not been very profitable. They normally are seen as an annoyance by users leading to a negative attitude towards the publisher and a loss of audience [38,40].

Search advertising networks offer to publishers the possibility of monetizing all their pages [31]. These networks account for thousands of advertisers interested in appearing associated with content related to their business. For example, Google's network is composed by more than one hundred thousand advertisers in several business areas [30]. These networks have the additional advantage of allowing the participation of small publishers. In traditional Web marketing, it would be very hard for a small publisher to sell ad space to larger advertisers [23,26,36,38,48]. Information gatekeepers have the extra benefit of showing ads in their database anywhere, anytime [15]. More important, since search-related ads are more targeted and passive, the user attitude towards the publisher and its content is positive [40].

3 Search Advertising Systems

Search Advertising applications can be seen as a particular kind of retrieval system, called Best Bets Systems [3]. Such systems are those in which content providers play a more active role in content delivery. In particular, the content providers bet on which the users will consider interesting and compete with each other by the user attention. For this, the advertisers (the content providers) devise queries (for example, the keywords) to select consumers for their ads (the content to be delivered). Notice that, in this case, the direction of the query is reversed since that, in a traditional IR system, the users play an active role and determine through their queries the documents they want to receive.

Retrieval systems can be described in terms of the retrieval model and the retrieval functions they provide. A retrieval model consists of an abstract description of the indexing process, the representations used for documents and queries, the matching process between them and the ranking criteria used to sort results. In the following, we formally define a general retrieval model R and the retrieval functions of a search advertising application expressed in terms of R, according to [3].

Let the retrieval model R be the tuple $\langle D, Q, match, rank\rangle$, where D is a document collection, Q is a query collection, $match\colon Q \times D \rightarrow \{0,1\}$ is a query matching function, and $rank\colon Q \times D \rightarrow [0,1]$ is a ranking function. In a traditional IR system the task is to get the best k ranking documents satisfying query q. Thus, the document retrieval functions can be defined by (1) and (2):

$$search(q, D) = \{d \in D | match(q, d)\} \tag{1}$$

$$searchTop(k, q, D) = \text{top } k \text{ docs in } sort(search(q, D), rank(q, .)) \tag{2}$$

where $sort(S, rank(q, .))$ sorts the documents in S according to $rank(q, .)$, the ranking function applied to query q. Now let $P = Q \times A$ be a set composed of pairs $\langle q, a\rangle$, where $q \in Q$ is a query that defines the criterium used to select consumers interested in products advertised by means of ads $a \in A$. In a search advertising system the task is to get the best k ads in which the selection criteria match the user query or the content of the page browsed by the user. Since a search advertising system can be seen as a reversed version of an IR system, its retrieval function can be defined by (3) and (4):

$$adSearch(d, P) = \{a \in A | \langle q, a\rangle \in search(d, Q) \ \wedge \langle q, a\rangle \in P\} \tag{3}$$

$$adSearchTop(k, d, A) = \text{top } k \text{ ads in } sort(adSearch(d, P), rank(., d)) \tag{4}$$

In this model the advertisers define the selection criteria that will indicate the users they want to deliver their ads by using complex queries. Note that the nature of these queries can be slightly different from that in traditional IR. In search advertising systems a query may be totally unrelated to the terms used in the ads. For example, an advertiser can select the keyword *wine* to associate with its ad about *cheese* if it considers that a consumer interested in wine could be also interested in cheese.

In an abstract retrieval model, the direction of search is not taken into account. However, techniques for implementing a specific retrieval function are tailored to optimize one direction. For example, traditional document search, which retrieves document from queries, explores inverted lists. Given the semi-static nature of the target collections used in these traditional applications, query optimization techniques such as those based on index pruning [42] are common. On the other hand, a search advertising model is characterized by dynamic ad collections and ranking functions based on parameters that vary dynamically and must be updated frequently.

To cope with these characteristics, authors in [3] proposed a search advertising system in which the employed optimization techniques aim to achieve efficient query search, incremental ad updates, and dynamic ranking. Efficient query search is essencial particularly in keyword-targeted advertising where the ads have to be shown along with the results of the user query. The possibility of incremental ad updates makes possible on-line modification of the ad collection. Finally, a dynamic ranking is necessary to model the competition between advertisers continuously updating the ranking parameters that they control. Figure 5 illustrates their system.

As we can see in this Fig. 5, ads and their associated selection criteria are stored in a collection as well as the parameters related to each advertiser. For each user request all ads that give a positive match are selected and ordered

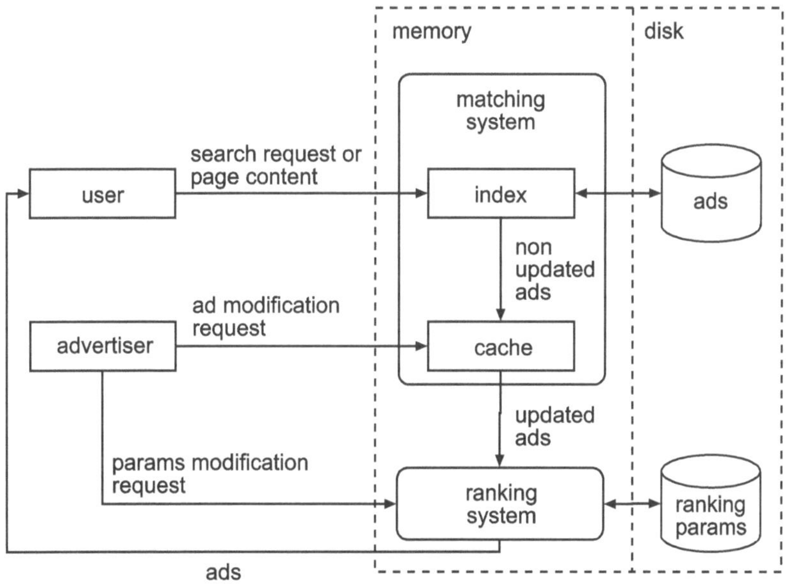

Fig. 5. Search advertising system.

according to a ranking measure. A positive matching indicates that the ad is relevant to the user request. The ranking measure indicates how useful is the matching for the actors in the search advertising network. Advertisers can modify their ads and the ranking parameters that they control such as, for example, the amount that they are willing to pay by a user click given a certain query. A caching system is used to support real time updates of queries and ranking parameters.

In the following sections we present some findings and research opportunities related to the matching and ranking systems and to the services provided by the brokers regarding to fraud detection, keyword suggestion and feedback.

3.1 Relevance Matching

Many works in advertising research have stressed the importance of relevant associations for consumers [28, 32]. For example, the studies in [40] point out that the user perception of Web content can be more positive whenever there is a strong relationship between this content and the advertised products. According to the authors in [28], even forced exposure is considered less intrusive if it is editorially congruent. Studies in [52] show that by ensuring the relevance of the ads, information gatekeepers are investing in the reinforcement of a positive user attitude towards the advertisers and their ads. The authors in [9] also notice that irrelevant or offensive ads can turn off users and that relevant ads are more likely to be clicked on, which leads to better performance since click-through is still a key element in evaluation.

As a consequence, sophisticated matching strategies have been developed to ensure that relevant ads will be shown to the users. This implies the support to exact and approximate matching methods by the match function. In the exact method, an ad keyword and a user query are matched if they are identical. This excludes, for instance, the case in which the keyword is *part* of the query. In the approximate method, the keywords are matched to the user query regardless of the order and of the distance of their component words. For example, the keyword "canon camera" has a degree of relevance to any query which contains the word "canon", the word "camera", or both words, such as "digital camera of canon". Further the words can be automatically expanded to include synonyms, related terms and plural forms. In many systems, the advertisers can still define terms that should not occur in the user query or even define sub-phrases that should occur in it [12, 31].

Studies on keyword matching have also shown that the nature and size of the keywords have impact on the likelihood of an ad to be clicked [19, 46]. These studies were motivated by the following intuition. A user in a buying cycle could start a search by employing short and generic keywords. In a later phase, however, he tends to be more focused. Consequently, his query evolves to include more words and these words are more specific. In such phase, it is common the use of brand names. To illustrate, a query that starts as "camera" can evolve to "canon powershot a95". Thus, conversion rates probably tend

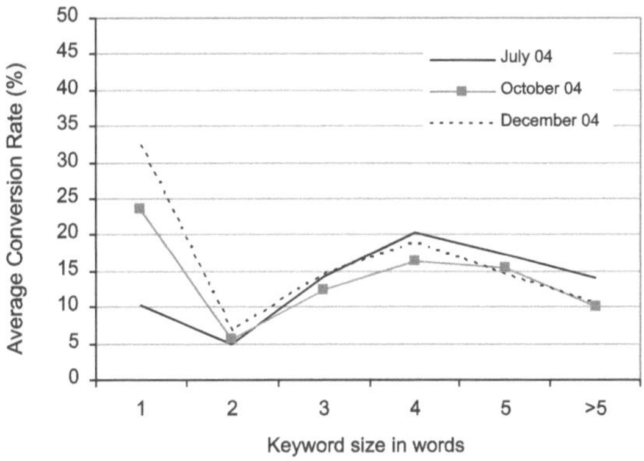

Fig. 6. Conversion rates per keyword size for a full database. Source: OneUpWeb, 2005.

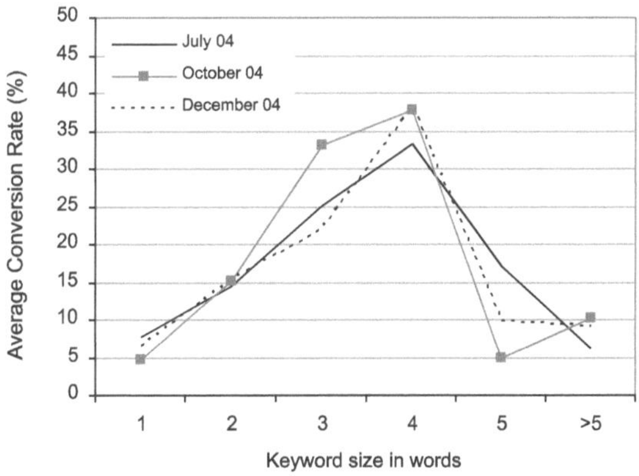

Fig. 7. Conversion Rates for the same database of Fig. 6 with two modifications: (a) only high traffic Keywords considered and (b) corporate names removed. Source: OneUpWeb, 2005.

to increase as the number of words in a search query increases. In fact, in order to test this intuition, the authors in [37] tracked conversion rates per query in 2004. These data were divided by month into categories based on the number of words in the queries and were used to build two collections. The first one included all the searches. The second one included only the high-traffic keywords. Additionally, all the company names were deleted from the second collection. Figures 6 and 7 show the results obtained in their experiments.

As we can see in these Figs. 6 and 7, discounted the single word keywords, they found that conversion rates peak at four-word keywords, dropping with larger keywords. The authors argued that this dropping occurred because users were searching for very specific information, i.e., products or services that were not found in the search engine databases. In the case of single word keywords (keyword size = 1), they found very high conversion rates for the collection in which brand names were not deleted (see Fig. 6). However, after removing these brand names, the conversion rates for single word keywords dropped to values lower than those obtained by keywords with size 2 to 4 words (see Fig. 7). This indicates that people searching for a particular company's name are more predisposed to make a conversion. In general, results can be explained by the fact that large keywords and corporate names tend to be less ambiguous, which leads to a more precise matching. Thus, these keywords could indicate a more qualified user, that is, a user in buying mode [46].

Besides the simple matching of terms, matching systems can also filter associations between ads and documents using evidence such as geographic information and categories of documents and ads.

By using browser information or IP identification, search advertising systems are able to determine the geographic location of the user. With this information, it is possible to show ad of products or services in user neighborhood. By means of geographic databases, these systems are able to recognize that some terms in the keywords represent specific locations. To illustrate, in the keyword "car rent in new york", such systems could infer that "new york" is a reference to the city of New York. In some cases, the advertisers can explicitly target specific countries and languages [22]. This is particularly interesting for products and services whose scope is geographically restricted as, for example, movie rent services, and sales promotions. In fact, the availability of such services in search advertising can lead to a great growth of the overall Web marketing business, since most of the advertisers are small companies interested in targeting consumers in certain regions.

Finally, it is important to notice that there are situations in which even relevant association between ads and page contents could lead to improper advertising [15,50]. For example, it is possible that an ad could be relevant to a page's content and damaging to the brand. This is the case of pages about catastrophes. Such pages hardly offer good opportunities to the announcement of products and services. As a consequence, matching systems normally have to be able to recognize and filter these contents. Such systems will also try to recognize unethical and illegal advertising to avoid placing ad of competitors, violating trademarks, and supporting pornography, drugs, and gambling games. For that brokers use categorization and filtering lists [33, 39, 49]. In many cases, however, manual editorial control is employed [9].

In the case of content-targeted advertising, matching systems have to be even more elaborated since the mapping technology has to extract the key topics from the full text of a Web page [15,21]. In fact, the authors in [44] have shown that great accuracy can be obtained in content-targeted advertising if the content of the pages and the selection criteria associated with the ads are enriched with terms from other sources of evidence. They started by testing straightforward methods in which the content of the triggering pages was matched to (a) the keywords related to the ads, (b) the content of the ad creative, that is, the content of its fields title and description, (c) the content

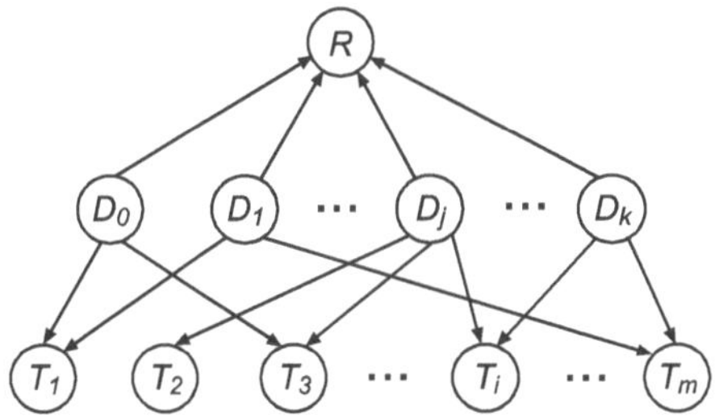

Fig. 8. Bayesian network model for selecting the terms that best represent the topics of a triggering page.

of the ad creative and its keywords, and (d) the content of the ad creative, its keywords, and the content of the ad's landing page.

The authors found that the best of these straightforward methods is matching the content of the triggering page to the content of the ad creative together its keywords, with the additional restriction that all the words in the keywords are required to occur in the triggering pages. Additional gains in average precision were reported through the expansion of the vocabulary of the triggering page p. For this, it is proposed a method that consists in determining the set $\mathcal{N}$ of the k nearest neighbors of the page p in a Web collection using a classic vector space model [4]. Given the k nearest neighbor pages of p, the Bayesian Network [41] depicted in Fig. 8 is used to select a subset from them that can be used to additionally represent the topics of the page p.

In this model, the nodes represent pieces of information in the domain. With each node is associated a binary random variable, which takes the value 1 to mean that the corresponding entity (a page or terms) is *observed* and, thus, relevant in the computations. In this case, we say that the information is *observed.* Node R represents the page r, a new representation for the triggering page p. Root nodes D_0 through D_k represent the documents in $\mathcal{N}$, that is, the triggering page D_0 and its k nearest neighbors. There is an edge from node D_j to node R if document d_j is in $\mathcal{N}$. Nodes T_1 through T_m represent the terms in the vocabulary of $\mathcal{N}$. There is an edge from node D_j to a node T_i if term t_i occurs in document d_j. In the model, the observation of the pages in $\mathcal{N}$ leads to the observation of a new representation of the triggering page p and to a set of terms describing the main topics associated with p and its neighbors. Let $\mathbf{d_j}$ be the state of the document nodes in which *only* document d_j is observed and probabilities $P(T_i|\mathbf{d_j})$ and $P(R|\mathbf{d_j})$ be defined as follows:

$$P(T_i|\mathbf{d_j}) = \eta \; w_{ij} \tag{5}$$

$$P(R|\mathbf{d_j}) = \begin{cases} (1-\alpha) & j = 0 \\ \alpha \; sim(p, d_j) & 1 \le j \le k \end{cases} \tag{6}$$

Table 2. Number of relevant ads per ad slot according to several matching methods. Columns labelled #1, #2, and #3 indicate total of relevant ads in first, second, and third ad slots in the triggering pages. These ads were selected from a collection of about 94,000 ads and matched to a collection of 100 triggering pages obtained from a newspaper. The term "keyword required" indicates the methods in which all the keyword words were required to occur in the triggering pages.

Methods	Relevant ads per slot			
	#1	#2	#3	total
Method (b)	41	32	13	86
Method (c)	51	28	17	96
Method (a)	46	34	28	108
Method (c) + keyword required	51	48	39	138
Method (d) + keyword required	52	50	46	148
Expansion + keyword required	70	52	53	175
Expansion + keyword required + landing page	64	61	51	176

where η is a normalizing constant, w_{ij} is the weight associated with term t_i in the document d_j, $sim(p, d_j)$ is a value of how related are the triggering page p and the document d_j, and α is a constant used to determine how important should be the influence of the triggering page p to its new representation. Notice that $P(\overline{T_i}|\mathbf{d_j}) = 1 - P(T_i|\mathbf{d_j})$ and $P(\overline{R}|\mathbf{d_j}) = 1 - P(R|\mathbf{d_j})$. Given these definitions, we can now use the network to determine the probability that a term t_i is a good term for representing a topic of the triggering page p. In other words, we are interested in the probability of observing the final evidence regarding a term t_i, given that the new representation of the page p has been observed, $P(T_i = 1|R = 1)$. This translates into the following equation[5]:

$$P(T_i|R) = \eta \left((1 - \alpha) \; w_{i0} + \alpha \sum_{j=1}^{k} w_{ij} \; sim(p, d_j) \right) \tag{7}$$

Equation (7) can be used to determine a set of terms that can be used as a new representation of p or as expansion terms for p. This expansion method allowed revealing new terms related to the main topics of the triggering page p, as well as emphasizing important terms, which led to improved precision (see Table 2). Notice that to implement such expansion strategies without affecting user experience, the ads should be assigned to the pages in off-line mode. In the case of content-targeted advertising this is possible because the ads do not have to be assigned at query time as in keyword-targeted advertising. These results suggest that great accuracy in content-targeted advertising can be attained with appropriate algorithms – a very interesting research avenue.

[5] To simplify our notation we represent the probabilities $P(X = 1)$ as $P(X)$ and $P(X = 0)$ as $P(\overline{X})$.

3.2 Ranking

To maximize the utility of a search network to all its participants, the ranking system has to satisfy different interests [21]: (a) the users want to receive relevant information, (b) the advertisers want to receive quality traffic at a minimum cost and with a minimum risk of negative user attitude towards them, and (c) the brokers and publishers want to maximize their revenues at the minimum risk of negative user attitude towards their brands, contents, and services.

A good ranking system has to meet these interests in a fair and transparent way in order to be useful to all participants. For example, by picking up an irrelevant ad, the system not necessarily poses a serious threat to the advertiser since ads are passive and probably the user will not click on irrelevant ads. However, this same ad could represent a problem for the publisher if it occupies a precious slot space in the triggering page [50]. Another interesting tradeoff is pointed out in [5]. Higher is the rank position in a paid list, higher are the click-through rates received by the advertiser and, by extent, higher are the revenues received by the brokers. This could lead to a very simple ranking strategy where user interests are not taken into account. In this ranking, paid list positions would be defined only by how much advertisers pay. However, one such ranking can lead the users to perceive the whole service as unreliable. That is, in the long run, it can diminish users' loyalty towards the publisher. So, such biased ranking can lead to less traffic. A similar result has been observed in the case of Information Gatekeepers, like search engines, that favor pages in their ranking that carry their ads. They risk their credibility with severe consequences in the long term [51]. These examples clearly show the necessity of fair and clever ranking algorithms.

A detailed study on such ranking algorithms is provided in [9]. In that work, the authors analyze paid placement strategies for keyword-targeted advertising. They start by defining a revenue model for a single keyword K. Suppose that companies interested in advertising for keyword K compete for n slots in a triggering page. Let v_j be the value that advertiser j is willing to pay for K. Let α_j be the true relevance score of ad j. This score is not observable to the brokers and users but can be approximated over time by the click-through rate received by this ad. However, since users are more likely to click on high-ranked items, the expected click-through rate of an item j at position i has to consider an exponentially decaying attention model. Thus, the expected click-through rate is given by $\frac{\alpha_j}{\delta^{i-1}}$, with $\delta > 1$. Let r be the ranking function which allocates position i to ad j and P_i be the payment for position i. In this model the overall quality of the paid placement portion of a site is given by $(\frac{\sum_i \alpha_{r(i)}}{n})^\lambda$, where λ represents the different user sensitivity towards advertising under different search scenarios ($\lambda \geq 1$, higher λ leads to a greater reduction in demand). Thus, the broker's revenue is given by 8.

$$\left(\frac{\sum_i \alpha_{r(i)}}{n}\right)^{\lambda} \times E\left(\sum_i^n P_i\right) \tag{8}$$

By using this revenue model, they compare several ranking strategies. Amongst them we cite:

- Rank by willingness to pay (WTP ranking): in this strategy, the ads with highest bids are selected and ranked according to their bids (v). The payments are made according to a first price auction system, that is, the advertisers pay what they bid.
- Rank by willingness to pay times click-through rate (WTPC ranking): in this strategy, the ads with highest bids are selected and ranked according to the product between their bids and their expected click-through rate ($v \times \alpha$). The payments are made according to a variant of a second price auction system. So, the winner advertiser pays the bid just below its own bid. In the case of a draw, it pays an amount that just exceeds the bid of its immediate opponent according to the ranking.

Notice that these ranking strategies are stylized versions of those employed by Overture and Google, respectively, in their keyword-targeted advertising systems. These strategies ignore, however, many factors present in real systems, including editorial control, inexact query matching, new pricing models like payment per conversion, marketing efforts, brand awareness, legal controls, and strategic alliances.

Amongst several strategies considered, the WTPC ranking performed best in almost all cases. This corroborates the intuition that click-through rates can be taken as a useful relevance judgement about the ads. To evaluate the impact of the editorial control over the ranking strategies, a variation of WTP strategy was developed to simulate the filtering of irrelevant or objectionable ads before the ranking process. This new strategy was significantly better than WTP ranking. In fact, it was even better than WTPC ranking. This suggests that a careful review of the mappings made by the ranking algorithms could be worthy if it ensures that ads are properly targeted [48, 50]. As pointed out by [9], brokers can therefore choose a suitable level of investment in the editorial process by trading the costs with the consequent increases in revenues.

Another interesting characteristic of WTPC ranking is its dynamic nature. In WTPC, ads can be promoted or demoted according to the number of clicks received. Thus, the ranking of an ad will change over the time, as its click-through rate itself changes. In a basic implementation of WTPC, each click increases the ad click-score by 1. Thus, each click has an equal impact on the likelihood of promotion. The authors in [9] suggest a different revision mechanism, in which the reward for a click is larger if it is received in a lower rank. Such clicks indicate situations in which it is more probable that the ranking algorithm has failed. Thus, by promoting more quickly those ads, the weighted WTPC is better able to recognize and correct errors in the ranking. This new mechanism has two important advantages when compared to the unweighted approach: it converges faster to the *optimal* ranking and is more stable. Convergence to the optimal ranking is important because any deviation leads to lower revenues. In its turn, speed of convergence is important because of the dynamic nature of demand for various search terms. For example, around the release date of the last movie in the series "Star Wars", advertisers might intensify the bid on keywords related to the movie. In this way, they update their campaigns to adapt them to changing environments. Thus, if a mechanism takes too long to converge for a certain keyword, it will implement the optimal policy for only a very short period.

The authors in [3] have tested the performance of their system using an augmented version of the WTPC ranking. Besides the value of the bid and the click-through rate, their ranking takes into consideration the maximum amount of money that an advertising is willing to spend in a day. In a test using a collection of one million ads running on a single PC and in which 20 concurrent threads continuously submitted queries and update requests, the system was able to achieve a steady performance of 180 searches per second. Tests using different collection sizes, cache sizes, and query formulation showed near linear performance scalability on these parameters.

The authors in [9] also conducted experiments to evaluate how many ads should be placed in a paid list. They concluded that when willingness to pay and relevance are positively correlated, the broker's expected revenue is approximately concave in the number of ads it enrolled. In their simulation, the optimal revenue values were obtained for lists with sizes varying from 3 to 7 ads. This is due to the tradeoff between direct revenue increases and indirect revenue losses due to consumer defection. This tradeoff can be explained as follows. Intuitively, the more ads a publisher shows, the more revenue it will receive from advertisers. However, large paid lists are more likely to enroll irrelevant ads. This has a negative impact on the overall quality of the publisher. Consequently, total traffic of the publisher and click-through rates will reduce lowering revenue from paid placement.

Overall, ranking is a key component of search advertising. Thus, designing and experimenting with new ranking algorithms provides another very interesting and promising research avenue.

3.3 Fraud Detection

The revenues in search advertising are directly associated with the user traffic in the network. Thus, the more publishers have users clicking in the ads shown in their pages, the more advertisers will pay to them. Clearly, there is a potential for frauds since publishers could simulate that traffic [8,28,34]. This is in fact a problem in the industry. For example, CompUSA[6] spent more than U$10 millions in 2004 due to fake traffic [28]. This is serious because

[6] `http://www.compusa.com`

advertisers could lose the confidence in the network, which could profoundly harm all the business [19].

To deal with this, research has concentrated on trying to characterize the fake traffic. For that, common strategies consist in analyzing the distribution of the clicks along the time. If an unusual number of clicks from a same user or group of users is detected in a certain window of time, or a pattern of invalid clicks is found, these clicks are considered illegal and are ignored. A more sophisticated treatment for the problem was suggested by the author in [8]. It considers this a classification problem in which real and fake traffic have to be distinguished with maximum precision. This is necessary to protect advertisers from paying for excess clicks and, at same time, to avoid penalizing the broker by discharging valid clicks. The author suggests the use of unlabelled data in the training phase since a huge amount of clicks is generated continuously and it would be impossible to label all of them.

3.4 Measurements and Feedback

In search advertising, advertisers have the possibility of getting detailed feedback about their performance. This can be used to determine how much they have to pay for the received traffic and what can be done to improve their campaigns. To help advertisers in such tasks, research, metrics, and tools have been developed. In fact, studies have shown that a careful analysis of feedback information is rewarding for advertisers [18, 19]. In general, companies that excel on biding, selecting keywords, selecting ad text, and preparing landing pages obtain higher revenues with search advertising.

Pay-per-performance strategies have been used since the beginning of the Web, pioneered by companies like CDNow[7] and Amazon[8]. However such strategies were exception in the early days of Web advertising when the traditional pricing model based on cost per thousand impressions (CPM) was dominant. In this model, payments were measured mainly based on the quantity of impressions of the ads. But since Internet is an accountable medium and there is uncertainty on the benefit of the advertisers in having their ads shown, there was a pressure for the adoption of metrics in which the payment is a function of the desired market response [14]. An example of such metrics is the cost per click-through (CPC) where the advertiser pays only when the user clicks on its ad. In fact, it has been shown that, in the case of uncertainty about the product's performance, performance-based metrics are more effective in Web marketing [6].

Such pricing models had fed many debates in the industry. If by one hand, performance-based models allow a direct estimate of advertiser's return of investments, by other hand, it is hard to measure brand awareness, brand recall, and purchase intent by means of clicks [2]. More important, CPM pricing models pose low risk to publishers since they cannot control many factors

[7] `http://www.cdnow.com`

[8] `http://www.amazon.com`

that affect the performance of an ad such as its design and offer. In contrast, CPC pricing models are better for the advertisers since they pay mainly by the visitors that are more likely to generate a conversion. Many studies have been made about the choice of these metrics. For example, the authors in [2] compared CPM and CPC metrics and studied the situations in which their employment would be more appropriate. They concluded that, in some cases, it would be worthy to use hybrid models. On their part, the authors in [16] used contract theory to explain why performance-based metrics are more desirable and what could be the form of a contract that aligns interests and gains of publishers and advertisers. As the debates go on, performance-based metrics have grown in the advertisers' preference and have become widely used in the industry, as we can see in Table 3.

Table 3. Internet advertising revenues, in percents, by pricing model – 1998–2004. Source: IAB, 1998–2004

	1998	1999	2000	2001	2002	2003	2004
Hybrid	47	53	47	40	34	20	17
CPM	49	40	43	48	45	43	42
Performance	4	7	10	12	21	37	41
Total	100	100	100	100	100	100	100

Advertisers know however that clicks will not necessarily be converted. Thus, they are also concerned with aspects such as brand awareness and perception, and are always pressuring for more detailed information that can provide them with a reliable estimate of the return of their investments (ROI). To meet this demand, brokers and third-party companies have made available several tools to help advertisers analyze their conversion rates and improve their campaigns accordingly [11]. These tools normally include tracking mechanisms that work in the advertiser's site. With these tools, advertisers can track users during their buying cycles, which makes possible to estimate impact of brand awareness, to determine which actions the user takes, how many leads, orders or sales transactions are generated, etc [12].

The problem previously cited of click conversion is particularly remarkable in the case of content-targeted advertising. Many studies [15,26,30,38,45,48, 50] have reported different performance for keyword-targeted and content-targeted advertising with worse performance for the second approach. This is normally attributed to the fact that users are less likely to generate a conversion while surfing. To deal with this problem, many tools are provided by third party companies to allow advertisers to evaluate and optimize separately keyword-targeted and content-targeted advertising.

Further, in order to provide a measure more indicative of the ROI, companies have developed alternative metrics to CPC [9,25,39]. In such metrics, lower CPC values are charged to sites that generate lower traffic quality, which encourages publisher to produce better content.

Using such tools, advertisers can also infer the more profitable keywords and auction strategies. For that, brokers can provide the advertisers with detailed information on their bids that includes, in some cases, demographic information about the users. For example, the MSN[9] adCenter system is able to inform the gender, age, lifestyle, and income of the users who search for a certain keyword [27]. In fact, a whole industry has flourished by selling consultancy services related to tasks as keyword selection and definition of auction strategies. For an example of such tools, consider a keyword suggestion tool that, for any given keyword, provides a sorted list of correlated keywords and suggests them to the advertiser.

Notice that by identifying a cluster to which a certain keyword belongs, it is possible to take the other keywords in that cluster as suggestions. The authors in [7] evaluated two clustering methods for determining (a) groups of keywords belonging to the same marketplace and (b) submarkets of advertisers showing common bidding behavior. The first clustering method was based on the idea that advertisers with common interests will bid on the same subset of keywords forming a submarket. Thus, if the relationship advertisers-keywords is represented as a graph, the nodes of this graph that represent advertisers and bided keywords in a submarket will be more strongly connected with each other than to the other graph nodes. Thus, the problem of finding these submarkets can be approached through partitioning a graph to find strongly connected subgraphs. For this, they employed a flow-based graph partitioning method.

Their second clustering method was based on the idea that related keywords present similar bidding patterns. Thus, these keywords can be represented as graph nodes connected by edges whose weights are proportional to the amount of overlapping between their set of bidders. Their method consisted in clustering these nodes using an agglomerative technique. After comparing the proposed methods against other commonly employed in the literature, they found that their first approach is better for providing a small number of larger clusters while their second approach is better for providing a large number of small clusters.

The authors in [10] also presented a keyword suggestion tool. Since they were interested in controlling the level of generality of the keywords suggested, they used a vector space model with a singular value decomposition (SVD) approach. Differently from the methods previously described, their strategy allows each keyword to potentially form soft clustering of related keywords. In their method, keywords are represented as vectors of advertisers. A nonzero entry in these vectors corresponds to an advertiser bidding on the keyword. The similarity between keywords is calculated as the cosine of the angle between their corresponding vectors. By projecting these vectors in an SVD

[9] http://www.msn.com

subspace, they can perform a conceptual match instead of a simple exact match. In other words, they can match keywords globally or conceptually without the need of explicit bidding associations.

4 Conclusions

In this chapter, we discussed the key concepts of search advertising. We first presented its main categories, the keyword-targeted advertising and the content-targeted advertising, emphasizing the differences between them. After that, we showed the search advertising network and described how their main actors interact. Specifically, we described the benefits earned by the broker, the publishers, the advertisers, and the users. Then, we showed important aspects of a search advertising system and research opportunities related to them.

In the discussion on relevance matching, we showed sophisticated matching algorithms that have been used in search advertising systems and recent findings on matching algorithm research. Later, we presented different ranking algorithms studied in the literature conceived to meet the diverse interests of the search advertising actors. We also presented studies and research opportunities related to topics like fraud detection, and metrics and feedback.

Search advertising has established itself as the most important format in Web marketing and the main source of monetization for Web services. Some reasons for this are the expanded marketing capacities provided by the Internet as well as its flexibility which facilitates accurate performance tracking and adaptation to change environments. However, the primary reason for such success is the increasing number of people which rely on the Internet and use it as their main source of information. Search advertising providers are aware of this and have hardly invested in the core technologies necessary to satisfy the user needs and sustain their credibility. In spite of this, there is little academic research on these subjects. Our primary aim in this work was to stimulate further research since much more work remains to be done.

References

1. Interactive Advertising Bureau. `http://www.iab.net/`.
2. Kursad Asdemir, Nanda Kumar, and Varghese Jacob. Internet Advertising pricing models. Technical report, School of Management, The University of Texas at Dallas, November 2002.
3. Giuseppe Attardi, Andrea Esuli, and Maria Simi. Best bets: thousands of queries in search of a client. In *WWW Alt. '04: Proceedings of the 13th international World Wide Web conference on Alternate track papers & posters*, pages 422–423, New York, NY, USA, 2004. ACM Press.
4. Ricardo Baeza-Yates and Berthier Ribeiro-Neto. *Modern Information Retrieval.* Addison-Wesley-Longman, 1st edition, 1999.

5. Hemant K. Bhargava and Juan Feng. Preferential placement in Internet search engines. In *Proceedings of the eleventh international conference on World Wide Web*, pages 117–123. ACM Press, 2002.
6. Hemant K. Bhargava and Shankar Sundaresan. Optimal design of contingency pricing in IT-intensive commerce. In *Proceedings of the Twenty-Third International Conference on Information Systems*, Barcelona, Espanha, December 2002.
7. John Joseph Carrasco, Daniel Fain, Kevin Lang, and Leonid Zhukov. Clustering of bipartite advertiser-keyword graph. In *Workshop on Clustering Large Datasets, 3th IEEE International Conference on Data Mining*, Melbourne, Florida, USA, November 2003. IEEE Computer Society Press. Available at `http://research.yahoo.com/publications.xml`.
8. Elena Eneva. Detecting invalid clicks in online paid search listings: a problem description for the use of unlabeled data. In Tom Fawcett and Nina Mishra, editors, *Workshop on the Continuum from Labeled to Unlabeled Data, Twentieth International Conference on Machine Learning*, Washington DC, USA, August 2003. AAAI Press.
9. Juan Feng, Hemant Bhargava, and David Pennock. Implementing paid placement in Web search engines: computational evaluation of alternative mechanisms. *INFORMS Journal on Computing*, 2005. To be published.
10. David Gleich and Leonid Zhukov. SVD based term suggestion and ranking system. In *Proceedings of the 4th IEEE International Conference on Data Mining*, pages 391–394, Brighton, UK, November 2004. IEEE Computer Society.
11. David Green. Search Engine Marketing: why it benefits us all. *Business Information Review*, 20(4):195–202, December 2003.
12. Robyn Greenspan. How to cheat Google AdWords select. SitePoint, December 2003. Available at `http://www.sitepoint.com/article/adwords-select-parts-1-4`.
13. Robyn Greenspan. Paid search listings unrecognizable to some. ClickZ Experts, December 2003. Available at `http://www.clickz.com/stats/sectors/search_tools/article.php/3290431`.
14. Donna L. Hoffman and Thomas P. Novak. How to acquire customers on the Web. *Harvard Business Review*, 106(31):179–188, 2002.
15. Gord Hotchkiss. Contextual text ads: the next big thing? SEO Today, August 2003. Available at `http://www.searchengineposition.com/info/netprofit/contextads.asp`.
16. Yu Hu. Performance-based pricing models in online Advertising. Discussion paper, Sloan School of Management, MIT, January 2004. Available at `http://ssrn.com/abstract=501082`.
17. IAB and PricewaterhouseCoopers. IAB internet advertising revenue report, April 2005. Available at `http://www.iab.net/2004adrevenues`.
18. Greg Jarboe. Report: Most search marketers are unsophisticated. Search Engine Watch, February 2005. Available at `http://searchenginewatch.com/searchday/article.php/3469301`.
19. Rob Kaiser. Web words become a lucrative market. Chicago Tribune, February 2005. Available at `www.chicagotribune.com/business/chi-0501240002-jan24,1,2431400.story?coll=chi-business-hed&ctrack=1&cset=true`.
20. Carol Krol. Zeroing in on content-targeted ads. BtoB Online, February 2005. Available at `http://www.btobonline.com/article.cms?articleId=23413`.

21. Kevin Lee. Context is king, or is it? ClickZ Experts, March 2003. Available at http://www.clickz.com/experts/search/strat/article.php/2077801.
22. Kevin Lee. Google regional targeting power. ClickZ Experts, October 2003. Available at http://www.clickz.com/experts/search/strat/article.php/3101731.
23. Kevin Lee. Making sense of AdSense... et al. ClickZ Experts, September 2003. Available at http://www.clickz.com/experts/search/strat/article.php/3076061.
24. Kevin Lee. The SEM content conundrum. ClickZ Experts, July 2003. Available at http://www.clickz.com/experts/search/strat/article.php/2233821.
25. Kevin Lee. AdSense makes much more sense. ClickZ Experts, May 2004. Available at http://www.clickz.com/experts/search/strat/article.php/3350001.
26. Kevin Lee. Separating search and contextual inventory. ClickZ Experts, January 2004. Available at http://www.clickz.com/experts/search/strat/article.php/3305651.
27. Kevin Lee. MSN's adCenter: more control and better results. ClickZ Experts, May 2005. Available at http://www.clickz.com/experts/search/strat/article.php/3490876.
28. Hairong Li and John Leckenby. Internet Advertising formats and effectiveness. Center for Interactive Advertising, 2004. Available at http://www.ciadvertising.org/studies/reports/measurement/ad_format_print.pdf.
29. Kate Maddox. Forrester reports advertising shift to online, May 2005. Available at http://www.btobonline.com/article.cms?articleId=24191.
30. Janis Mara. Overture breaks up contextual, search listings. ClickZ Experts, January 2004. Available at http://www.clickz.com/news/article.php/3296211.
31. Fredrick Marckini. Contextual Advertising. ClickZ Experts, October 2003. Available at http://www.clickz.com/experts/search/results/article.php/3087311.
32. Sally McMillan. Internet Advertising: one face or many? In editors David W. Schumann & Esther Thorson, editor, *Internet Advertising: Theory and Research*. Lawrence Erlbaum, 2005.
33. Kevin Newcomb. FindWhat unveils answer to AdSense. ClickZ Experts, September 2004. Available at http://www.clickz.com/news/article.php/3415431.
34. Kevin Newcomb. Google sues AdSense publisher for click fraud. ClickZ Experts, November 2004. Available at http://www.clickz.com/news/article.php/3440341.
35. Kevin Newcomb. Google ads B2B sites to AdSense. ClickZ Experts, February 2005. Available at http://www.clickz.com/news/article.php/3483586.
36. Kevin Newcomb. Marketers react to Yahoo! AdSense alternative. ClickZ Experts, March 2005. Available at http://www.clickz.com/news/article.php/3489851.
37. OneUpWeb. How keyword length affects conversion rates, January 2005. Available at http://www.oneupweb.com/landing/keywordstudy_landing.htm.
38. Pamela Parker. Google extends AdSense overseas. ClickZ Experts, December 2003. Available at http://clickz.com/experts/brand/buzz/article.php/3291591.
39. Pamela Parker. The next context. ClickZ Experts, April 2004. Available at http://www.clickz.com/experts/brand/buzz/article.php/3341161.

40. Jeffrey Parsons, Katherine Gallagher, and K. Dale Foster. Messages in the medium: An experimental investigation of Web Advertising effectiveness and attitudes toward Web content. In Jr. Ralph H. Sprague, editor, *Proceedings of the 33rd Hawaii International Conference on System Sciences-Volume 6*, page 6050, Washington, DC, USA, 2000. IEEE Computer Society.
41. Judea Pearl. *Probabilistic Reasoning in Intelligent Systems: Networks of plausible inference.* Morgan Kaufmann Publishers, 2nd edition, 1988.
42. Michael Persin. Document filtering for fast ranking. In *SIGIR '94: Proceedings of the 17th annual international ACM SIGIR conference on Research and development in information retrieval*, pages 339–348, New York, NY, USA, 1994. Springer-Verlag New York, Inc.
43. Andy Reinhardt. And you thought that Web ad market was dead. Business Week, May 2003. Available at `http://www.businessweek.com/magazine/content/03_18/b3831134_mz034.htm`.
44. Berthier Ribeiro-Neto, Marco Cristo, Paulo B. Golgher, and Edleno Silva de Moura. Impedance coupling in content-targeted advertising. In *SIGIR '05: Proceedings of the 28th annual international ACM SIGIR conference on Research and development in information retrieval*, pages 496–503, New York, NY, USA, 2005. ACM Press.
45. Catherine Seda. Contextual ads: vital to a search marketing campaign? Search Engine Watch, October 2004. Available at `http://searchenginewatch.com/searchday/article.php/3418151`.
46. Mike Shields. Multi-term searches get high conversion. Media Week, February 2005. Available at `http://www.mediaweek.com/mw/search/article_display.jsp?schema=&vnu_content_id=1000816071`.
47. Mike Shields. Online publishers foresee dynamic ad spending. Adweek, February 2005. Available at `http://www.adweek.com/aw/search/article_display.jsp?schema=&vnu_content_id=1000797161`.
48. Michael Singer. Contextual ad debate rouses critics. ClickZ Experts, August 2003. Available at `http://www.clickz.com/news/article.php/3066971`.
49. Richard Stern. Is gatoring unfair or illegal? *IEEE Micro*, 22(1):6–7, 92–93, 2002.
50. Danny Sullivan. The content context contest. ClickZ Experts, June 2003. Available at `http://www.clickz.com/experts/search/opt/article.php/2241231`.
51. Danny Sullivan. Contextual ads and the little guy. ClickZ Experts, June 2003. Available at `http://www.clickz.com/resources/search_reference/contextual_advertising/article.php/2224611`.
52. Chingning Wang, Ping Zhang, Risook Choi, and Michael Daeredita. Understanding consumers attitude toward Advertising. In *Eighth Americas Conference on Information Systems*, pages 1143–1148, August 2002.
53. Tessa Wegert. Contextual ads: a consumer's point of view. AtNetwork, April 2003. Available at `http://www.atnewyork.com/news/article.php/2196831`.
54. Tessa Wegert. Contextual ads: a media buyer's point of view. ClickZ Experts, April 2003. Available at `http://www.clickz.com/experts/media/media_buy/article.php/2192031`.
55. Melius Weideman. Ethical issues on content distribution to digital consumers via paid placement as opposed to website visibility in search engine results. In *The Seventh ETHICOMP International Conference on the Social and Ethical Impacts of Information and Communication Technologies*, pages 904–915. Troubador Publishing Ltd, April 2004.

56. Melius Weideman and Timothy Haig-Smith. An investigation into search engines as a form of targeted advert delivery. In *Proceedings of the 2002 annual research conference of the South African institute of computer scientists and information technologists on Enablement through technology*, pages 258–258. South African Institute for Computer Scientists and Information Technologists, 2002.

Information Loss in Continuous Hybrid Microdata: Subdomain-Level Probabilistic Measures

Josep Domingo-Ferrer, Josep Maria Mateo-Sanz, and Francesc Sebé

Rovira i Virgili University of Tarragona, Av. Països Catalans 26, E-43007 Tarragona, Catalonia
{josep.domingo,josepmaria.mateo,francesc.sebe}@urv.net

Summary. The goal of privacy protection in statistical databases is to balance the social right to know and the individual right to privacy. When microdata (*i.e.* data on individual respondents) are released, they should stay analytically useful but should be protected so that it cannot be decided whether a published record matches a specific individual. However, there is some uncertainty in the assessment of data utility, since the specific data uses of the released data cannot always be anticipated by the data protector. Also, there is uncertainty in assessing disclosure risk, because the data protector cannot foresee what will be the information context of potential intruders. Generating synthetic microdata is an alternative to the usual approach based on distorting the original data. The main advantage is that no original data are released, so no disclosure can happen. However, subdomains (*i.e.* subsets of records) of synthetic datasets do not resemble the corresponding subdomains of the original dataset. Hybrid microdata mixing original and synthetic microdata overcome this lack of analytical validity. We present a fast method for generating numerical hybrid microdata in a way that preserves attribute means, variances and covariances, as well as (to some extent) record similarity and subdomain analyses. We also overcome the uncertainty in assessing data utility by using newly defined probabilistic information loss measures.

1 Introduction

Statistical databases come in two flavors: tabular data (aggregated data) and microdata (records on individual persons or entities). Microdata can be continuous, *e.g.* salary or weight, or categorical, for instance sex, hair color or instruction level. Releasing microdata, or any statistical data for that matter, must face the tension between respondent privacy and data utility. In the case of microdata, providing respondent privacy means that an intruder should be unable to make a decision whether a published record corresponds to a specific respondent. On the other hand, providing data utility means that the published set of data should preserve as many statistical properties as possible from the original set.

J. Domingo-Ferrer et al.: *Information Loss in Continuous Hybrid Microdata: Subdomain-Level Probabilistic Measures*, StudFuzz **197**, 287–298 (2006)
www.springerlink.com

However, assessing respondent privacy and data utility is inherently uncertain:

- The privacy of respondents is inversely proportional to the risk of disclosure of their responses, *i.e.* the risk that an intruder can link specific released data to specific respondents. But the ability of an intruder to do so depends on how many and how good are the external identified information sources he can gather (*e.g.* censuses, yearbooks, phonebooks, etc.). Unfortunately, the data protector cannot exactly anticipate how much external information intruders will be able to link to the data being released in anonymized form.
- The utility of the released data depends on the specific data uses it has to serve. But the data protector is very often unable to foresee how the released data will be used by legitimate users.

One possibility for protecting a microdata set is to use a *masking method* (*e.g.* additive noise, microaggregation, etc., cf. [3]) to transform original data into protected, publishable data. An alternative to masking the original data is to generate a a *synthetic* data set not from the original data, but from a set of random values that are adjusted in order to fulfill certain statistical requirements [14, 11, 13, 12]. A third possibility is to build a *hybrid* data set as a mixture of the masked original values and a synthetic data set [6, 1, 2].

The advantage of synthetic data over masking is that deciding whether a synthetic data record corresponds to a particular respondent does not make sense any more. So the problem of assessing the privacy of respondents is circumvented. The drawback of synthetic data is that they have limited data utility: they only preserve the statistics that were taken into account by the data protector when generating the synthetic dataset. This raises an interesting question: why not directly publish the statistics that should be preserved rather than a synthetic dataset preserving them?

For synthetic microdata to be really useful, they ought to preserve record-level similarity to some extent, so that subdomains of the protected dataset still yield acceptably accurate analyses. Hybrid microdata are better than purely synthetic microdata at preserving record-level similarity.

1.1 Contribution and Plan of This Paper

In this paper, a method for generating continuous hybrid microdata is proposed such that:

- It is non-iterative and fast, because its running time grows *linearly* with the number of records in the original dataset;
- It *exactly* reproduces the means and the covariance matrix of the original dataset, which implies that variances and Pearson correlations are also exactly preserved;
- It allows preservation of record-level similarity to some extent, so that subdomains of the protected dataset can still be analytically valid (this is

especially true if the relevant subdomains can be decided before applying the method).

When assessing the utility of the data output by the proposed method, we propose to deal with the uncertainty about data uses using probabilistic information measures.

Section 2 describes our proposal for generating hybrid data. Section 3 deals with the complexity and the data utility properties of the proposed method. Probabilistic information loss measures are described in Section 4. Empirical results are reported in Section 5. Finally, Section 6 contains some conclusions and suggestions for future research.

2 A Low-cost Method for Hybrid Microdata Generation

Let X be an original microdata set, with n records and m attributes. Let X'' be a hybrid microdata set to be generated, also with n records and m attributes. In fact, both X and X'' can be viewed as $n \times m$ matrices. The method presented is a hybrid evolution of the synthetic data generation method in [8]. Like [8], it exactly preserves both univariate and multivariate statistical properties of X, such as means, covariances and correlations. As we show below, the improvement with respect to [8] is that, since X'' is hybrid rather than synthetic, a fair amount of record-level similarity between X and X'' is preserved, which allows for subdomain analysis. The algorithm below constructs X'' from X:

Algorithm 1 (Basic Procedure)

1. Use a masking method to transform the original dataset X into a masked dataset X'.
2. Compute the covariance matrix C of the original microdata matrix X.
3. Use Cholesky's decomposition on C to obtain
$$C = U^t \times U$$
where U is an upper triangular matrix and U^t is the transposed version of U.
4. Compute an $n \times m$ matrix A as
$$A := X' \cdot U^{-1}$$
5. Use Algorithm 2 to modify matrix A so that its covariance matrix is the $m \times m$ identity matrix.
6. Obtain the hybrid microdata set as
$$X'' = A \cdot U$$
By construction, the covariance matrix of X'' equals the covariance matrix of X (see [15]).
7. Due to the construction of matrix A, the mean of each attribute in X'' is 0. In order to preserve the mean of attributes in X, a last adjustment is performed. If $\bar{x}_j$ is the mean of the j-th attribute in X, then $\bar{x}_j$ is added to the j-th column (attribute) of X'':
$$x''_{ij} := x''_{ij} + \bar{x}_j \quad \text{for } i = 1, \cdots, n \text{ and } j = 1, \cdots, m \tag{1}$$

Note that modification of matrix A is needed for two reasons:

- To preserve covariances;
- To prevent the hybrid data X'' from being exactly the masked data X'.

We now need to specify how to modify an $n \times m$ matrix A so that its covariance matrix is the $m \times m$ identity matrix.

Algorithm 2 (Modification of Matrix A)

1. Given an $n \times m$ matrix A with elements $a_{i,j}$, view the m columns of A as samples of attributes $A_1, \cdots, A_m$. If $Cov(A_j, A_{j'})$ is the covariance between attributes A_j and $A_{j'}$, the objective of the algorithm is that
$$Cov(A_j, A_{j'}) = \begin{cases} 1 \text{ if } j = j' \\ 0 \text{ otherwise} \end{cases}$$
for $j, j' \in \{1, \cdots, m\}$.
2. Let $\bar{a}_1$ be the mean of A_1. Let us adjust A_1 as follows:
$$a_{i,1} := a_{i,1} - \bar{a}_1 \quad i = 1, \ldots, n$$
The mean of the adjusted A_1 is 0.
3. In order to reach the desired identity covariance matrix, some values of attributes $A_2, \cdots, A_m$ must change. For $v = 2$ to m do:
   a) Let $\bar{a}_v$ be the mean of attribute A_v.
   b) For $j = 1$ to $v - 1$, the covariance between attributes A_j and A_v is
   $$Cov(A_j, A_v) = \frac{\sum_{i=1}^{n} a_{i,j} \cdot a_{i,v}}{n} - 0 \cdot \bar{a}_v = \frac{\sum_{i=1}^{n} a_{i,j} \cdot a_{i,v}}{n}$$
   c) In order to obtain $Cov(A_j, A_v) = 0, \quad j = 1 \ldots v - 1$, some elements $a_{i,v}$ in the v-th column of A are assigned a new value. Let $x_1, \ldots, x_{v-1}$ the unknowns for the following linear system of $v - 1$ equations:
   $$\frac{\sum_{i=1}^{n-v+1} a_{i,j} \cdot a_{i,v} + \sum_{i=1}^{v-1} a_{n-v+1+i,j} \cdot x_i}{n} = 0 \text{ for } j = 1 \ldots v - 1$$
   that is
   $$\sum_{i=1}^{n-v+1} a_{i,j} \cdot a_{i,v} + \sum_{i=1}^{v-1} a_{n-v+1+i,j} \cdot x_i = 0 \text{ for } j = 1 \ldots v - 1$$
   Once the aforementioned linear system is solved, the new values are assigned:
   $$a_{n-v+1+i,v} := x_i \text{ for } i = 1 \ldots v - 1$$
   d) Let $\bar{a}_v$ be the mean of attribute A_v. A final adjustment on A_v is performed to make its mean 0:
   $$a_{i,v} = a_{i,v} - \bar{a}_v \text{ for } i = 1 \ldots n$$
4. In the last step, values in A are adjusted in order to reach $Cov(A_j, A_j) = 1$ for $j = 1 \ldots m$. If σ_j is the standard deviation of attribute A_j, the adjustment is computed as:
$$a_{i,j} := \frac{a_{i,j}}{\sigma_j}, \quad i = 1 \ldots n, j = 1 \ldots m$$

3 Properties of the Proposed Scheme

3.1 Performance and Complexity

The computational complexity for the proposed method will next be estimated; we exclude the masking step, because it is not part of the method itself. Let n be the number of records and m the number of attributes. Then the complexities of the various operations are as follows:

- Calculation of the covariance matrix: $\mathcal{O}(n + m^2)$;
- Cholesky's decomposition: $\mathcal{O}(m^3/6)$ (see [10]);
- Inversion of the triangular matrix U: $\mathcal{O}(m^2/2)$;
- Calculation of A: $\mathcal{O}(2nm + 2m^3 + 2m^4/3)$ where the term $2m^4/3$ is the cost of solving a Gauss system m times [10];
- Matrix product: $\mathcal{O}(nm^2)$;
- Mean adjustment: $\mathcal{O}(nm)$.

In summary, the overall complexity is $\mathcal{O}(nm + 2m^4/3) = \mathcal{O}(n + m^4)$. To understand this complexity, one should realize that, in general, the number of records n is much larger than the number of attributes m, *i.e.* $n \gg m$. Thus this proposal maintains the strong point of [8] that *its complexity is linear in the number of records.*

The method has been tested with several data set sizes and execution times to generate X'' from X' (that is, excluding the running time for masking X into X') are shown in Table 1.

Table 1. Running time (in seconds) on a 1.7 GHz desktop Intel PC under a Linux OS. Note that time for random matrix generation is included

	Number of attributes m			
Number of records n	5	10	25	50
1,000	0.00	0.01	0.06	0.32
10,000	0.06	0.20	1.28	5.33
100,000	0.50	1.95	12.43	51.19

3.2 Data Utility

As stated above, the proposed scheme exactly reproduces the statistical properties of the original data set.

- The means of attributes in the original data set X are exactly preserved in the synthetic data set X''.
- The covariance matrix of X is exactly preserved in X'' (see [15]). Thus, in particular:

- The variance of each attribute in X is preserved in X'';
- The Pearson correlation coefficient matrix of X is also exactly preserved in X'', because correlations are obtained from the covariance matrix.

The difference between this method and the one in [8] is that record-level similarity is preserved to some extent, as shown in Section 5 below on empirical results. Unlike [8], which used a random matrix A, the method in our paper uses a matrix A which is derived from a masked version X' of the original data X.

4 A Generic Information Loss Measure

To measure information loss, we assume that the original dataset X is a population and the hybrid dataset X'' is a sample from the population. Given a population parameter θ on X, we can compute the corresponding sample statistic $\hat{\Theta}$ on X''. Let us assume that $\hat{\theta}$ is the value taken by $\hat{\Theta}$ in a specific instance of sample X''. The more different is $\hat{\theta}$ from θ, the more information is lost when publishing the sample X'' instead of the population X. We show next how to express that loss of information through probability.

If the sample size n is large, the distribution of $\hat{\Theta}$ tends to normality with mean θ and variance $Var(\hat{\Theta})$. According to [5], values of n greater than 100 are often large enough for normality of all sample statistics to be acceptable. Fortunately, most protected datasets released in official statistics consist of $n > 100$ records, so that assuming normality is safe. Thus, the standardized sample discrepancy

$$Z = \frac{\hat{\Theta} - \theta}{\sqrt{Var(\hat{\Theta})}}$$

can be assumed to follow a $N(0,1)$ distribution.

Therefore, in [7] we defined a generic probabilistic information loss measure $pil(\theta)$ referred to parameter θ as the probability that the absolute value of the discrepancy Z is less than or equal to the actual discrepancy we have got in our specific sample X', that is

$$pil(\hat{\Theta}) = 2 \cdot P\left(0 \leq Z \leq \frac{|\hat{\theta} - \theta|}{\sqrt{Var(\hat{\Theta})}}\right) \qquad (2)$$

Being a probability, the above measure is bounded in the interval $[0,1]$, which facilitates comparison and tradeoff against disclosure risk (which is also bounded). This is a major advantage over previous non-probabilistic information loss measures [4,16], which are unbounded.

5 Empirical Work

5.1 Information Loss and Disclosure Risk Measures

To assess data utility, we will compute the above generic probabilistic information loss measure for specific population parameters θ and sample statistics $\hat{\Theta}$: quantiles, means, variances, covariances and Pearson correlations. We write below PIL rather than pil when the probabilistic measure has been averaged over all attributes or pairs of attributes, rather than being computed for a single attribute or attribute pair. Also, the measures below have been multiplied by 100 to scale them within $[0, 100]$ rather than within $[0, 1]$.

1. $PIL(Q)$ is 100 times the average $pil(Q_q)$ for all attributes and quantiles Q_q from $q = 5\%$ to $q = 95\%$ in 5% increments over all attributes; this is the average impact on quantiles;
2. $PIL(m_1^0)$ is the average impact on means over all attributes;
3. $PIL(m_2)$ is the average impact on variances over all attributes;
4. $PIL(m_{11})$ is the average impact on covariances over all attribute pairs;
5. $PIL(r)$ is the average impact on Pearson's correlation coefficients over all attribute pairs.

Regarding disclosure risk, it is measured using the following three measures defined in [3, 9], which also take values in $[0, 100]$ as follows:

- DLD (Distance-based Linkage Disclosure) is the average of $DLD - 1$ to $DLD - 10$, where $DLD - i$ is the percentage of correctly linked pairs of original-hybrid records using distance-based record linkage through i attributes;
- RID (Rank Interval Disclosure) is the average of $RID - 1$ to $RID - 10$, where $RID - i$ is the percentage of original values of the i-th attribute that fall within a rank interval centered around their corresponding hybrid value with width $p\%$ of the total number of records;
- $SDID$ (Standard Deviation Interval Disclosure) is analogous to RID but using intervals whose width is a percentage of each attribute's standard deviation.

5.2 The Data Set

The microdata set X used for testing was obtained from the U.S. Energy Information Authority and contains 4092 records[1]. Initially, the data file contained 15 attributes from which the first 5 were removed as they corresponded to identifiers. We have worked with the attributes: RESREVENUE, RESSALES, COMREVENUE, COMSALES, INDREVENUE, INDSALES, OTHREVENUE, OTHRSALES, TOTREVENUE, TOTSALES. This dataset was also used in [9].

[1] `http://www.eia.doe.gov/cneaf/electricity/page/eia826.html`

Table 2. Information loss and disclosure risk measures for the overall masked and hybrid datasets

	$PIL(Q)$	$PIL(m_1^0)$	$PIL(m^2)$	$PIL(m_{11})$	$PIL(r)$	DLD	RID	$SDID$
$X'(mic)$	5.3	0	6.6	2.0	27.0	19.3	93.0	84.5
$X''(mic)$	49.4	0	0	0	0	2.0	41.1	41.4
$X'(sw)$	0	0	0	99.9	100.0	3.8	71.3	62.7
$X''(sw)$	53.3	0	0	0	0	0.1	31.7	32.7
$X'(rn)$	52.6	5.2	29.2	3.4	66.5	7.7	39.1	26.6
$X''(rn)$	62.7	0	0	0	0	0.3	28.9	20.5

5.3 The Results

Results on the Overall Dataset

Three different masking methods have been used in Step 1 of Algorithm 1: microaggregation with parameter 3 to obtain $X'(mic)$, rank swapping with parameter 7 to obtain $X'(sw)$ and random noise with parameter 0.16 to obtain $X'(rn)$ (see [3] for details on those methods). Call $X''(mic)$, $X''(sw)$ and $X''(rn)$ the hybrid datasets obtained with Algorithm 1 using $X'(mic)$, $X'(sw)$ and $X'(rn)$, respectively. By construction, the hybrid datasets preserve means, variances, covariances and Pearson correlations of the original dataset X. Table 2 gives the above information loss measures for each masked and hybrid dataset.

It can be seen from Table 2 that X'' have higher $PIL(Q)$ than X' for all three masking methods (only slightly higher for random noise); however, X'' have substantially lower values than X' for the remaining information loss measures. Disclosure risk measures are also much lower for X'' than for X' for all masking methods. Thus, all in all, the hybrid datasets X'' are much better than the masked X.

Top-down Generation: Posterior Subdomains

The next question is how does the proposed hybrid data generation method behave for subdomains that were not predictable at the moment of generating the hybrid data (posterior subdomains). This is relevant if the user is interested in a subset of records selected according to the values of a (maybe external) categorical attribute (*e.g.* the records corresponding to women).

To answer that question, X and X'' have been partitioned into d corresponding subdomains, for $d = 2, 3, 4, 5$. The various information loss measures have been computed between each subdomain in X and its corresponding subdomain in X''. Table 3 gives the average results, *i.e.* for each d the measures averaged over the d pairs of $X - X''$ subdomains.

At least two things can be observed from Table 3, regardless of the masking method used:

Table 3. Top-down information loss and disclosure risk measures (induced in subdomains)

	d	$PIL(Q)$	$PIL(m_1^0)$	$PIL(m^2)$	$PIL(m_{11})$	$PIL(r)$	DLD	RID	$SDID$
$X''(mic)$	2	54.2	78.0	72.1	71.5	96.5	2.2	32.2	28.8
$X''(mic)$	3	56.3	84.1	80.5	86.4	97.5	2.5	28.2	23.1
$X''(mic)$	4	57.9	81.6	84.7	90.3	96.9	2.6	25.7	20.5
$X''(mic)$	5	58.8	87.2	86.2	93.5	96.1	2.8	24.4	19.2
$X''(sw)$	2	57.9	85.9	80.5	75.3	95.5	0.2	20.5	20.6
$X''(sw)$	3	59.5	88.1	86.9	89.9	96.8	0.3	16.7	16.9
$X''(sw)$	4	60.6	89.0	92.1	94.1	97.0	0.3	14.6	15.5
$X''(sw)$	5	62.5	92.4	94.7	96.2	96.0	0.4	13.9	14.5
$X''(rn)$	2	63.9	88.1	79.3	77.0	95.2	0.4	21.4	15.8
$X''(rn)$	3	67.3	92.0	86.9	88.3	96.0	0.5	18.2	14.0
$X''(rn)$	4	66.9	85.9	92.8	93.3	95.5	0.5	16.6	12.8
$X''(rn)$	5	70.2	91.1	94.3	95.5	96.2	0.5	15.9	12.3

- Means, variances, covariances and Pearson correlations of the original dataset are *not* preserved in the subdomains; further, the discrepancy of these statistics between the subdomains in X and X'' is remarkable;
- The more subdomains are made, the larger the information loss and the smaller the interval disclosure risks RID and $SDID$.

Bottom-up Generation: Prior Subdomains

We now turn to the performance of the proposed hybrid data generation when directly applied to subdomains that can be decided *a priori*. By construction, means, variances, covariances and Pearson correlations of subdomains of X are preserved in subdomains of X''. The interesting point is that the overall X''_t obtained as union of d hybrid subdomains also preserves means, variances, covariances and Pearson correlations of X. Table 4 shows average information loss and disclosure risk measures for subdomains in a way analogous to Table 3; in addition, for each partition into d subdomains, it gives the measures between the overall X and X''_t.

We can see from Table 4 that, regardless of the masking method used:

- When assembling the hybrid subdomains into X''_t, $PIL(Q)$ and DLD decrease (even if the latter is already very low). At the same time, the interval disclosure measures RID and $SDID$ increase substantially;
- As the number d of subdomains increases, interval disclosure measures decrease for each subdomain, but they increase for the overall dataset X''_t.

6 Conclusions and Future Research

In this paper, we have presented a new method for generating numerical hybrid microdata, based on combining a masking method with Cholesky's decompo-

Table 4. Bottom-up information loss and disclosure risk measures

	d	$PIL(Q)$	$PIL(m_1^0)$	$PIL(m^2)$	$PIL(m_{11})$	$PIL(r)$	DLD	RID	$SDID$
$X''(mic)$	2	45.0	0	0	0	0	2.5	36.6	22.1
$X_t''(mic)$	2	37.8	0	0	0	0	2.3	50.0	54.6
$X''(mic)$	3	45.5	0	0	0	0	5.6	38.8	21.7
$X_t''(mic)$	3	37.3	0	0	0	0	5.1	57.3	56.9
$X''(mic)$	4	44.5	0	0	0	0	6.4	37.3	18.7
$X_t''(mic)$	4	37.5	0	0	0	0	5.3	58.7	58.5
$X''(mic)$	5	47.0	0	0	0	0	5.4	34.6	15.1
$X_t''(mic)$	5	37.2	0	0	0	0	4.2	61.1	59.4
$X''(sw)$	2	50.9	0	0	0	0	0.3	23.1	12.7
$X_t''(sw)$	2	46.5	0	0	0	0	0.2	37.7	46.8
$X''(sw)$	3	54.0	0	0	0	0	0.4	20.6	9.6
$X_t''(sw)$	3	45.0	0	0	0	0	0.3	41.5	48.1
$X''(sw)$	4	54.4	0	0	0	0	0.5	17.9	7.3
$X_t''(sw)$	4	41.5	0	0	0	0	0.3	42.8	49.7
$X''(sw)$	5	53.5	0	0	0	0	0.5	17.7	6.8
$X_t''(sw)$	5	40.5	0	0	0	0	0.4	45.2	51.0
$X''(rn)$	2	53.8	0	0	0	0	0.4	21.1	11.1
$X_t''(rn)$	2	47.3	0	0	0	0	0.4	32.1	40.6
$X''(rn)$	3	57.0	0	0	0	0	0.5	18.4	8.5
$X_t''(rn)$	3	45.2	0	0	0	0	0.6	34.6	42.0
$X''(rn)$	4	55.6	0	0	0	0	0.7	16.5	7.4
$X_t''(rn)$	4	42.6	0	0	0	0	0.6	37.0	46.1
$X''(rn)$	5	55.4	0	0	0	0	0.6	15.4	6.3
$X_t''(rn)$	5	42.8	0	0	0	0	0.5	38.7	47.0

sition. Excluding masking computation (dependent on the particular masking method used), the method is very fast, because its running time is linear in the number of records. The method preserves a fair amount of record-level similarity, which allows for subdomain analysis. The best results are obtained when the relevant subdomains can be decided in advance by the data protector: in that case, the method can be applied independently for each subdomain and the information loss and disclosure risk measures for the overall dataset are still very good (bottom-up approach). If the relevant subdomains cannot be anticipated, then the only option is to apply the method to the overall dataset and hope for the best when subdomains are analyzed (top-down approach); in this case, the method does not guarantee intra-subdomain preservation of any statistics, but it distorts those statistics less than sheer synthetic data generation.

A first line for future research is to extend the notion of subdomain from a subset of records to a subset of records *and* attributes. The idea is to evaluate the bottom-up and the top-down approaches for subdomains comprising only a subset of attributes for each record.

A second line for future research is to give some guidance to help the data protector anticipate the relevant subdomains in the bottom-up approach (*e.g.* use categorical variables to partition records in subdomains, etc.).

Acknowledgments

The authors are partly supported by the Spanish Ministry of Science and Education through project SEG2004-04352-C04-01 "PROPRIETAS", by the Government of Catalonia under grant 2002 SGR 00170 and by Cornell University under contract no. 47632-10043.

References

1. J. M. Abowd and S. D. Woodcock (2004) Multiply-imputing confidential characteristics and file links in longitudinal linked data. In J. Domingo-Ferrer and V. Torra, editors, *Privacy in Statistical Databases*, volume 3050 of *LNCS*, pages 290–297, Berlin Heidelberg: Springer.
2. R. Dandekar, J. Domingo-Ferrer, and F. Sebé (2002) LHS-based hybrid microdata vs rank swapping and microaggregation for numeric microdata protection. In J. Domingo-Ferrer, editor, *Inference Control in Statistical Databases*, volume 2316 of *LNCS*, pages 153–162, Berlin Heidelberg: Springer.
3. J. Domingo-Ferrer and V. Torra (2001) Disclosure protection methods and information loss for microdata. In P. Doyle, J. I. Lane, J. J. M. Theeuwes, and L. Zayatz, editors, *Confidentiality, Disclosure and Data Access: Theory and Practical Applications for Statistical Agencies*, pages 91–110, Amsterdam: North-Holland.
4. J. Domingo-Ferrer and V. Torra (2001) A quantitative comparison of disclosure control methods for microdata. In P. Doyle, J. I. Lane, J. J. M. Theeuwes, and L. Zayatz, editors, *Confidentiality, Disclosure and Data Access: Theory and Practical Applications for Statistical Agencies*, pages 111–134, Amsterdam: North-Holland.
5. M. G. Kendall, A. Stuart, S. F. Arnold J. K. Ord, and A. O'Hagan (1994) *Kendall's Advanced Theory of Statistics, Volume 1: Distribution Theory (6th Edition)*. London: Arnold.
6. A. B. Kennickell (1999) Multiple imputation and disclosure protection: the case of the 1995 survey of consumer finances. In J. Domingo-Ferrer, editor, *Statistical Data Protection*, pages 248–267, Luxemburg: Office for Official Publications of the European Communities.
7. J. M. Mateo-Sanz, J. Domingo-Ferrer, and F. Sebé (2005) Probabilistic information loss measures for continuous microdata. *Data Mining and Knowledge Discovery*, to appear.
8. J. M. Mateo-Sanz, A. Martínez-Ballesté, and J. Domingo-Ferrer (2004) Fast generation of accurate synthetic microdata. In J. Domingo-Ferrer and V. Torra, editors, *Privacy in Statistical Databases*, volume 3050 of *LNCS*, pages 298–306, Berlin Heidelberg: Springer.

9. J. M. Mateo-Sanz, F. Sebé, and J. Domingo-Ferrer (2004) Outlier protection in continuous microdata masking. In J. Domingo-Ferrer and V. Torra, editors, *Privacy in Statistical Databases*, volume 3050 of *LNCS*, pages 201–215, Berlin Heidelberg: Springer.
10. W. Press, W. T. Teukolsky, S. A. Vetterling, and B. Flannery (1993) *Numerical Recipes in C: The Art of Scientific Computing.* Cambridge, UK: Cambridge University Press.
11. T. J. Raghunathan, J. P. Reiter, and D. Rubin (2003) Multiple imputation for statistical disclosure limitation. *Journal of Official Statistics*, 19(1):1–16.
12. J. P. Reiter (2005) Releasing multiply-imputed, synthetic public use microdata: An illustration and empirical study. *Journal of the Royal Statistical Society, Series A*, 131(2):365–377.
13. J. P. Reiter (2005) Significance tests for multi-component estimands from multiply-imputed, synthetic microdata. *Journal of Statistical Planning and Inference*, 168:185–205.
14. D. B. Rubin (1993) Discussion of statistical disclosure limitation. *Journal of Official Statistics*, 9(2):461–468.
15. E. M. Scheuer and D. S. Stoller (1962) On the generation of normal random vectors. *Technometrics*, 4:278–281.
16. W. E. Yancey, W. E. Winkler, and R. H. Creecy (2002) Disclosure risk assessment in perturbative microdata protection. In J. Domingo-Ferrer, editor, *Inference Control in Statistical Databases*, volume 2316 of *LNCS*, pages 135–152, Berlin Heidelberg: Springer.

Access to a Large Dictionary of Spanish Synonyms: A Tool for Fuzzy Information Retrieval*

Alejandro Sobrino-Cerdeiriña[1], Santiago Fernández-Lanza[1], and Jorge Graña-Gil[2]

[1] Departamento de Lógica y Filosofía Moral, Univ. de Santiago de Compostela, Campus Sur s/n, 15782 – Santiago de Compostela, Spain
{lfgalex, sflanza}@usc.es

[2] Departamento de Computación, Universidad de La Coruña, Campus de Elviña s/n, 15071 – La Coruña, Spain
grana@udc.es

Summary. We start by analyzing the role of imprecision in information retrieval in the Web, some theoretical contributions for managing this problem and its presence in search engines, with special emphasis on the use of thesaurus in order to increase the number and relevance of the documents retrieved. We then present a general architecture for implementing large dictionaries in natural language processing applications which is able to store a considerable amount of data relating to the words contained in these dictionaries. In this modelling, efficient access to this information is guaranteed by the use of minimal deterministic acyclic finite-state automata. In addition, we implement a Spanish dictionary of synonyms and illustrate how our general model helps to transform the original dictionary into a computational framework capable of representing semantic relations between words. This process allows us to define synonymy as a gradual relation, which makes the final tool more suitable for word sense disambiguation tasks or for information retrieval applications than other traditional approaches. Moreover, our electronic dictionary, called FDSA, will be freely available very soon for stand-alone use.

1 Introduction

Nowadays, when there is more and more digital information available, to ask questions and recover their corresponding relevant answers is of great importance. In this task, Web searchers play a fundamental role. Until now, the criteria preferred by search engines were of syntactic nature: a document was

* This work was partially supported by the projects TIN2004-07246-C03-02 of the Spanish Ministry of Education and Science, and PGIDTI02PXIB30501PR of the Autonomous Government of Galicia (Spain).

A. Sobrino-Cerdeiriña et al.: *Access to a Large Dictionary of Spanish Synonyms: A Tool for Fuzzy Information Retrieval*, StudFuzz **197**, 299–316 (2006)
www.springerlink.com

recovered as a relevant answer when the words used by the engine to index that document fitted with the words introduced by the user in the search box.

However, a search method based on simple syntactic matchings is not enough, and it should include some kind of weighting for the terms to match. Simple matching of words does not ensure relevance of documents respect to queries, since relevance is of a semantic nature. Moreover, relevance is hardly absolute, but a gradual subject. In fact, a document is seldom completely informative or completely devoid of any interest. Usually, documents inform of a theme in a partial way, i.e. to a certain degree. Weighting of words according to their field (in semi-structured texts) or to their position in the document (title, abstract, etc.) has been the most commonly applied method for automatically inferring relevance, as shown in a great number of proposed models for information retrieval, e.g. the classic vectorial model [13] or the TF×IDF weights [9]. Based on the vectorial space model, the relevance score of a document is the sum of weights of the query terms that appear in the document, normalized by the Euclidean vector length of the document. The weight of a term is a function of the word's occurrence frequency (also known as TF, the term frequency) and the number of documents containing the word in collection (also known as IDF, the inverse document frequency).

Nevertheless, weighting of terms is not enough in many cases for an appropriate retrieval. If we need, for instance, information about `powerful cars`, there will be relevant pages about `fast cars`, and this relevance is not given by any weighting of the word `powerful`, but is given by an identification of `powerful` and `fast` as synonymous terms in this context. This intuitive idea arises from our linguistic knowledge, which allows us to consider as relevant pages those including terms explicitly present in the query, and also those including terms semantically related with the query through the synonymy process. In practice, the application of this idea implies expanding the query over a word with its *synset* or set of synonymous terms semantically related.

The use of associative methods for recovering information is not a recent proposal in the field of information retrieval. Query expansion through synonymy is a traditional resort for improving performances, and, since similarity of meanings is a gradual subject, fuzzy logic plays an important role in this task. In fact, there exist various fuzzy thesauri [11, 12] in which, if two words are related, this link presents a degree which points out the strength of their semantic association. This degree is determined by means of statistical data and by validation processes performed by users. Fuzzy thesauri have been used for recovering textual information with good results, even though they show three main problems:

- With the use of this kind of thesauri, the coverage is increased, but this often diminishes precision.
- These thesauri do not distinguish meanings or contexts, and therefore they cause wrong recoveries, increase the latency of the search, and make the list of results redundant or useless.

- For queries with sentences or complex terms, these thesauri do not integrate an appropriate semantic to combine the corresponding simple terms.

In this work, we propose an alternative solution for avoiding these problems. This solution is given by our electronic dictionary of synonyms, which presents the following features:

- Our dictionary mechanizes a Spanish dictionary of synonyms considering, for each word, its possible meanings and, for each meaning, its synset or list of synonymous terms. To consider different meanings of a word is crucial for avoiding erroneous searches. In some way, these meanings associated to the words provide a (partial) ontology, because a meaning indicates the contextual use of a word, and separates this use from the other possible uses. Moreover, our dictionary assigns a degree of synonymy to all the words of the synset. This degree can be calculated by using different definitions of similarity. The use of these different criteria for measuring similarity is important because vagueness of words is plural and can require different models (fuzzy logics) for its formalization, particularly in the case of searches involving sentences or complex terms. Currently, in our dictionary, this functionality must be used manually. In the future, it will be convenient to integrate an automatic method to select a specific similarity measure, depending on the different situations in which vagueness appears. Nevertheless, for this last task, it would be necessary to have a very wide ontology available.
- Our dictionary is also efficient from a computational point of view. With regard to this aspect, to handle an electronic dictionary for the synonymous words of a given language, considering their corresponding meanings, is in fact a complex task and involves a great computational cost. The last part of this work attend to this problem by designing a method to build an acyclic deterministic automaton, which allows us to complete the printed dictionary and to access its contents rapidly.

Since synonymy is a linguistic feature modeled with a gradual logic, such as fuzzy logic [8], it should be a resort used by the well-known searchers. The query `fuzzy search engines` in GOOGLE provides, among others, the following results:

- NORTHERN LIGHT (`www.northernlight.com`). It is said that it has a fuzzy *and* because it uses singulars and plurals in the search, giving wider results than a precise searcher (as is described in its technical manual).
- DISCOUNT ENGINE (`www.discountdomainsuk.com/glossary/3/499/0`). It defines fuzzy search as a search that succeeds when matching words that are partially or wrongly spelled. The same definition can be found in `www.search-marketing.info/search-glossary/engine-terms/fuzzy-search.htm`.
- NETSCAPE (`www.netscape.com`). It uses the fuzzy operator ~, which must be typed after some symbols. Its mission is to force the system to guess

what comes after it: for instance, with `acetil~` the searcher should provide all the pages including terms that are compatible with this prefix, even those related with `acetilsalicilic`, the complete name of the term perhaps partially forgotten.

An overview to the rest of URLs provided by this query shows that the word *fuzzy* is applied, in most cases, when the searcher has algorithms to correct wrongly typed strings (GOOGLE itself has this functionality) and, in any case, when the searcher handles linguistic resources such as synonymy.

To correct wrongly spelled words is obviously very useful. However, this approximate matching is only fuzzy in a formal aspect, but not from a semantic point of view. A search is said to be genuinely approximate when it is handled by considering the meaning of the terms involved in it, i.e. by using their representation in the index of the search engine, and the relations established between them attending to their meanings (and, for instance, to their synonyms), and not to the way in which they are written. Our dictionary can be an appropriate tool for performing fuzzy searches in the Web because it seems plausible that:

- To distinguish meanings will reduce the number of pages provided as relevant answers by the searcher.
- To have an efficient implementation does not penalize the latency of the searching proccess.
- To implement different similarity measures will permit progress to be made in the selection of the appropriate operators to combine the terms of the queries.

We are sure that a fuzzy semantic search is a necessary step on the way to obtaining more robust information retrieval systems. However, its implementation in a commercial search engine and an evaluation of the results is a challenge that we are obliged to undertake in the future. For the moment, we restrict the objective of this work to building an electronic dictionary of synonyms for Spanish.

In Sect. 2, we include a brief discussion on synonymy. Section 3 gives our way to treat synonymy and specifies how to calculate the degree of synonymy between two entries of the dictionary. Section 4 describes our general model of electronic dictionary and allows us to understand the role of the finite-state automata in this context. In Sect. 5, we describe the Spanish dictionary of synonyms [2] and detail all the transformations performed on it with the help of our automata-based architecture for dictionaries. As we have seen, our main aim is to integrate this dictionary in natural language processing applications of a more general nature, as a tool able to provide greater precision in the analysis of synonymy relations. Nevertheless, our electronic dictionary, called FDSA[1], will be available in the very near future for stand-alone use. In Sect. 6,

[1] Fuzzy Dictionary of Synonyms and Antonyms.

we present its main features and functionalities. Finally, Section 7 presents our conclusions after analysing the data contained in this new dictionary.

2 A Short Historical Introduction to Synonymy

Synonymy, or the relationship of similarity of meaning, has long been a subject of interest. The first recorded mention of the concept was made by the Ancient Greek philosopher Prodicus of Keos (465–399? B. C.), and Aristotle refers to it in his *Topics* (I 7:103a6-32):

> "... from the outset one should clearly state, with regard to that which is identical, in how many ways it can be said."

This description led to continued interest in the subject of synonymy, and influenced the aim of grouping together as synonyms those words whose meanings, although coinciding, showed certain differences.

The Romans took up this tradition, as is shown by Seleucus' treatise *On the difference between synonyms* and a rudimentary dictionary by Ammonius, *On similar and different expressions*. Both Greeks and Romans alluded to two fundamental traits of synonymy:

1. it is a characteristic of the meaning of words, and deals with the plurality of signifiers of a single reference;
2. it is a relationship of the similarity of meanings.

But the first attempt at a systematic study of synonymy as a lexical relationship was made by the Frenchman Gabriel Girard at the beginning of the 18th century. In his work *La Justesse de la Langue Française, ou les Differents Significations de Mots qui Passent pour Synonymies* (1718), he stated that:

> "In order to obtain propriety, one does not have to be demanding with words; one does not have to imagine at all that so-called synonyms are so with all the rigorousness of perfect resemblance; since this only consists of a principal idea which they all enunciate, rather each one is made different in its own way by an accessorial idea which gives it its own singular character. The similarity bought about by the principal idea thus makes the words synonyms; the difference that stems from the particular idea, which accompanies the general one, means that they are not perfectly so, and that they can be distinguished in the same manner as the different shades of the same colour." [6, pp. VIII ff.]

Girard clearly considers synonymy more as an approximate relation than a matter of perfect resemblance. This conception still prevails today where a common definition of synonymy is that two expressions are synonyms if they have the same, or approximately the same, meaning. In natural language there are few examples of words that have exactly the same meaning, nevertheless in dictionaries of synonyms there are many examples of words that have approximately the same meaning. The imprecise characteristics of word synonymy in natural language will be studied in this work in terms of the automatisation of a dictionary of synonyms.

As we have seen, this definition of synonymy lies in the concept of meaning. Although it may seem appropriate to analyze this concept, which has been the subject of long-standing controversy in the fields of philosophy of language and linguistics, such an analysis lies outside the scope of the present work, in which we only try to look for a computational way to represent the meaning of a word. Our proposal is to consider the set of words that a dictionary of synonyms gives for an entry as a computational way to represent the meaning of that entry. This is not to say that the meaning of a word is the set of synonym words that a dictionary of synonyms associate with it. Our approach is strictly empirical and practical, but could be helpful for those that analyse meaning from a theoretical point of view in order to test their theories.

This empirical point of view is not free of problems. There is always an excesive dependence on the particular published dictionary that we use, and dictionaries are man-made tools. In consequence, dictionaries may contain mistakes, may not be complete, or may give a slanted view of synonymy. Moreover it is usual to find relations other than that of synonymy between the words appearing in dictionaries of synonyms. This difficulty was perceived by John Lyons when he stated that, strictly speaking, the relation that holds between the words in dictionaries of synonyms is quasi-synonymy more than synonymy (see [10]). We assume these risks on behalf of the practical applications and utility of our resulting tool: FDSA. Some of these applications and utilities will be shown in Sects. 6 and 7.

3 A Computational View of Synonymy

In the previous section, we have discussed three main ideas. Firstly, we have seen that it is usual to conceive synonymy as a relation between two expressions with identical or similar meaning. Secondly, we were also able to infer that the controversy of understanding synonymy as a precise question or as an approximate question, i.e. as a question of identity or as a question of similarity, has always been present since the beginnings of the study of this semantic relation. And finally, in order to provide a method to apply synonymy in practice, we have stated that, in this work, synonymy is understood as a gradual relation between words.

In order to calculate the degree of synonymy, we use measures of similarity applied on the sets of synonyms provided by a dictionary of synonyms for each of its entries. In the examples shown in this work, we will use as our measure of similarity *Jaccard's coefficient*, which is defined as follows. Given two sets X and Y, their *similarity* is measured as:

$$sm(X,Y) = \frac{|X \cap Y|}{|X \cup Y|}$$

This similarity measure yields values ranging between 0 (the words are not synonymous at all) and 1 (the words are completely synonymous).

On the other hand, let us consider a word w with M possible meanings m_i, where $1 \le i \le M$, and another word w' with M' possible meanings m_j, where $1 \le j \le M'$. By $dc(w, m_i)$, we will represent the function that gives us the set of synonyms provided by the dictionary for every entry w in the concrete meaning m_i. Then, the degree of synonymy of w and w' in the meaning m_i of w is calculated as follows [3, 4]:

$$dg(w, m_i, w') = \max_{1 \le j \le M'} sm[dc(w, m_i), dc(w', m_j)]$$

Furthermore, by calculating

$$k = \arg \max_{1 \le j \le M'} sm[dc(w, m_i), dc(w', m_j)]$$

we obtain in m_k the meaning of w' closest to the meaning m_i of w.

Let us consider this example[2], extracted from the dictionary we will use in this work:

$$w = \texttt{abandonado} \qquad m_i = m_2 \qquad w' = \texttt{sucio}$$

$$dc(w, m_2) = \{\texttt{abandonado}, \texttt{desaseado}, \texttt{desaliñado}, \texttt{sucio}\}$$

Case $m_j = m_1$:

$$\begin{aligned} dc(w', m_1) &= \{\texttt{sucio}, \texttt{impuro}, \texttt{sórdido}\} \\ dc(w, m_2) \cap dc(w', m_1) &= \{\texttt{sucio}\} \\ dc(w, m_2) \cup dc(w', m_1) &= \{\texttt{abandonado}, \texttt{desaseado}, \texttt{desaliñado}, \texttt{sucio}, \\ &\qquad \texttt{impuro}, \texttt{sórdido}\} \\ sm[dc(w, m_2), dc(w', m_1)] &= \frac{|dc(w, m_2) \cap dc(w', m_1)|}{|dc(w, m_2) \cup dc(w', m_1)|} = \frac{1}{6} = 0.16666667 \end{aligned}$$

Case $m_j = m_2$:

$$\begin{aligned} dc(w', m_2) &= \{\texttt{sucio}, \texttt{inmundo}, \texttt{puerco}, \texttt{cochino}, \texttt{desaseado}\} \\ dc(w, m_2) \cap dc(w', m_2) &= \{\texttt{sucio}, \texttt{desaseado}\} \\ dc(w, m_2) \cup dc(w', m_2) &= \{\texttt{abandonado}, \texttt{desaseado}, \texttt{desaliñado}, \texttt{sucio}, \\ &\qquad \texttt{inmundo}, \texttt{puerco}, \texttt{cochino}\} \\ sm[dc(w, m_2), dc(w', m_2)] &= \frac{|dc(w, m_2) \cap dc(w', m_2)|}{|dc(w, m_2) \cup dc(w', m_2)|} = \frac{2}{7} = 0.28571429 \end{aligned}$$

[2] The Spanish words involved in this example (and one of their corresponding translations into English) are: abandonado (*slovenly*), sucio (*dirty*), desaseado (*untidy*), desaliñado (*down-at-heel*), impuro (*impure*), sórdido (*squalid*), inmundo (*foul*), puerco (*nasty*), cochino (*filthy*), obsceno (*obscene*) and deshonesto (*indecent*).

Case $m_j = m_3$:

$$
\begin{aligned}
dc(w', m_3) &= \{\texttt{sucio}, \texttt{obsceno}, \texttt{deshonesto}\} \\
dc(w, m_2) \cap dc(w', m_3) &= \{\texttt{sucio}\} \\
dc(w, m_2) \cup dc(w', m_3) &= \{\texttt{abandonado}, \texttt{desaseado}, \texttt{desaliñado}, \texttt{sucio}, \\
&\quad\ \texttt{obsceno}, \texttt{deshonesto}\} \\
sm[dc(w, m_2), dc(w', m_3)] &= \frac{|dc(w, m_2) \cap dc(w', m_3)|}{|dc(w, m_2) \cup dc(w', m_3)|} = \frac{1}{6} = 0.16666667
\end{aligned}
$$

Finally, we have:

$$dg(w, m_2, w') = \max_{1 \leq j \leq 3} sm[dc(w, m_2), dc(w', m_j)] = 0.28571429$$

$$k = \arg \max_{1 \leq j \leq 3} sm[dc(w, m_2), dc(w', m_j)] = 2$$

That is, the degree of synonymy of the second meaning of the word `abandonado` with respect to `sucio` is 0.28571429 and the meaning of `sucio` that is more similar to `abandonado` is m_2.

The conception of synonymy as a gradual relation implies a distancing from the idea that considers it as a relation of perfect equivalence. This is coherent with the behaviour of synonymy in the printed dictionary, since it is possible to find cases in which the reflexive, symmetrical and transitive properties do not hold:

- The reflexive relation is usually omitted in dictionaries in order to reduce the size of the corresponding implementations, since it is obvious that any word is a synonym of itself in each one of its individual meanings.
- The lack of symmetry can be due to several factors. In certain cases, the relation between two words can not be considered as one of synonymy. This is the case of the words `granito` (*granite*) and `piedra` (*stone*), where the relation is a hyponymy. This phenomenon also occurs with some expressions: for instance, the expression `ser uña y carne` (*to be inseparable* or, in literal translation, *to be nail and flesh*) and the word `uña` (*nail*) appear as synonyms. In other cases, symmetry is not present because a word can have a synonym which is not an entry in the dictionary. One reason for this is that the lemmas of the words are not used when these words are provided as synonyms. Another possible reason is an omission by the lexicographer who compiled the dictionary, but, in general, all these problems support the claim of Lyons when he talks of quasi-synonymy to define the relation between words appearing in dictionaries of synonyms ([10]).
- Finally, if synonymy has been understood as similarity of meanings, it is reasonable that transitivity does not always hold.

The dictionary used also includes antonyms of each entry. The main problem with antonyms in most of the Spanish published dictionaries of synonyms and antonyms is that the sets of antonyms for an entry are frequently incomplete. For example, the first meaning of the word `abandonado` has a set of associated antonyms formed by the words `diligente` (*diligent*) and `amparado` (*protected*). Neither word appears as a dictionary entry, but only as a synonym of other entries. Most synonyms of `diligente` and `amparado` are antonyms of `abandonado` and must be included in the set of antonyms of this entry. In other words, a synonym of an antonym of `abandonado` is an antonym of this word. The inclusion of new antonyms in the sets of antonyms under this criterion can be performed automatically by FDSA. It is known that synonymy and antonymy are distinct semantic relations that do not work in exactly the same way. It is not the purpose of this work to deal with the computational treatment of antonymy but our proposal for synonymy could be helpful for the aforementioned lack of antonyms in published dictionaries. Once again, it is necessary to state that our approach is fundamentally guided by applied and practical criteria and the results with regard to antonymy were positive.

The use of finite-state automata to implement dictionaries efficiently is a well-established technique [1]. The main reasons for compressing a very large dictionary of words into a finite-state automaton are that its representation of the set of words is compact, and that the process of looking up a word in the dictionary is proportional to the length of the word, and therefore very fast [7]. Of particular interest for natural language processing applications are minimal acyclic finite-state automata, which recognize finite sets of words, and which can be constructed in various ways [15, 5]. The aim of the present work was to build a general architecture to handle a large Spanish dictionary of synonyms [2].

In the following sections, we will describe a general architecture that uses minimal deterministic acyclic finite-state automata in order to implement large dictionaries of synonyms, and how this general architecture has allowed us to modify an initial dictionary with the purpose of letting the relations between the entries and the expressions provided as answers satisfy the reflexive and symmetrical properties, but not the transitive one.

4 General Architecture of an Electronic Dictionary of Synonyms

Words in a dictionary of synonyms are manually inserted by linguists. Therefore, our first view of a dictionary is simply a text file, with the following line format:

```
word meaning homograph synonym
```

Words with several meanings, homographs or synonyms use a different line for each possible relation. With no loss of generality, these relations could be alphabetically ordered. Then, in the case of Blecua's dictionary, the point at which the word `concesión` (*concession*) appears could have this aspect:

```
concesión 1 1 gracia      (grace)
concesión 1 1 licencia    (licence)
concesión 1 1 permiso     (permission)
concesión 1 1 privilegio  (privilege)
concesión 2 1 epítrope    (a figure of speech)
```

For a later discussion, we say that the initial version of the dictionary had $M = 27,029$ different words, with $R = 87,762$ possible synonymy relations. This last number is precisely the number of lines in the text file. The first relation of `concesión` appears in line 25,312, but the word takes the position 6,419 in the set of the M different words ordered lexicographically.

Of course, this is not an operative version for a dictionary. It is therefore necessary to provide a compiled version to compact this large amount of data, and also to guarantee an efficient access to it with the help of automata. The compiled version is shown in Fig. 1, and its main elements are:

- The `Word_to_Index` function changes a word into its relative position in the set of different words (e.g. `concesión` into $6,419$).
- In a *mapping* array of size $M+1$, this number is changed into the absolute position of the word (e.g. 6,419 into 25,312). This new number is used to access the rest of arrays, all of them of size R. The lexicographical ordering guarantees that the relations of a given word are adjacent, but we need to know how many they are. For this, it is enough to subtract the absolute position of the word from the value of the next cell (e.g. 25,317–25,312 = 5 relations).
- The arrays *m1* and *h1* store numbers which represent the meanings and homographs, respectively, of a given word. The arrays *m2* and *h2* have the same purpose for each of its synonyms.
- The array *w2* is devoted to synonyms and also stores numbers. A synonym is a word that also has to appear in the dictionary. The number obtained by the `Word_to_Index` function for this word is the number stored here, since it is more compact than the synonym itself. The original synonym can be recovered by the `Index_to_Word` function.
- The array *dg* directly stores the degrees of every possible synonymy relation. In this case, no reduction is possible.

Note that the arrays *m2*, *h2* and *dg* store data that are not present in the original version of the dictionary. This new information was easily calculated from the rest of arrays with the formulas explained in Sect. 3, once the dictionary had been compiled into this general model and those initial data could be efficiently accessed. The specific transformations performed on the initial dictionary are detailed in Sect. 5.

This is the most compact architecture for storing all the information of the words present in a dictionary, when this information involves specific features of each word, such as the degree of a synonymy relation. Furthermore, this

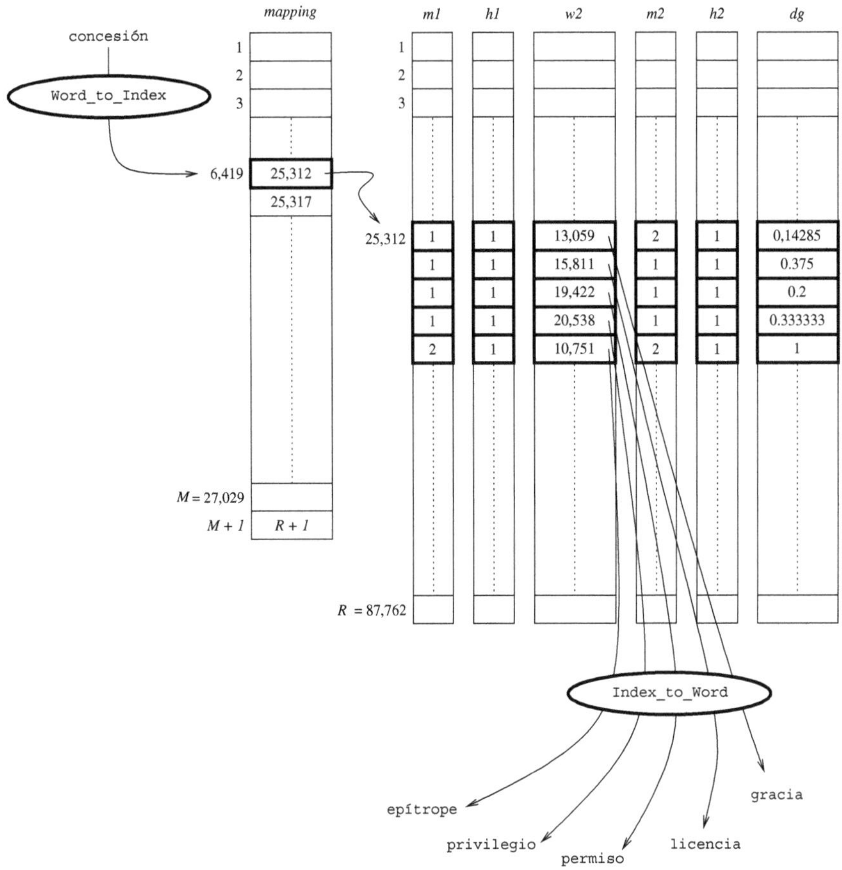

Fig. 1. Compact modeling of an electronic dictionary of synonyms.

architecture is very flexible: it is easy to incorporate new arrays for other additional data (such as part-of-speech tags), or to remove unused arrays (thus saving the corresponding space). To complete this model, we only need the implementation of the functions `Word_to_Index` and `Index_to_Word`. Both functions operate over a special type of automata, the numbered minimal acyclic finite-state automata described in [5], allowing us to efficiently perform perfect hashing between numbers and words.

5 Improving the Dictionary

The implementation of the dictionary of synonyms [2] was carried out in several steps, some of which required manual processes whereas others could be made automatically. The initial version of the dictionary had 21,098 entries; however it also included 5,931 expressions that appear as synonyms of others but were not entries by themselves (from now, no-entries). This version had 87,762 pairs of synonyms and our first goal was to fill the information corresponding to the *m2*, *h2* and *dg* arrays described in Sect. 4. With respect to *dg* and *m2*, this could be done mechanically by using the formulas of Sect. 3. The automatic detection of homographs *h2* was carried out by including all the homographs of the second word in the calculation of the degree, but only 300 entries of dictionary proved to have homographs. From these initial data we made further modifications related to the properties that the synonymy relation satisfies in the dictionary:

- Symmetry: One of the reasons why symmetry does not hold is the existence of 11,596 pairs involving no-entries. This means that there exist pairs of the form (entry, no-entry) but not the converse pairs. The next improvement was to add as entries all no-entry expressions. In order to do so, we had to associate a set of synonyms to each no-entry. This set was made up of all entries in which the no-entry expression appears as a synonym. In this way, the number of entries and the number of pairs were increased to 27,029 and 99,358 respectively. At this moment, all the expressions involved in the dictionary appear as entries. Nevertheless, the synonymy relation is still non-symmetric. Since we use a measure of similarity, in this case Jaccard's coefficient, two meanings of two different entries will be non zero synonyms (i.e. will be synonyms) if their associated sets of synonyms have some element in common. Following this criterion, if an entry x has synonyms in common with another y given two respective meanings of them, y will have the same synonyms in common with x for those meanings. We have improved the dictionary again by adding to each set of synonyms X the new entries that had meanings whose associated sets of synonyms had elements in common with X. This second step does not further modify the number of entries of the dictionary, but the number of pairs is modified increasing to 621,265. We obtained a symmetric relation of synonymy and we transformed the initial dictionary into a richer one.
- Reflexivity: The final improvement was to incorporate the reflexive pairs in the synonymy relation by adding for each entry of the dictionary the entry word itself in all the corresponding sets for each meaning of it. This is useful in order to avoid some problems in the calculation of the degrees of synonymy. For instance, when a word x appears as a unique synonym of another y and y as a unique synonym of x for two specific meanings of them, the corresponding sets of these meanings have no elements in common. In this case, the degree of synonymy is 0; therefore x and y will not be considered as synonyms, which is not very intuitive. By adding the reflexive case in the sets of synonyms we avoid this problem. For example, let us consider the Spanish words `carrete` and `bobina`[3]. The only meaning of the word `carrete` had as its set of associated synonyms {`bobina`} and the only meaning of the word `bobina` had as its set of associated synonyms

[3] Both words can be translated into English as *bobbin*, *reel* or *spool*.

{carrete}. If we calculate the degree of synonymy between both words using the similarity measure that we have presented in Sect. 3, we can see that the corresponding sets of synonyms are not similar at all, resulting in a degree of synonymy equal to 0. But if we include the reflexive case in the set of synonyms, we will have the associated set {carrete, bobina} for both words, which results in a degree of synonymy equal to 1 (i.e. the maximum degree). This second option is more coherent with our intuitions about the synonymy of carrete and bobina. After this modification the number of pairs increases to 655,583 and the relation is now reflexive and symmetric.

- Transitivity: Since the criterion followed indicates that two entries are synonyms if their corresponding sets of synonyms have elements in common, it is reasonable to think that the synonymy relation is not necessarily a transitive one. This is because, in general, from the fact that a set of synonyms X has elements in common with Y and Y has elements in common with Z it can not be inferred that X has elements in common with Z. Although there exist some dictionaries of synonyms whose synonymy relation is transitive, the dictionary we have used includes a considerable number of examples showing the non-existence of this property.

With regard to the time figures involved in this final configuration of Blecua's dictionary, the time needed to build the automaton (27,029 words, 27,049 states and 49,239 transitions) is 0.63 seconds, in a Pentium Centrino M715 1.5 GHz. under Linux operating system. A further 1.65 seconds are needed to incorporate the information regarding meanings, homographs, synonyms and degrees, thus giving us a total compilation time of 2.28 seconds. Finally, it should be noted that the recognition speed of our automata is around 180,000 words per second. This figure makes it possible to access the information very rapidly, thus proving the suitability of our general architecture for both the process of improvement and the use of this Spanish dictionary of synonyms.

6 Stand-alone Use of FDSA

FDSA (Fuzzy Dictionary of Synonyms and Antonyms) is the generic name of our electronic dictionary of synonyms, although we usually reserve this name for its stand-alone version. As we have seen, we propose a general architecture for this kind of linguistic resource, it being possible to use this model to implement an electronic dictionary of synonyms and antonyms for any language. In this work, we build a dictionary for Spanish from all the information that appears in Blecua's printed dictionary of synonyms [2].

Since FDSA is able to calculate the degree of synonymy between two entries, it presents some advantages with respect to the printed dictionary, amongst which are:

- It provides, using automatic procedures, the meaning of synonyms and antonyms that it gives as answers.

- It provides, by automatic procedures too, the homograph of synonyms and antonyms when these have various homographs.
- It orders the synonyms by the degree of synonymy with respect to the entry.
- It includes in the answer words that do not appear in the dictionary as synonyms but which could be synonyms because they have a non null degree of synonymy.
- It provides more antonyms than the printed dictionary using the criterion that a synonym of an antonym of the entry may be an antonym of the entry.
- It allows the user to reduce the number of answers by the use of thresholds.

Moreover, FDSA offers the user the possibility of implementing all the improvement processes described in this work. The main components of this software are: the electronic dictionaries, the algorithms that calculate the degrees of synonymy and antonymy, and the graphical user interface.

The electronic dictionaries. Initially, they include all the information contained in the printed dictionary. This information is stored in three different electronic dictionaries, `Syn`, `Ant` and `Inf`:

- `Syn` contains all the information about synonyms. To each entry we can associate one or more homographs, to each homograph one or more meanings, and to each meaning a set of synonyms.
- `Ant` contains all the information about antonyms. Its structure is similar to `Syn` but now the sets associated to each meaning are sets of antonyms. In other words, the information is classified in the same way for synonyms and antonyms. Of course, this is not to say that synonymy and antonymy have the same structure.
- `Inf` contains notes on style and usage, such as information about inflexion suffixes, grammatical issues, technical terms, dialectalisms, loanwords, pragmatic issues, diachronic issues, etc. The structure is similar to `Syn` and `Ant` but now the sets associated to each meaning consist of this additional information.

Modifications can be made to these initial versions of the electronic dictionaries by applying various improvement techniques carried out by the algorithms that calculate the degrees of synonymy and antonymy.

The algorithms that calculate the degrees of synonymy and antonymy. These algorithms have been used to apply the improvements described in this work, thus adding new information to the electronic dictionaries, such as:

- degrees of synonymy and antonymy,
- meanings and homographs of the synonymous and antonymous words,
- no-entries,

- reflexive cases,
- additional synonyms (with respect to the previous version of the dictionary).

Each of these improvements can be incorporated in an independent step, thus providing different versions of the electronic dictionaries. Moreover, the last of the improvements cited above (additional synonyms) can be implemented repeatedly, as a recurrent process. In each iteration, we will also obtain a different version of the electronic dictionaries. However, care must be taken when exercising this option, since it may lead to distortion of our starting point, i.e. the representation of the meaning of a word using the sets of synonyms with which it is associated in the printed dictionary of synonyms. This is particularly critical in the case of polysemic and imprecise words.

The graphical user interface. The graphical user interface of FDSA has three components (see Fig. 2): a main window and two dialog boxes (one of them for synonyms and the other for antonyms). In the dialog for synonyms, a user can introduce a Spanish word and will obtain, among other things, the synonyms that the printed dictionary gives, the words that do not appear in the printed dictionary as synonyms but could be synonyms because they have a non null degree of synonymy, the corresponding degree of synonymy, the

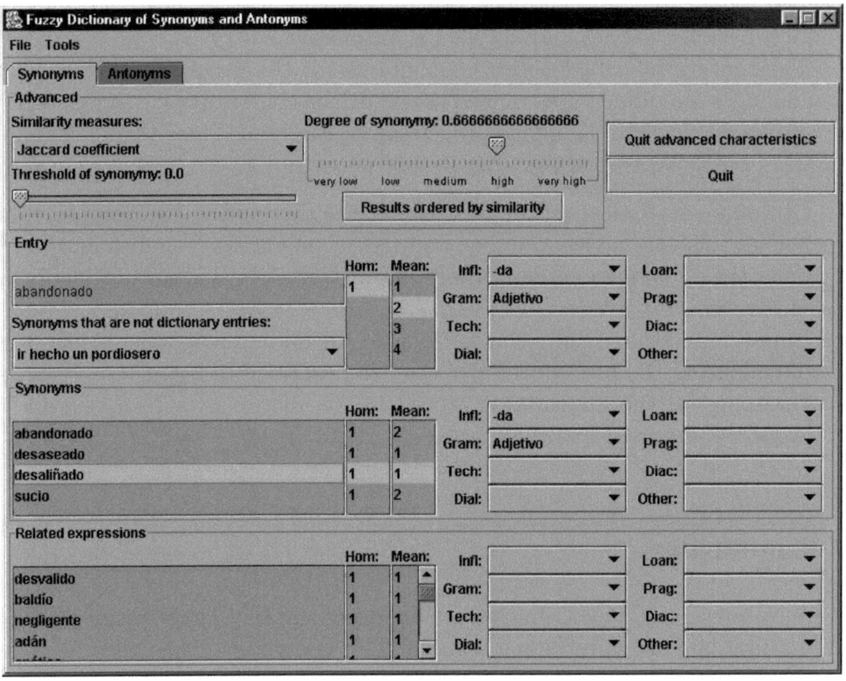

Fig. 2. The graphical user interface of FDSA.

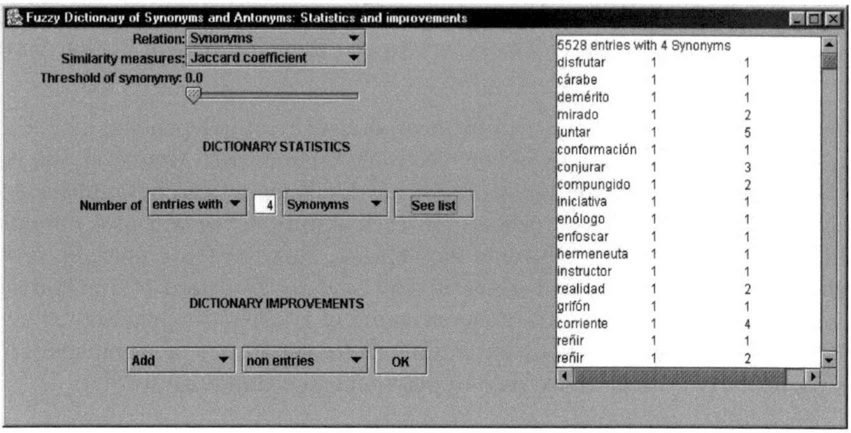

Fig. 3. The graphical user interface of the module Statistics and improvements.

verbalization of the degree indicating whether the synonymy is low, medium, or high, and the information about inflection suffixes, grammatical issues, technical terms, etc. Moreover, the user can list the synonyms ordered by degree of synonymy, can fix a threshold that reduces the number of answers, and can select one of various similarity measures.

The dialog for antonyms has the same structure but now the semantic relation between entries and answers is antonymy.

The tool as described above is of interest to the general user, but FDSA also includes a module for advanced users such as computational linguists, lexicographers or natural language processing researchers. This module is named Statistics and improvements (see Fig. 3) and is useful for:

- Obtaining statistical information about the dictionary, for example, number and list of entries, number and list of no-entries, number and list of pairs that hold symmetry, number and list of pairs that do not hold symmetry, etc.
- Improving automatically the dictionaries by adding no-entries or new synonyms using the procedures and criteria described in Sect. 5.
- Dumping all the information stored in the dictionaries to a text file in order to be reused by other tools.

7 Conclusions

We have presented a contribution for handling suitably large sets of words in the natural language processing domain. This contribution has been to design a general architecture for dictionaries which is able to store large amounts of data related to the words contained in them. We have shown that it is the most compact representation when we need to deal with very specific information about these words such as degrees of synonymy.

We have described how our general model has helped to implement and transform a Spanish dictionary of synonyms into a computational framework able to represent relations of synonymy between words. This framework, characterized by the conception of synonymy as a gradual relation, could be useful in order to improve the efficiency of some natural language processing tasks such as word sense disambiguation or query expansion in information retrieval systems.

With respect to this last task, one of the main problems of using synonymy to increase recall is the loss of precision. The information about degrees of synonymy, not present in other classical approaches such as the Spanish version of EuroWordNet [14] but included in the improved version of the Spanish dictionary of synonyms described in this work, makes it possible to use thresholds to control this loss of precision. If we consider synonymy as an approximate relation between words, we can obtain a greater or smaller number of answers depending on the user specifications of precision. In some cases a high degree of synonymy of the answers with respect to the entries could be necessary, but in other cases we do not need to be so strict with this requirement. Furthermore, Spanish EuroWordNet does not detect the meaning of the words that it gives as an answer.

Therefore, these features of our dictionary lead to conjecture that its use will increase recall without diminishing too much precision and latency in a fuzzy information retrieval system which is still at the experimental stage.

References

1. Aho, A., Sethi, R., Ullman, J. *Compilers: Principles, Techniques and Tools.* Addison-Wesley, Reading, MA, 1985.
2. Blecua-Perdices, J. *Diccionario Avanzado de Sinónimos y Antónimos de la Lengua Española.* Bibliograf, 1997.
3. Fernández-Lanza, S., Sobrino-Cerdeiriña, A. Hacia un Tratamiento Computacional de la Sinonimia. *Revista de la Sociedad Española para el Procesamiento del Lenguaje Natural*, 26:89–95, 2000.
4. Fernández Lanza, S. *Una Contribución al Procesamiento Automático de la Sinonimia Utilizando Prolog.* Ph. D. Thesis, University of Santiago de Compostela, 2001.
5. Graña-Gil, J., Barcala-Rodríguez, F. M., Alonso-Pardo, M. Compilation Methods of Minimal Acyclic Finite-State Automata for Large Dictionaries. *Lecture Notes in Computer Science*, 2494:135–148, 2001. Springer-Verlag.
6. Girard, G. *Synonymes François et leur Différents Significations et le Choix qu'il Faut Faire pour Parler avec Justesse.* Veuve d'Houry, Paris, 9th edition, 1749.
7. Hopcroft, J., Ullman, J. *Introduction to Automata Theory, Languages and Computation.* Addison-Wesley, Reading, MA, 1979.

8. López de Mántaras, R., Trillas, E. Towards a Measure of the Degree of Synonymy. In Sánchez, E. (ed.), *Fuzzy Information, Knowledge Representation and Decision Analysis*, Pergamon Press, 1984.
9. Lee, D.L., Chuang, H., Seamons, K.E. Document Ranking and the Vector-Space Model. *IEEE Software*, 14(2):67–75, 1997.
10. Lyons, J. *Linguistic Semantics. An Introduction.* Cambridge University Press, Cambridge, 1995.
11. Miyamoto, S. Information Retrieval based on Fuzzy Associations. *Fuzzy Sets and Systems*, 38(2):191–205, 1990.
12. Neuwirth, E., Reisinger, L. Dissimilarity and Distance Coefficients in Automation-Supported Thesauri. *Information Systems*, 7(1):47–52, 1982.
13. Salton, G., McGill, M. *Introduction to Modern Information Retrieval.* McGraw-Hill, New York, 1983.
14. Vossen, P. *EuroWordNet. A Multilingual Database with Lexical Semantic Networks.* Kluwer Academic Publishers, Dordrecht, The Netherlands, 1998.
15. Watson, B. *A Taxonomy of Finite Automata Construction Algorithms.* Computing Science Note 93/43, 1993, Eindhoven University of Technology, The Netherlands.

MIX
Papier aus verantwortungsvollen Quellen
Paper from responsible sources
FSC® C105338

If you have any concerns about our products,
you can contact us on
ProductSafety@springernature.com

In case Publisher is established outside the EU,
the EU authorized representative is:
Springer Nature Customer Service Center GmbH
Europaplatz 3, 69115 Heidelberg, Germany

Printed by Libri Plureos GmbH
in Hamburg, Germany